GOOD SCIENCE

NSW Syllabus for the Australian Curriculum
Stage 5

9+10

Rebecca **Cashmere**
Emma **Craven**
Aaron **Elias**
Haris **Harbas**
Niru **Kumar**
Natalie **Stinson**

Series reviewer: Maria **de Lima**
Investigations reviewer: Margaret **Croucher**

Good Science Stage 5 NSW Syllabus for the Australian Curriculum
1st edition
Emma Craven
Rebecca Cashmere
Aaron Elias
Haris Harbas
Niru Kumar
Natalie Stinson

Publisher: Catherine Charles-Brown
Content development: Patrick O'Duffy
Publishing director: Olive McRae
Senior editor: Amy Nicholls-Diver
Copy editors: Catherine Greenwood and Kelly Robinson
Proofreader: Jane Fitzpatrick
Indexer: Max McMaster
Illustrator: QBS Learning
Design manager: Jo Groud
Cover and text designer: Regine Abos
Production controller: Nicole Ackland
Digital production: Erin Dowling
Permissions researcher: Annthea Lewis
Typesetters: Cath Pirret and Leigh Ashforth

This edition published in 2021 by

Matilda Education Australia, an imprint
of Meanwhile Education Pty Ltd
Level 1/274 Brunswick St
Fitzroy, Victoria Australia 3065
T: 1300 277 235
E: customersupport@matildaed.com.au
www.matildaeducation.com.au

First edition published in 2020 by Macmillan Science and Education Australia Pty Ltd

Publication data

Author: Emma Craven, Rebecca Cashmere, Aaron Elias, Haris Harbas, Niru Kumar, Natalie Stinson

Title: *Good Science Stage 5 NSW Syllabus for the Australian Curriculum*

ISBN: 9781420246155

A catalogue record for this book is available from the National Library of Australia

Warning: It is recommended that Aboriginal and Torres Strait Islander peoples exercise caution when viewing this publication as it may contain names or images of deceased persons.

Printed in Malaysia by Vivar Printing
1 2 3 4 5 6 7 25 24 23 22 21

MAKE EVERY LESSON A GOOD LESSON

CONTENTS

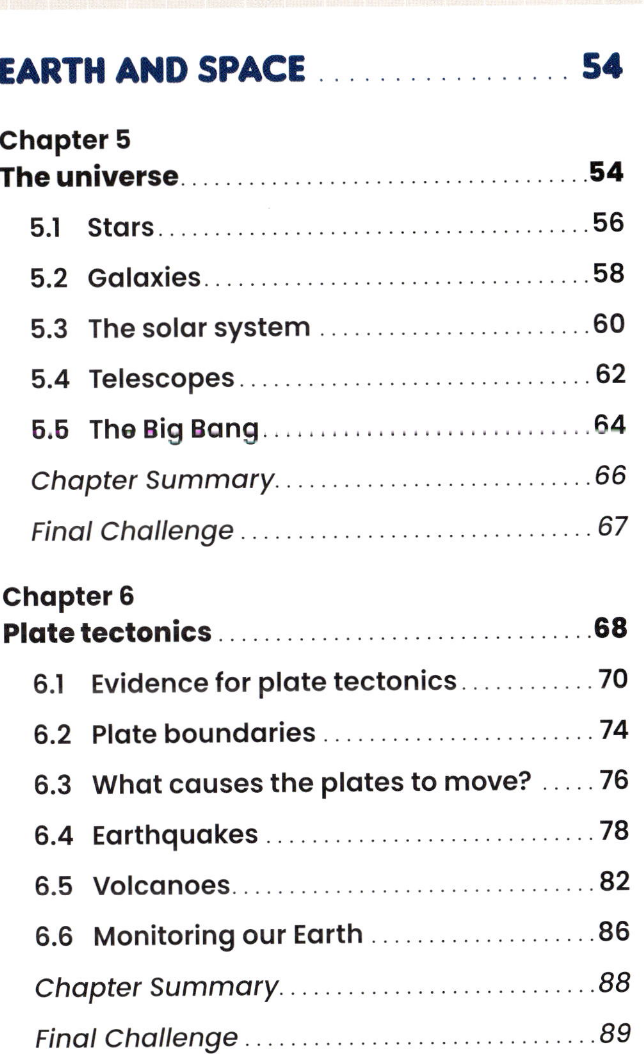

EARTH AND SPACE 54

CONTENTS

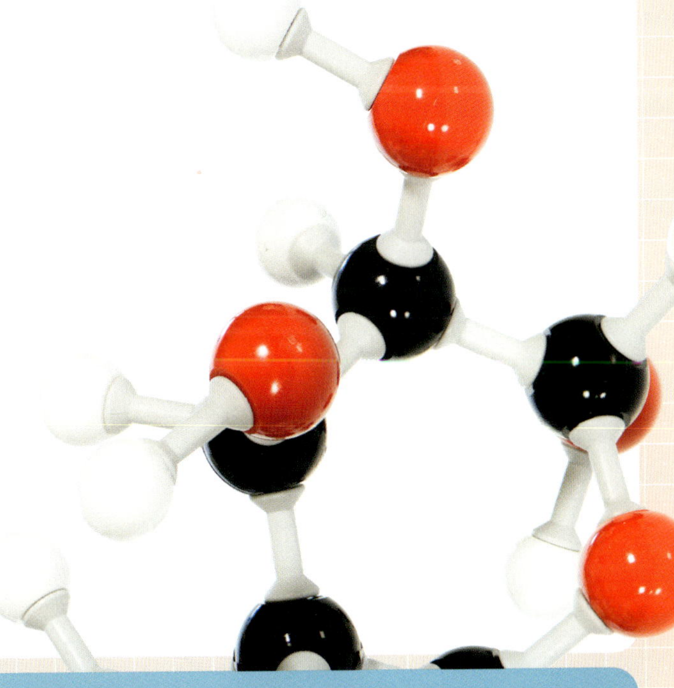

SYLLABUS CORRELATION GRID

	Stage 5 Outcomes		1 Energy transfer	2 Motion	3 Electricity	4 Energy usage	5 The universe	6 Plate tectonics	7 Global systems	8 Body systems	9 Energy and matter in ecosystems	10 Genetics	11 Evolution	12 Inside atoms	13 The periodic table	14 Chemical reactions	15 Rates of reactions	Investigations
Skills	SC5-4WS	A student develops questions or hypotheses to be investigated scientifically	✓	✓	✓	✓	✓	✓	✓	✓	✓	✓	✓	✓	✓	✓	✓	✓
Skills	SC5-5WS	A student produces a plan to investigate identified questions, hypotheses or problems, individually and collaboratively	✓	✓	✓	✓	✓	✓	✓	✓	✓	✓	✓	✓	✓	✓	✓	✓
Skills	SC5-6WS	A student undertakes first-hand investigations to collect valid and reliable data and information, individually and collaboratively	✓	✓	✓	✓	✓	✓	✓	✓	✓	✓	✓	✓	✓	✓	✓	✓
Skills	SC5-7WS	A student processes, analyses and evaluates data from first-hand investigations and secondary sources to develop evidence-based arguments and conclusions	✓	✓	✓	✓	✓	✓	✓	✓	✓	✓	✓	✓	✓	✓	✓	✓
Skills	SC5-8WS	A student applies scientific understanding and critical thinking skills to suggest possible solutions to identified problems	✓	✓	✓	✓	✓	✓	✓	✓	✓	✓	✓	✓	✓	✓	✓	✓
Skills	SC5-9WS	A student presents science ideas and evidence for a particular purpose and to a specific audience, using appropriate scientific language, conventions and representations	✓	✓	✓	✓	✓	✓	✓	✓	✓	✓	✓	✓	✓	✓	✓	✓
Physical World	SC5-10PW	A student applies models, theories and laws to explain situations involving energy, force and motion	✓	✓	✓	✓												
Physical World	SC5-11PW	A student explains how scientific understanding about energy conservation, transfers and transformations is applied in systems	✓	✓	✓	✓												
Physical World	PW1	Energy transfer through different mediums can be explained using wave and particle models. (ACSSU182)	✓															
Physical World	PW2	The motion of objects can be described and predicted using the laws of physics. (ACSSU229)		✓														

	Stage 5 Outcomes		1 Energy transfer	2 Motion	3 Electricity	4 Energy usage	5 The universe	6 Plate tectonics	7 Global systems	8 Body systems	9 Energy and matter in ecosystems	10 Genetics	11 Evolution	12 Inside atoms	13 The periodic table	14 Chemical reactions	15 Rates of reactions	Investigations
	PW3	Scientific understanding of current electricity has resulted in technological developments designed to improve the efficiency in generation and use of electricity.			✓													
	PW4	Energy conservation in a system can be explained by describing energy transfers and transformations (ACSSU190)				✓												
Earth and Space	SC5-12ES	A student describes changing ideas about the structure of the Earth and the universe to illustrate how models, theories and laws are refined over time by the scientific community					✓	✓	✓									
Earth and Space	SC5-13ES	A student explains how scientific knowledge about global patterns of geological activity and interactions involving global systems can be used to inform decisions related to contemporary issues						✓	✓									
Earth and Space	ES1	Scientific understanding, including models and theories, are contestable and are refined over time through a process of review by the scientific community. (ACSHE157, ACSHE191)					✓											
Earth and Space	ES2	The theory of plate tectonics explains global patterns of geological activity and continental movement. (ACSSU180)						✓										
Earth and Space	ES3	People use scientific knowledge to evaluate claims, explanations or predictions in relation to interactions involving the atmosphere, biosphere, hydrosphere and lithosphere. (ACSHE160, ACSHE194)							✓									

SYLLABUS CORRELATION GRID

	Stage 5 Outcomes		1 Energy transfer	2 Motion	3 Electricity	4 Energy usage	5 The universe	6 Plate tectonics	7 Global systems	8 Body systems	9 Energy and matter in ecosystems	10 Genetics	11 Evolution	12 Inside atoms	13 The periodic table	14 Chemical reactions	15 Rates of reactions	Investigations
Living World	SC5-14LW	A student analyses interactions between components and processes within biological systems								✓	✓	✓	✓					
Living World	SC5-15LW	A student explains how biological understanding has advanced through scientific discoveries, technological developments and the needs of society								✓	✓	✓	✓					
Living World	LW1	Multicellular organisms rely on coordinated and interdependent internal systems to respond to changes in their environment. (ACSSU175)								✓								
Living World	LW2	Conserving and maintaining the quality and sustainability of the environment requires scientific understanding of interactions within, the cycling of matter and the flow of energy through ecosystems.									✓							
Living World	LW3	Advances in scientific understanding often rely on developments in technology, and technological advances are often linked to scientific discoveries. (ACSHE158, ACSHE192)										✓						
Living World	LW4	The theory of evolution by natural selection explains the diversity of living things and is supported by a range of scientific evidence. (ACSSU185)											✓					

	Stage 5 Outcomes		1 Energy transfer	2 Motion	3 Electricity	4 Energy usage	5 The universe	6 Plate tectonics	7 Global systems	8 Body systems	9 Energy and matter in ecosystems	10 Genetics	11 Evolution	12 Inside atoms	13 The periodic table	14 Chemical reactions	15 Rates of reactions	Investigations
Chemical World	SC5-16CW	A student explains how models, theories and laws about matter have been refined as new scientific evidence becomes available												✓	✓	✓	✓	
Chemical World	SC5-17CW	A student discusses the importance of chemical reactions in the production of a range of substances, and the influence of society on the development of new materials.												✓	✓	✓	✓	
Chemical World	CW1	Scientific understanding changes and is refined over time through a process of review by the scientific community.												✓				
Chemical World	CW2	The atomic structure and properties of elements are used to organise them in the Periodic Table. (ACSSU186)													✓			
Chemical World	CW3	Chemical reactions involve rearranging atoms to form new substances; during a chemical reaction mass is not created or destroyed. (ACSSU178)														✓		
Chemical World	CW4	Different types of chemical reactions are used to produce a range of products and can occur at different rates and involve energy transfer. (ACSSU187)															✓	

1 Energy transfer

Did you know that the stars you look at in the night sky are a glimpse back in time? This is because by the time the light reaches your eyes, millions or even billions of years have passed, so the stars you are looking at are ancient. If an alien species with a powerful telescope looked at Earth from 65 million light-years away, they would see dinosaurs!

Light travels in wave form — as do sound, radio waves, microwaves and X-rays.

1 LEARNING LINKS

Brainstorm as much as you can remember about these topics.

Heat energy moves in liquids by convection, and in solids by conduction.

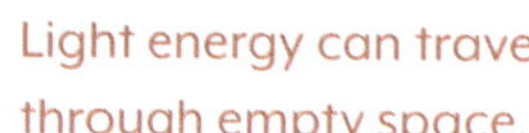

Light energy can travel through empty space.

Sound energy needs a medium to travel in.

2 SEE-KNOW-WONDER

List three things you can **see**, three things you **know** and three things you **wonder** about this image.

3 CRITICAL + CREATIVE THINKING

The question: Write five questions that have the answer 'wave'.

The disadvantages: List five disadvantages of sound.

The commonality: Outline some points of commonality between an X-ray machine and a television.

4 THE AUSSIE INVENTIONS!

Did you know that wi-fi is an Australian invention? In 1992, scientists at CSIRO accidentally discovered the technology after studying radio astronomy and looking for echoes of black holes.

Australians also developed ultrasound technology in 1976. Ever since, ultrasound technology has been used extensively to diagnose medical conditions. Ultrasound replaced the use of harmful X-rays to monitor foetal development.

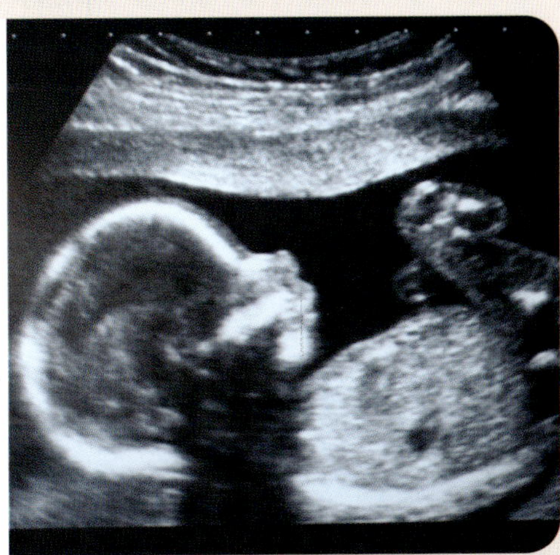

1.1 Conduction and convection

At the end of this lesson I will be able to:

- **explain**, in terms of the particle model, the processes underlying convection and conduction of heat energy.

KEY TERMS 

conduction
the transfer of heat through a substance

conductor
a substance that allows the transfer of heat

convection
the transfer of heat by movement of a liquid or gas

insulator
a substance that resists the transfer of heat

LITERACY LINK 

Create a mind map using the key terms and at least two additional words.

NUMERACY LINK 

Temperature can be measured in °C or Kelvins (K), where:

Temp (in K) = 273.15 + Temp (in °C)

What temperature is 25°C in K?

What temperature is 200 K in °C?

Heat is a form of thermal energy. In liquids and gases, heat is transferred by convection from a warm area to a cooler area. On a warm day at the beach, you may notice a cooling breeze. Cool air from above the ocean moves towards the land, where the air is warmer. The cold air quickly replaces the hot air, causing the colder breeze. In solids, heat is transferred by conduction. Your feet are warmed by conduction, by touching the solid hot sand.

Figure 1.1 The heat from the sand is transferred by conduction to your feet. Wearing thongs, which are made from an insulating material, protects your feet.

❶ Conduction is the transfer of heat in solids

Heat moves through solids by **conduction**. The particles in solids vibrate. When heated, the particles gain energy and vibrate even more. The particles bump into neighbouring particles and transfer some of their heat energy.

The best **conductors** of heat are metals. Non-metals and gases are poor conductors of heat and are known as **insulators**. Plastic, wood and rubber are good insulators, which is why rubber thongs protect your feet from hot sand.

How does conduction occur?

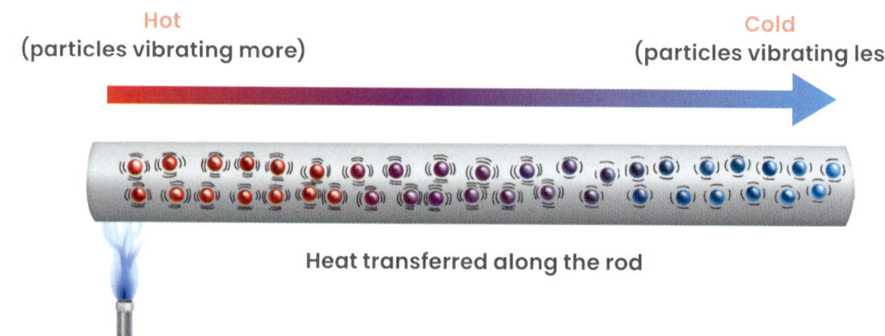

Figure 1.2 Heat transfer in a metal rod by conduction. Heat travels along the metal rod as the metal particles vibrate and bump into each other.

❷ Convection is the transfer of heat by movement of a liquid or gas

Heat moves through liquids and gases by **convection.** Particles in liquids and gases can move around more than particles in solids can. When a liquid or gas is heated, the particles move and take the place of particles with less heat energy. If a liquid is heated from underneath, the particles move from the bottom towards the top where the liquid is cooler.

When liquids and gases are heated, the particles move further apart from each other and the substance expands. The hot liquid or gas is less dense than the cooler liquid or gas, so it rises into the cold areas. The denser cold liquid or gas falls into the warm areas, producing convection currents that transfer heat from place to place. Similarly, hot air will rise and move to an area of colder air.

Convection explains why hot air balloons rise, and why it is often hotter in the upper floor of a two-storey house than downstairs.

Why does convection occur?

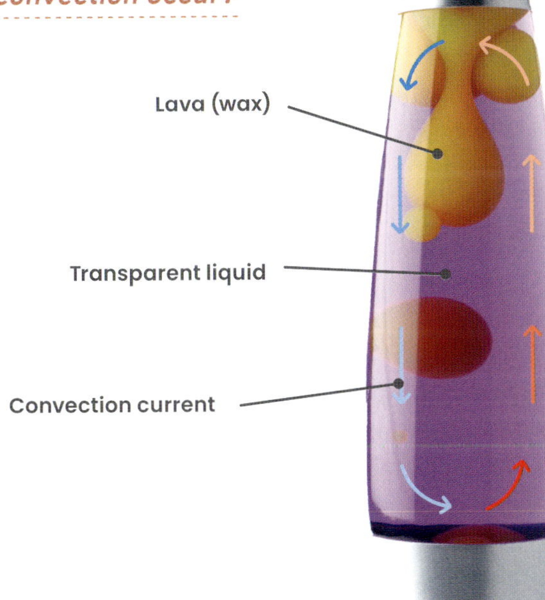

Figure 1.3 Convection currents can be seen in lava lamps, which contain a wax in a transparent liquid. The bulb in the base heats the wax, which expands and becomes less dense. The warm wax rises, then cools and becomes denser as it moves away from the heat. It sinks to the bottom where it is warmed again.

INVESTIGATION 1.1
Convection currents

CHECKPOINT 1.1

1 Explain what convection and conduction are.

2 Describe the difference between insulators and conductors.

3 Give two examples of materials that make good:
  a insulators
  b conductors.

4 Complete this sentence.
  Conduction occurs in ________, whereas convection occurs in ________ and ________.

5 Explain why convection cannot take place within a solid.

6 Explain how a convection current would work to heat up a pot of water on a stove. Include a diagram.

7 In designing a house, would it be best to place a heater on the floor or ceiling? Justify your response, using your knowledge of convection.

CHALLENGE

8 Design a simple experiment to compare the movement of heat by conduction through different solids.

SKILLS CHECK

- I can use examples to explain conduction.
- I can use examples to explain convection.
- I can describe how particles act during conduction and convection.

1.2 Waves transfer energy

At the end of this lesson I will be able to:

- **identify** situations where waves transfer energy
- **describe**, using the wave model, the features of waves, including wavelength, frequency and speed.

Figure 1.4 Slinkies are a fun way to observe waves.

A wave is a disturbance that travels through a substance. Waves are energy moving from one place to another without the movement of matter. Sound, light, surf and earthquakes are all examples of waves. If you have ever played with a slinky, you have probably observed waves and noticed how you can change their size and frequency. You can make waves in a slinky that go up and down like ocean waves, or back and forth like sound waves.

KEY TERMS 

amplitude
half the height of the wave

frequency
the number of waves passing a point every second

longitudinal
running lengthwise rather than across

transverse
running across rather than lengthwise

vacuum
empty space

wavelength
the distance from the peak of one wave to the next

1 Waves are energy carriers

Waves are moving energy. Some waves, called mechanical waves, require matter to move through. Energy is transferred from one place to another when a wave moves through a substance. Sound is a mechanical wave because it requires particles (such as air particles) for the sound to move through. Sound cannot travel through space because it is a **vacuum**, an area free of matter, and there is no medium for it to pass through. However, electromagnetic waves, such as light, can travel through a vacuum. This is how the heat and light from the Sun pass through empty space to reach Earth.

A wave has three properties: amplitude, frequency and wavelength.

- **Amplitude** is the distance from the peak (top) or trough (bottom) of a wave to the middle of its position. A wave with a high amplitude carries more energy than a wave with a low amplitude. In a sound wave, a high amplitude results in a louder sound.
- **Frequency** is the number of waves that pass a point every second. Frequency is measured in hertz (Hz). High-frequency waves carry more energy than low-frequency waves. In a sound wave, a higher frequency causes a higher pitch.

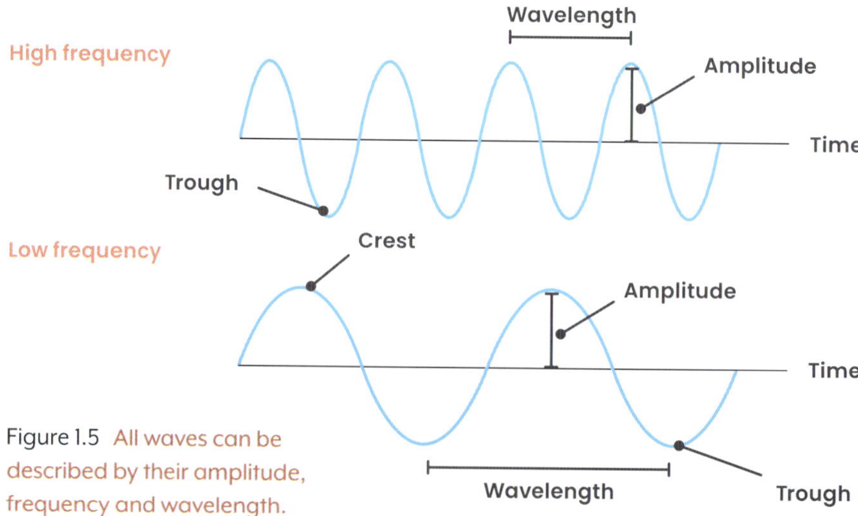

Figure 1.5 All waves can be described by their amplitude, frequency and wavelength.

LITERACY LINK 

Draw a Venn diagram to compare and contrast longitudinal and transverse waves.

NUMERACY LINK 

Speed of a wave (m/s) = wavelength (m) x frequency (Hz)
The speed of sound is 340 m/s. If the frequency of a guitar string is 490 Hz, calculate the wavelength.

- **Wavelength** is the distance from the peak of one wave to the next. Wavelength is measured in metres (m), centimetres (cm) or similar units. A wave with a shorter wavelength carries more energy than a wave with a longer wavelength.

What are three properties of waves?

2 Longitudinal waves vibrate in the same direction that they move

A **longitudinal** wave is a mechanical wave in which the particles of the medium vibrate in the same direction as the wave. Sound waves, tsunamis and earthquakes are longitudinal waves.

A longitudinal wave is made up of different regions called compressions and rarefactions. A compression is where the particles are bunched up. A rarefaction is where the particles are stretched apart. The wavelength is the distance from one compression to the next.

When sound waves travel through a medium, they make the particles of the material vibrate in the direction of the wave. The denser the material, the faster the sound travels. This is why sound travels better through water than it does through air.

What is a longitudinal wave?

3 Transverse waves vibrate at right angles to their movement

Transverse waves vibrate at right angles to their direction of motion. In other words, they vibrate 'up and down' rather than 'back and forth'. You can produce transverse waves in a rope by moving the ends of the rope up and down.

The strings in musical instruments, ocean waves and light are all types of transverse waves. Transverse waves may or may not require a medium to move through.

What is a transverse wave?

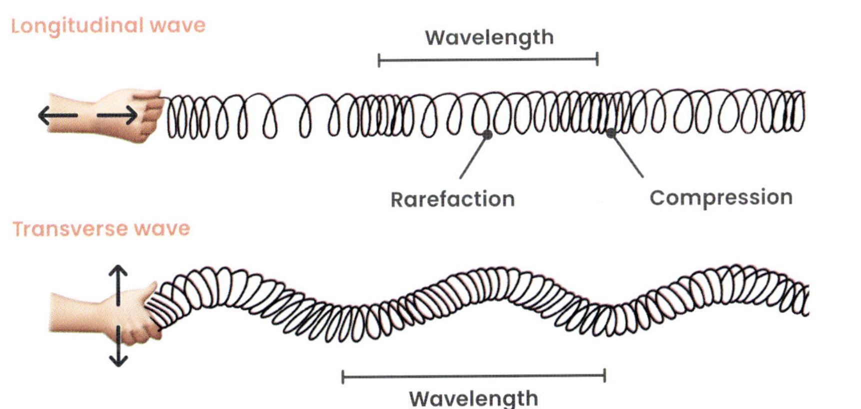

Figure 1.6 Longitudinal waves and transverse waves vibrate in different directions.

INVESTIGATION 1.2
Waves in a slinky

CHECKPOINT 1.2

1 Draw and label a:
 a longitudinal wave
 b transverse wave.
2 Explain the following terms:
 a *wavelength*
 b *frequency*
 c *amplitude.*
3 Describe how amplitude, frequency or wavelength can affect how much energy is transferred by a wave.
4 Describe the different regions of a longitudinal wave.
5 Give some examples of longitudinal and transverse waves.
6 A 'Mexican wave' at a sporting event is an example of a transverse wave. Explain why.
7 If you hear a very loud sound that has a very low pitch, what would this mean about the amplitude and frequency of the sound wave?

CHALLENGE

8 Research why earthquakes are hard to detect before they occur.

SKILLS CHECK

- I can identify at least two situations where waves transfer energy.
- I can use the wave model to discuss key features of waves.

1.3 Sound energy

At the end of this lesson I will be able to:

- **explain**, using the particle model, the transmission of sound in different mediums.

KEY TERMS 

reflect
send back sound or light without absorbing it

sonar
sound navigation and ranging

LITERACY LINK 

Use particle diagrams of solids, liquids and gases to write an explanation about what happens when sound passes through them.

NUMERACY LINK 

Use the data in Table 1.1 to construct a column graph comparing the speed of sound in air, lead, gold and glass.

Sound energy is transmitted through materials as longitudinal waves. The particles of the material vibrate as the sound travels through it. The more compact the material, the faster the sound travels. Sound cannot travel at all in a vacuum.

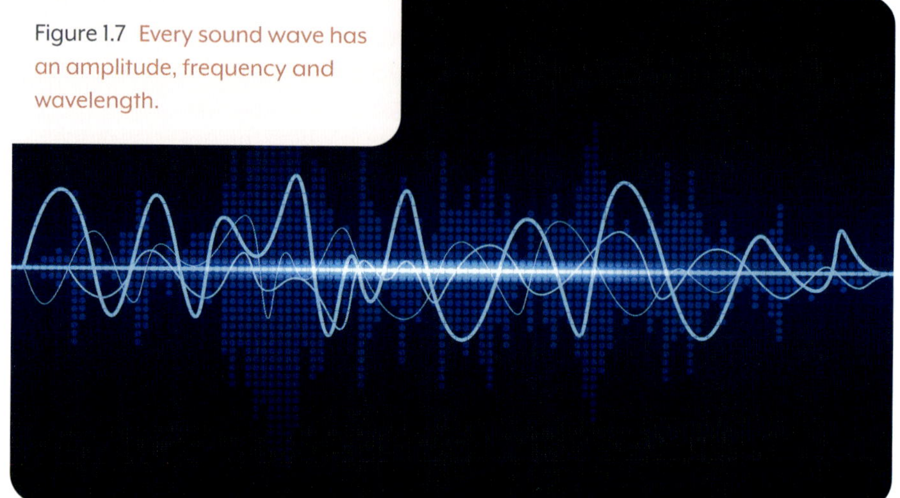

Figure 1.7 Every sound wave has an amplitude, frequency and wavelength.

❶ Sound travels quickly through many solids

Sound wave transmission relies on particles vibrating in different materials. Sounds travel very quickly though most solids.

If you put your ear on a metal rail, you could hear the sound of someone tapping on it before you could hear it through the air. This is because solid particles are bunched up, and sound wave vibrations are carried very quickly along the solid particles.

Sound travels faster through denser solids, such as metals, than through less dense solids, such as rubber.

How do sounds travel through solids?

Table 1.1 The speed of sound in different materials

Material	Speed of sound (m/s)
Rubber	60
Air (20°C)	343
Air (40°C)	355
Lead	1210
Gold	3240
Glass	4540

2 Sound also travels through liquids and gases

Sounds travel through liquids, such as water. Particles in liquids are more spread out than particles in a solid, so sound waves move more slowly through liquids. Dolphins use sound waves in water to navigate, and submarines use their **sonar** systems in a similar way.

Gas particles are very spread out, so sound waves travel the slowest through gases, such as air. It takes longer for vibrations to be passed from particle to particle. Heat can increase the vibrations of particles so sound travels faster through hot air than through cold air.

How do sounds travel through liquids and gases?

3 Sound can be reflected and absorbed

Hard surfaces such as concrete walls, caves and tiles **reflect** sound, creating an echo. Soft materials such as carpet and curtains absorb sound, converting it into heat energy. This explains why an unfurnished room not only echoes, but feels colder than a furnished room.

Spaces designed for plays and musical performances are built so that sounds from the stage are reflected towards the audience. The ancient Greek ampitheatres, which were built more than 2000 years ago, had limestone seats that absorbed low-frequency background noise and reflected the high-frequency sounds of performers. This allowed people at the back of the massive space to hear the plays.

What kind of materials relfect or absorb sound?

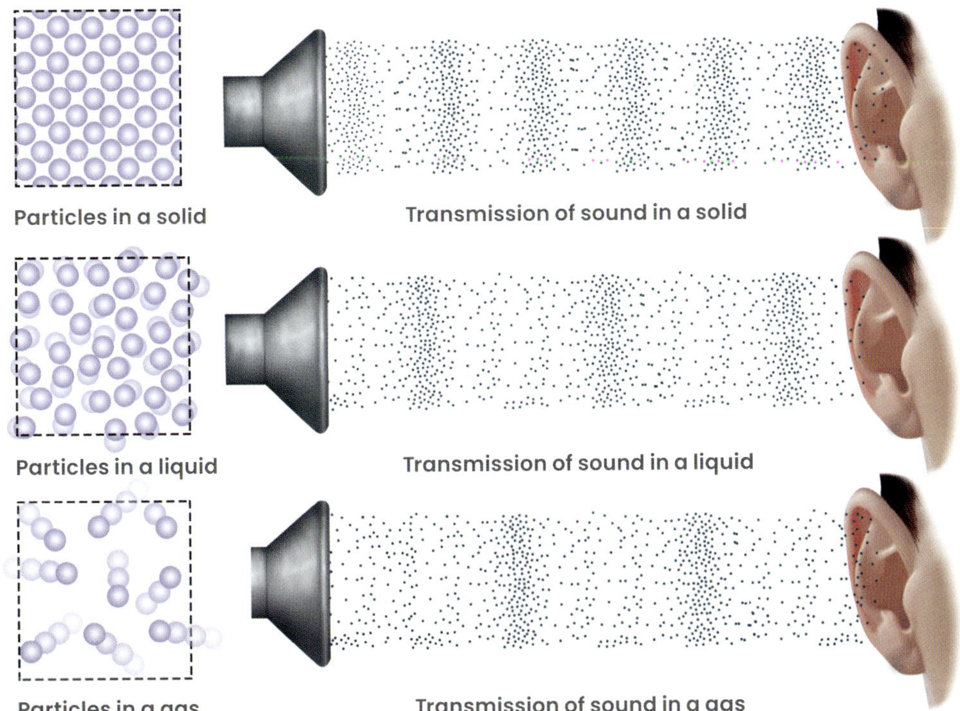

Figure 1.8 Sound travels fastest through solids and slowest through gases because of how densely the particles are packed.

CHECKPOINT 1.3

1 Explain why sound requires a medium to pass through.
2 Explain why sounds travel faster through solids than through liquids and gases.
3 Suggest why sounds travel the slowest in gases.
4 Explain why sound reflection and absorption occurs.
5 Give three everyday examples where sound is:
  a reflected
  b absorbed.
6 Use the particle theory to explain why sound travels faster as temperature increases.

CHALLENGE

7 Research sonar technology and how it can be used to calculate the depth of oceans and help fishing crews locate fish.

SKILLS CHECK

- I can describe how sounds travel in different mediums by using the particle model.
- I can explain why sound travels at different speeds in different mediums.

1.4 The electromagnetic spectrum

At the end of this lesson I will be able to:

- **relate** the properties of different types of electromagnetic radiation to their uses in everyday life, including communications technology.

KEY TERMS 

electromagnetic spectrum
all the different electromagnetic waves

gamma ray
radiation emitted by radioactive materials

infrared light
an electromagnetic wave with longer wavelength than red light

X-ray
a high-energy ray that can penetrate materials

LITERACY LINK 

Using the information obtained from the challenge question, create an information brochure for patients. Include information about the properties, safety and use of the specific technique, and refer to the electromagnetic waves used in the technique.

NUMERACY LINK 

The wavelengths on the electromagnetic spectrum are measured in nm, where $1 \text{ nm} = 1 \times 10^{-9}$ m.
Write this number out in decimal notation.

Have you ever wondered about the invisible waves around us? We use microwaves not just to heat up food, but also for mobile phone communication and satellite technology. We turn on our TVs using remote controls that use infrared rays. We apply sunscreen to protect us from UV rays that can burn us. Watching TV or listening to the radio uses radio waves. If you break your arm, you will get an X-ray. If you have a CT scan, you will be exposed to gamma rays, which help doctors to diagnose your medical conditions. You can see everything around you because of visible light rays reflecting from objects and reaching your eyes.

1 Electromagnetic radiation travels in waves

Electromagnetic waves originate from the Sun and other stars, and travel through the vacuum of space in different forms. The range of all the different types of electromagnetic radiation is called the **electromagnetic spectrum**. Electromagnetic waves are transverse waves made up of moving magnetic fields and electric fields travelling together. All electromagnetic waves are similar to each other. They all travel at 300 000 km/s – the speed of light.

Electromagnetic waves differ in wavelength and frequency. As wavelength increases, frequency decreases and the energy decreases. As wavelength decreases, frequency increases and the energy increases. In order of increasing energy, electromagnetic waves include radio waves, microwaves, infrared waves, visible light waves, ultraviolet (UV) waves, X-rays and gamma rays.

What is the electromagnetic spectrum?

Figure 1.9 The electromagnetic spectrum. Frequency is measured in hertz (Hz), which equals one cycle per second.

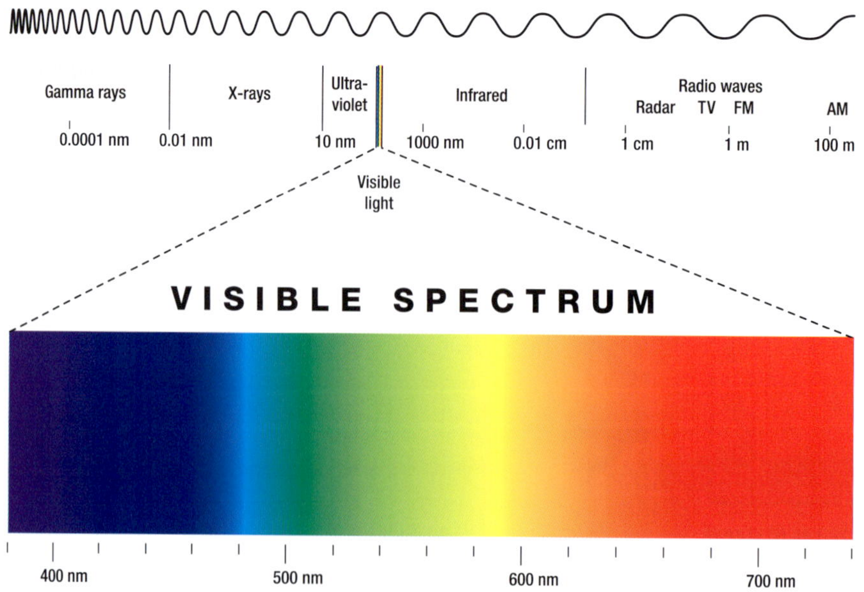

2 Gamma rays, X-rays and UV waves are high-energy rays

Gamma rays can penetrate materials such as lead and concrete. They are dangerous, because they can damage the cells in your body. Gamma rays are used in industry for detecting cracks in metal structures and underground pipes. They are also used for sterilising medical equipment, radiation therapy for the treatment of cancer, and CT scans.

X-rays are high-frequency rays that are produced when high-energy electrons hit a metal surface. You cannot see X-rays directly, but they affect photographic film. Denser materials absorb more X-rays than less dense materials do. X-rays are used to produce images of bones. Soft tissue, such as skin and organs, cannot absorb the high-energy rays, and the beam passes through them. X-rays are also used to kill cancer cells, and in the transport industry to check baggage.

Ultraviolet means 'beyond violet'. This is the part of the electromagnetic spectrum that has shorter wavelengths and higher frequencies than violet rays. You cannot see UV light, but insects such as bees can. UV light is used often in forensic science, to detect blood or other substances, as well as checking signatures for forgeries.

Figure 1.10 Bees can see UV light, which means they can see patterns on flowers that humans can't.

What is ultraviolet light?

3 Visible light, infrared light, radio waves and microwaves are low-energy rays

Light is sometimes called visible light because it is the only part of the spectrum that we can see. Our eyes contain special cells that detect only this type of electromagnetic radiation.

Infrared light has a lower frequency and longer wavelength than red light. People give off infrared light in the form of heat. Infrared light is used in electrical heaters, short-range communications (i.e. remote controls), and thermal imaging cameras, which can detect people in the dark.

Microwaves have shorter wavelengths than radio waves, and are given off by mobile phones. Microwaves are used in telecommunications; for example, with satellite telephone towers. Shorter microwaves are used in cooking food.

Radio waves are used for broadcasting television and radio, and in communications and satellite transmissions. The radio waves are produced by vibrating electrons, which cause antennas to vibrate. This is then converted into images on the TV, or sound on the radio.

How are radio waves produced?

CHECKPOINT 1.4

1 Explain what is meant by electromagnetic radiation and the electromagnetic spectrum.
2 Identify the origin of electromagnetic waves.
3 Define *electromagnetic waves*.
4 What is the relationship between wavelength and frequency?
5 Identify a high-energy wave and a low-energy wave in the electromagnetic spectrum.
6 Describe one property and one use for:
   a gamma rays
   b X-rays
   c UV light
   d infrared light
   e visible light
   f microwaves
   g radio waves.

CHALLENGE

7 Research one medical technique that uses electromagnetic radiation for either treatment or diagnostic purposes.

SKILLS CHECK

- I can identify the different waves of the electromagnetic spectrum, including their properties.
- I can relate the properties of different waves of the electromagnetic spectrum to their everyday uses.

1.5 Absorption, reflection and refraction

At the end of this lesson I will be able to:

- **describe** the occurrence and some applications of absorption, reflection and refraction in everyday situations.

KEY TERMS 

angle of incidence
the angle at which light hits a surface

angle of reflection
the angle at which light reflects from a surface

opaque
a material that doesn't allow any light to pass through

refract
bend (light)

translucent
a material that allows some light to pass through

transparent
a material that allows light to pass through

NUMERACY LINK 

This table shows data for a light ray entering a glass prism at 20°C.

Speed of light ×106 (m/s)	Angle of refraction
300	20
226	14.9
200	13.2
125	8.2

Draw a line graph for this data and state the relationship between speed of light and angle of refraction.

Light waves travel very fast, and can pass though transparent objects such as glass, but will usually refract (or bend) as they do. They can also bounce off mirrors or scatter when they strike rough materials. Why does light behave in such a manner? Why does it bounce or bend, and where does the light energy go? What is light energy transformed into when it hits rough surfaces?

Figure 1.11 Light can be reflected.

1 Light is an electromagnetic wave

Light is an electromagnetic wave made up of alternating magnetic and electric fields. Light travels extremely fast, at 300 000 km/s. Light can also be reflected, absorbed or refracted when it strikes a surface.

Transparent materials (such as cling wrap) allow light to pass through, and you can see a clear image through them. **Translucent** materials (such as wax paper) allow only some light to pass through, so you see a blurred image. **Opaque** materials (such as aluminium foil and cardboard) either reflect or absorb light, but do not allow any light to pass through.

What happens to light when it strikes a surface?

2 Light can be reflected and absorbed

When light hits a smooth shiny surface such as water or a mirror, it reflects off at the same angle (**angle of reflection**) as the angle at which it strikes the surface (**angle of incidence**). This is known as the law of reflection. Complete reflection results in a clear image.

Opaque materials that have very rough surfaces, such as brick and concrete, do not reflect light in a regular way. Also, some of the light is absorbed by the material and converted into heat.

When is light reflected?

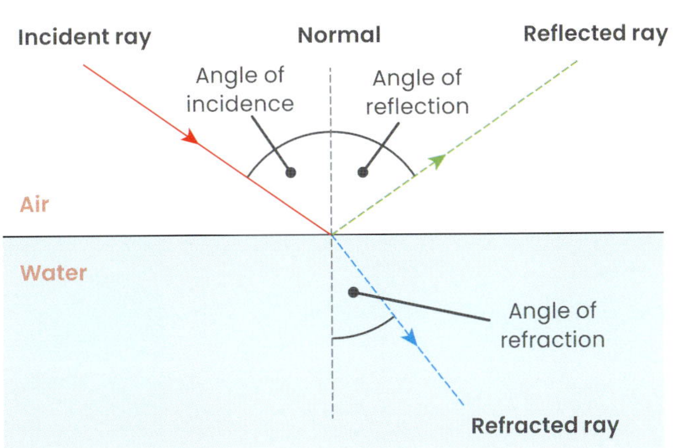

Figure 1.12 When light hits a smooth surface such as water, it is reflected back at the same angle. When light is refracted, it bends towards the normal (perpendicular) line.

INVESTIGATION 1.5A
The law of reflection

INVESTIGATION 1.5B
Refraction

3 Light bends when it travels through different mediums

When light passes from one medium to another (for example, from air to water), it will usually **refract**, or bend. This occurs because light travels at different speeds in different mediums. When the light moves from air to water, it slows down. This change in speed makes the light bend inwards, causing objects to appear to be in a different position.

Why does light refract?

4 Convex and concave lenses make objects look bigger and smaller

Lenses are very useful for bending light. Convex lenses (which bulge in the middle) make light rays meet at a point (converge). Concave lenses (which curve inwards) make light rays spread out (diverge).

A magnifying glass uses a convex lens to make objects look bigger than they really are. If you look at an object with a concave lens, it will look smaller than it is. People who are short-sighted (cannot see long distances) wear glasses with concave lenses. People who are long-sighted (cannot focus on close objects) wear glasses with convex lenses.

What is the difference between concave and convex lenses?

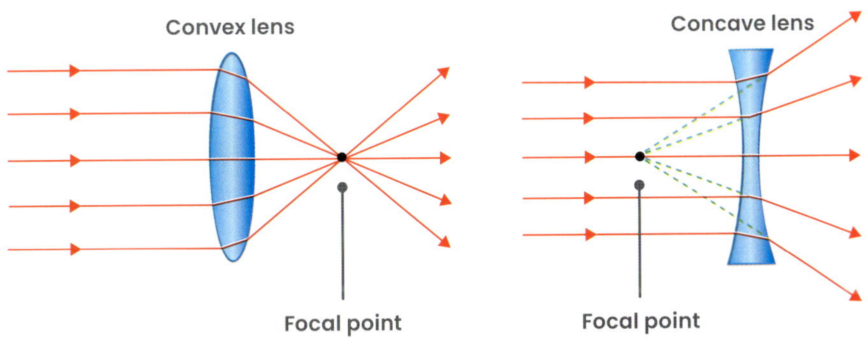

Figure 1.13 A convex lens makes light rays converge at a single point, the focal point. A concave lens makes rays diverge.

CHECKPOINT 1.5 

1 Explain the difference between reflection, refraction and absorption.

2 Identify three things that can happen when light hits different materials.

3 What happens to light when it hits a:
   a smooth surface?
   b rough surface?

4 Identify everyday examples of light reflecting and light absorbing.

5 Why does light refract? Give an example of this.

6 What is the difference between how concave and convex lenses refract light?

CHALLENGE

7 Research concave and convex mirrors and their everyday uses.

SKILLS CHECK

- I can describe reflection, refraction and absorption.
- I can identify everyday examples of reflection, refraction and absorption.

CHAPTER SUMMARY

The electromagnetic spectrum includes all the different types of electromagnetic radiation, from radio waves to gamma rays.

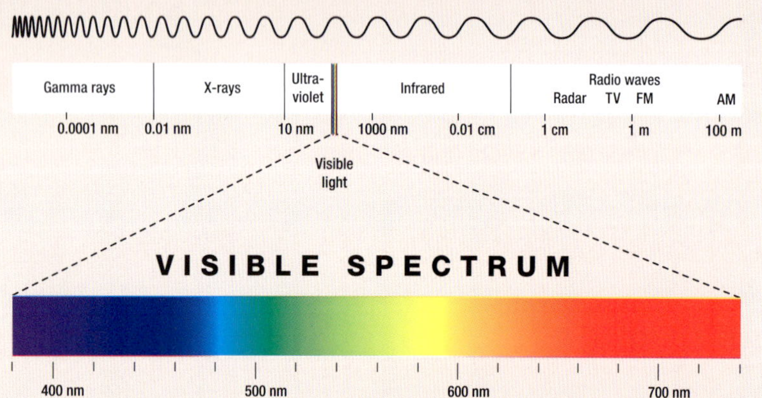

Light energy is the only part of the electromagnetic spectrum that we can see.

Longitudinal waves vibrate back and forth along their path.

Transverse waves vibrate at right angles to the wave path.

Light can be reflected and absorbed.

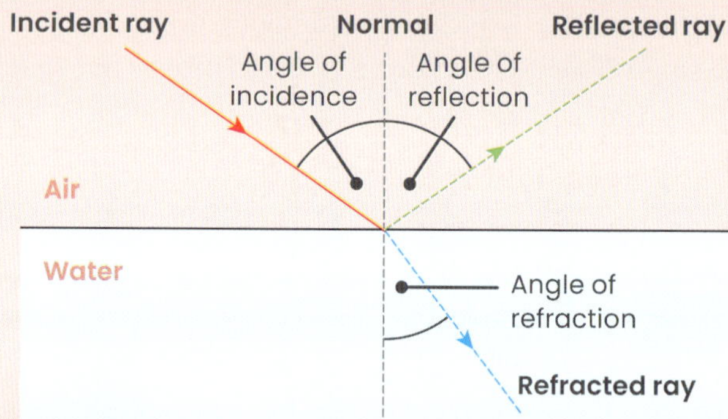

Sound waves are longitudinal waves and can be transmitted through matter.

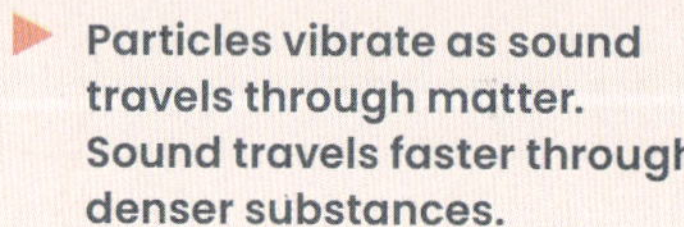

Particles vibrate as sound travels through matter. Sound travels faster through denser substances.

Convection is the movement of heat in liquids and gases, in which the material's particles carry the heat energy around.

Conduction is the transfer of heat from one material to another through the vibration of particles.

Concave lenses make objects appear smaller.

Convex lenses make objects appear larger.

★ FINAL CHALLENGE ★

1. Identify the similarities and differences between conduction and convection in terms of the particle model.
2. Describe an everyday example of conduction and convection other than what is discussed in the chapter.

LEVEL 1 ★☆☆☆☆☆ 50XP LEVEL UP!

3. Draw, label and describe the properties of a wave.
4. Compare and contrast longitudinal and transverse waves.
5. Sound travels slower through a liquid than through a solid. Suggest why.

LEVEL 2 ★★☆☆☆☆ 100XP LEVEL UP!

6. Explain, using the particle model, the transmission of sound in different mediums.
7. Why can sound be absorbed or reflected by different materials?

LEVEL 3 ★★★☆☆☆ 150XP LEVEL UP!

8. Recreate this table, correctly matching the type of wave, its properties and where it is used.

Waves	Property	Use
Gamma ray	Given off by mobile phones	Optical fibres for high-speed internet
X-ray	Shortest wavelength, highest frequency	Broadcasting television and radio
UV ray	Longest wavelength, shortest frequency	Satellite telephone towers
Light	Lower frequency than red light	Radiation therapy in treating cancer
Infrared light	High-frequency rays, produced when electrons hit a metal	TV remote controls
Microwaves	Beyond ultraviolet	Detecting forgeries and checking signatures

LEVEL 4 ★★★★☆☆ 200XP LEVEL UP!

9. Describe some everyday examples and applications of absorption, reflection and refraction.
10. Why do shadows form? Use at least three key terms from this chapter in your answer. Provide a supporting diagram with labels.

LEVEL 5 ★★★★★★ 300XP LEVEL UP!

2 Motion

How coordinated are you? Do you find it easy or hard to catch a ball? When you catch a ball, your brain does some very rapid calculations so that you can quickly work out the ball's speed, angle and position. The motion of cars, satellites and bullets is more complex than the motion of a thrown ball. However, you can still predict how complex objects will move by using the laws of motion. These laws govern how all sorts of things move under different conditions and forces.

1 LEARNING LINKS

Brainstorm as much as you can remember about these topics.

In order to change an object's motion, you need to apply an unbalanced force.

Some forces require contact while others can act over a distance.

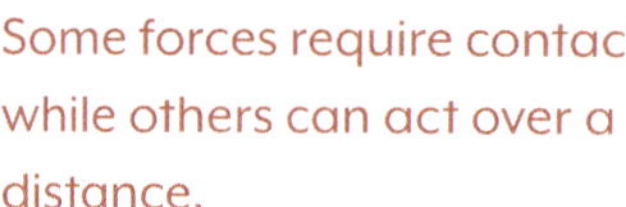

Many pieces of safety equipment are designed to reduce the impact of forces.

2 SEE-KNOW-WONDER

List three things you can **see**, three things you **know** and three things you **wonder** about this image.

3 CRITICAL + CREATIVE THINKING

What if ... Earth and the Moon suddenly swapped gravitational fields?

Ten things: List 10 things that cannot be accelerated.

Alternatives: List all the ways you could change the speed or direction of a car without touching the pedals or steering wheel.

4 THE STRONGEST!

The strongest force in the universe is called the strong nuclear force. This force keeps the smallest particles in an atom (called quarks) stuck together. Luckily, the strong nuclear force doesn't work over distances greater than the nucleus of an atom. Otherwise, the force of attraction between all matter would be so great that everything on Earth would be squashed together into the size of a house, and a tennis ball would weigh as much as 3 million suns!

2.1 Force, mass and acceleration

At the end of this lesson I will be able to:

- **describe** the relationship between force, mass and acceleration.

KEY TERMS 

acceleration
a change in speed or direction of motion over time

force
a push, pull or twist

inversely proportional
when one value decreases at the same rate that the other increases

mass
the amount of matter that something consists of

newton
the unit of measurement of force

LITERACY LINK 

Write a short paragraph explaining the relationship between force, mass and acceleration to a primary school student.

NUMERACY LINK 

Rearrange the $F = m \times a$ equation to:

- make m the subject
- make a the subject.

Have you ever tried to push a car? It requires a lot of force to get a car moving. On the other hand, you can get a skateboard moving with just your foot. The difference is mass: a car is much heavier than a skateboard, and so requires a much greater force to make it accelerate.

This tells us that there must be a relationship between force, mass and acceleration.

Figure 2.1 A golf ball can be accelerated rapidly because it has a small mass.

1 Acceleration depends on mass and the size of the force being applied

The relationship between force, mass and acceleration is summed up by Newton's second law, which says that the acceleration of an object is proportional to the force applied, and **inversely proportional** to the mass of the object.

Newton's second law can also be written as an equation:

$$\text{force} = \text{mass} \times \text{acceleration}$$

If you increase the force on an object of fixed mass, the acceleration will increase. If you use the same force but increase the mass, the acceleration will decrease.

Acceleration is a measure of how quickly an object's speed or direction of motion changes. Acceleration is usually measured in metres per second squared (m/s^2). If you drop a ball, it accelerates towards the ground – that is, its speed increases downwards. If you press the accelerator pedal in a car, the car's speed increases in the direction it is moving.

As long as speed or direction is changing, the object is accelerating. However, if an object isn't moving or is moving at a constant speed, then it is not accelerating.

What is acceleration a measure of?

Figure 2.2 Drag cars undergo rapid acceleration, and so their speed increases very quickly.

❷ Mass is the amount of matter in an object

The **mass** of an object is a measure of how much matter it contains and is usually measured in kilograms (kg). A heavy object contains more matter than a lighter object. Think about an empty cardboard box – it is quite light because it doesn't contain much matter. If you fill the box with heavy items, it becomes much heavier. Obviously, as the box gets heavier, it becomes more difficult to move.

In other words, the heavier an object is, the more force you need to apply to get it to accelerate.

What is mass?

Figure 2.3 The empty box has less mass than the full box because it contains less matter.

❸ Force is a push, pull or twist

A force is a push, pull or twist. Force is measured in **newtons** (N). The size of a force is important. A person pulling on a broken-down car with a rope won't generate a large pulling force, and the car will experience only a very small acceleration. If you attach the car to a tractor, the car will experience a much greater acceleration because the tractor can generate a much larger pulling force than a person can.

The amount of acceleration an object experiences depends on how much force is applied to the object. The greater the force, the greater the acceleration.

What is the relationship between force and acceleration?

Figure 2.4 A tractor can apply a large force and so is able to accelerate a stationary car easily.

INVESTIGATION 2.1
Acceleration changes due to mass

CHECKPOINT 2.1

1 What is the formula used to calculate force?
2 Which box would have the greatest acceleration? Explain why.
   a Box A and box B have identical masses, but box B is experiencing a smaller force.
   b Box C and box D are both experiencing the same force, but box C is heavier.
3 A motorcycle is travelling at a constant speed. Is the motorcycle accelerating?
4 Which has a greater mass – a bowling ball or a beach ball? Explain how you can tell.
5 Describe the relationship between acceleration, mass and force.

CHALLENGE

6 A rocket is very heavy and undergoes a large acceleration as it takes off. So, the force on a rocket must be extremely large. Use the internet to find out how big the force is, and where this force comes from.

SKILLS CHECK

- I can describe what is meant by acceleration, force and mass.
- I can predict how changing force and mass will change the acceleration of an object.

2.2 Speed, distance and time

At the end of this lesson I will be able to:

- **explain** the relationship between distance, speed and time.

KEY TERMS 

distance
the amount of space between two points

equation
a mathematical statement showing that one value is equal to another

speed
the distance an object travels divided by the time taken

LITERACY LINK 

Write a script for a video that explains the relationship between speed, distance and time.

NUMERACY LINK 

Convert 60 km to m, and 1 hour to seconds.

Then convert 60 km/h to m/s. Show your working.

Convert 8 m to km, and 1 second to hours.

Then convert 8 m/s to km/h. Show your working.

Figure 2.5 A fighter jet can travel large distances in very little time.

The fastest thing in the universe is light. Light can travel amazingly quickly – it can travel around Earth seven times per second! The speed of light in a vacuum is 300 000 000 m/s. This means that in one second, light can travel 300 000 000 metres. This is extraordinarily fast, and it is very difficult to picture just how quick this is.

However, you can see from this that there is a clear relationship between speed, time and distance.

1 Speed is distance divided by time

Speed is the distance an object travels divided by the time it takes to travel that distance. Speed is often measured in metres per second or kilometres per hour. You can rewrite the definition of speed as an **equation**:

$$\text{speed} = \frac{\text{distance}}{\text{time}}$$

So, if a car travels 60 kilometres in 1 hour, its speed would be equal to 60 km/h. From this equation, you can see that if object A travels a certain distance in less time than object B, then object A has a higher speed than object B.

What is speed?

2 A formula triangle allows you to calculate speed, distance and time

You can calculate speed from the formula: speed = distance ÷ time. Speed is usually measured in metres per second (m/s), distance is usually measured is metres (m) and time is usually measured in seconds (s). To calculate distance or time, you can use a formula triangle (Figure 2.6).

d
÷ ÷
t × s

Figure 2.6 A formula triangle for distance (*d*), speed (*s*) and time (*t*)

To use the formula triangle, cover up the quantity you wish to find. What you can then see is the formula you need.

For example, if you know speed and time and wish to find distance, cover up the *d* on the triangle. This leaves *t* × *s*. So, the formula is distance = time × speed. If you wish to find time, cover up the *t*, which leaves *d* ÷ *s*. So, the formula is time = distance ÷ speed.

How can you use a formula triangle?

3 Distance is how far you travel

Imagine that one person is jogging and another person is sprinting. Assuming they maintain their speed for the entire time, who do you think will travel further in 1 minute? The sprinter will travel further because they will have been moving at a greater speed. Remember, speed is just a measure of how far you travel in a set amount of time – if you have a greater speed, you will travel further in that time.

To practise calculating distance, imagine you are riding a bicycle at a speed of 4 metres per second. How far would you travel in 3 seconds? Check the formula triangle – you can see that the formula for distance is *d* = *t* × *s*. So, your **distance** travelled will be 3 s × 4 m/s = 12 m.

How is distance related to speed?

Figure 2.7 You can calculate the distance this cyclist travels if you know her speed and the time it takes her to travel that distance.

INVESTIGATION 2.2
Ticker timers

CHECKPOINT 2.2

1 Describe how distance and time are related to speed.
2 What are the three formulas for finding speed, time and distance?
3 Usain Bolt can run 100 metres in 9.58 seconds. Calculate his average speed.
4 How long would it take to walk from Melbourne to Brisbane (a distance of 1600 km) at an average speed of 5 km/h?
5 Sound can travel at 330 m/s. A spectator hears a starter's pistol 1.8 seconds after it was shot. How far away is the spectator from the pistol?
6 If you were riding on a roller coaster track, how would your speed change when you went:
   a up a hill?
   b down a hill?
7 Create a formula triangle for the equation: force = mass × acceleration, from section 2.1.

CHALLENGE

8 Use the internet to find out how long it takes light to travel from:
   a the Sun to Earth
   b our nearest star to Earth
   c the centre of the galaxy to Earth.

SKILLS CHECK

- I can explain the relationship between distance, speed and time.
- I can state the formulas to calculate distance, speed and time.

2.3 Graphing motion

At the end of this lesson I will be able to:

- **explain** the difference between speed and velocity
- **describe** the relationships between displacement, time, velocity and acceleration, using the equations of motion

KEY TERMS

displacement
the distance an object is from its starting position

distance
the total length an object travels

scalar
a quantity that has a magnitude but no direction

vector
a quantity that has both a magnitude and a direction

velocity
a measure of how quickly displacement changes

LITERACY LINK

Create flashcards for the key terms listed above by writing the term on one side of a small card and the definition on the other. Study them for five minutes and then test a friend using the cards.

NUMERACY LINK

State the formulas for both speed and velocity.

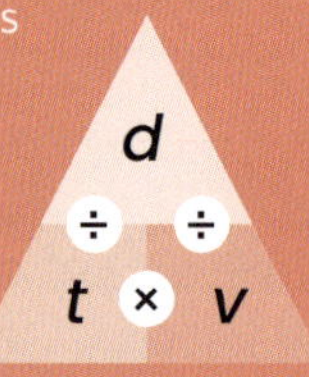

The world's fastest person is Jamaican sprinter Usain Bolt. Bolt is nearly 2 metres tall and hold world records in the 100-metres sprint, 200-metres sprint and 4 × 100-metres relay. You can use the distance of his races, as well as the time he ran them in, to calculate his speed. You can also figure out his velocity by using his displacement and time.

Figure 2.8 Usain Bolt can reach a speed of 44.72 km per hour.

1 Distance is different from displacement

When discussing motion, it's important to use the right terms. **Distance** is how far an object actually travels, whereas **displacement** is how far the object is from its initial position.

Consider this scenario. From a starting position, you walk 6 metres north, then 8 metres west. This means that you have travelled a total distance of 6 + 8 = 14 metres. However, you are only 10 metres from your starting position, so your displacement is 10 metres northwest. Figure 2.9 shows another example of distance and displacement.

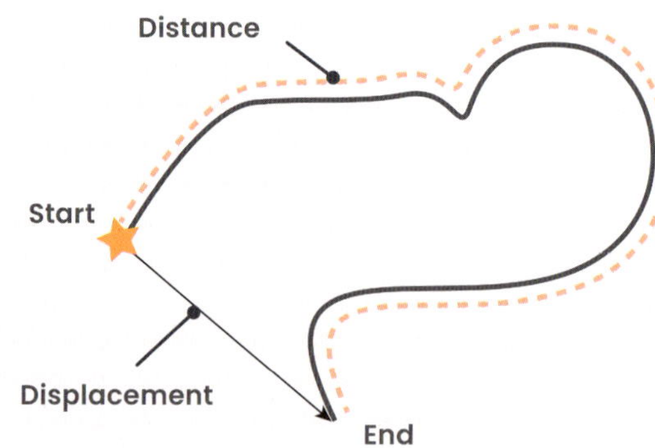

Figure 2.9 Distance and displacement can be very different.

Speed and velocity are both related to distance and displacement. Speed is a measure of how much distance changes in a certain time, while the **velocity** is a measure of how much displacement changes in a certain time.

In our example above, let's suppose you were walking for 10 seconds. Your average speed would be the distance travelled divided by the time taken, so 14/10 = 1.4 ms^{-1}. Your average velocity would be your displacement divided by the time taken, so 10/10 = 1.0 ms^{-1} northwest.

Displacement and velocity should always be given a direction as well as a magnitude (such as 6 ms^{-1} east). Quantities that have both a direction and a magnitude are called vector quantities. Other vector quantities include acceleration and force. Distance and speed do not require a direction, and are known as scalar quantities. Other scalar quantities include mass and energy.

What is the difference between displacement and distance?

2 Distance–time graphs represent motion

Consider a student walking to class. In the first 10 seconds, she walks 15 metres. She then realises she's running late and jogs for the next 10 seconds, travelling 30 metres. Finally, she sprints the last 25 metres to class in 5 seconds. A graph of the student's journey would look like Figure 2.10. This is a distance–time graph, and is a simple visual of the motion that occurred during the journey.

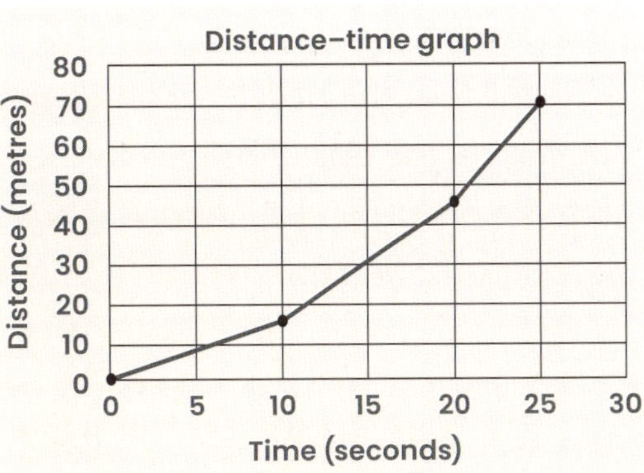

Figure 2.10 A distance–time graph of the student's journey

The average speed can be calculated by finding the gradient of the graph. For example, in the 0–10 second interval, the speed is 15/10 = 1.5 ms^{-1}. At each stage of this particular graph, the gradient increases. This tells us that the student is moving at a greater speed in each successive interval.

What does the gradient of a distance–time graph show?

3 Speed–time graphs provide more information

Consider the speed the student is travelling in our example. In the first interval, she's moving at 1.5 ms^{-1}; from 10 seconds to 20 seconds, she's travelling at 30/10 = 3.0 ms^{-1}; and from 20 seconds to 25 seconds, she's travelling at 25/5 = 5 ms^{-1}.

We can graph the speed she's travelling against the time, to create a speed–time graph.

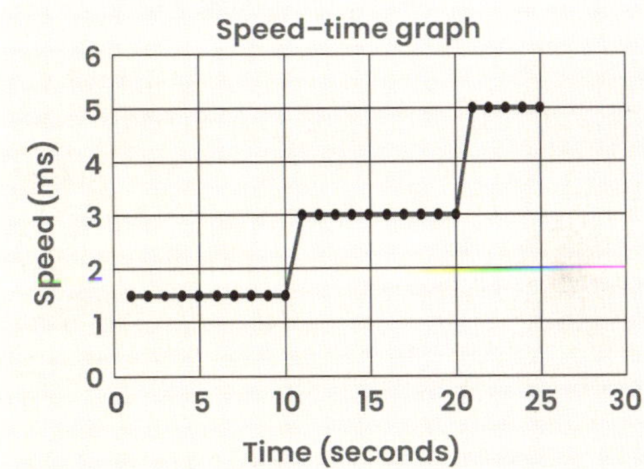

Figure 2.11 A speed–time graph of the student's journey

A speed–time graph gives us a little more information about the journey, as we can see the speed at each point. For example, between 0 and 10 seconds, the graph is flat, showing that the student was travelling at a constant speed of 1.5 ms^{-1}. Between 10 and 11 seconds, the graph increases, showing that the student is accelerating. In fact, the magnitude of the acceleration is given by the gradient of the line. In this case, the acceleration of the student is 1.5/1 = 1.5 ms^{-2}. Finally, the area of the underside of the graph will show the distance travelled. The area under the graph for the first interval in Figure 2.11 is 1.5 × 10 = 15 m, which is exactly how far the student travelled in Figure 2.10.

What does the gradient of a speed–time graph show?

INVESTIGATION 2.3
Vehicles and pedestrians

CHECKPOINT 2.3

1 Explain the difference between distance and displacement using an example.
2 What is velocity and how is it different to speed?
3 What does a decreasing gradient on a speed–time graph demonstrate?
4 a A person walks 8 metres north and then 2 metres south. Give both the distance and displacement for this scenario.
   b If it took this person 8 seconds to complete the walk, calculate both their speed and velocity.

CHALLENGE

5 Usain Bolt is the world's fastest man. Research his incredible 200-metre world record race in 2009. Calculate Bolt's average speed and velocity during this race, using a diagram of the track as well as the calculations for speed and velocity. Hint: You will need to find his displacement first.

SKILLS CHECK

- I can explain the difference between distance and displacement and provide an example diagram of each.
- I can explain both speed and velocity and give the formulas for both.
- I can interpret speed–time and distance–time graphs.

2.4 Net force and acceleration

At the end of this lesson I will be able to:

- **relate** acceleration to a change in speed and/or direction as a result of a net force.

KEY TERMS 

friction
a contact force that opposes motion, caused by objects rubbing against each other

net force
all the forces acting on an object added together

normal force
the force of the floor or ground pushing an object upwards

weight force
the force generated by gravity pulling an object downwards

LITERACY LINK 

List the different types of forces mentioned in this section. Write a definition of each force.

NUMERACY LINK 

The units for acceleration are m/s^2, which is the same as m/s/s or ms^{-2}.

If $a = \Delta v/t$, show how the units are expressed as m/s^2, m/s/s and ms^{-2}.

Think about the forces acting on you right now. The first one you will probably think of is gravity, the force that pulls you down to the ground.

Recall that the acceleration of an object is related to the force acting on the object as well as the mass of the object. So, if there is force acting on you, and you have a mass, why don't you accelerate? The answer is in a concept called 'net force'.

1 Net force is the sum of all forces acting on an object

The **net force** is all the forces acting on an object added together. Consider the forces acting on a cyclist. The driving force of their feet on the pedals is propelling them forward. Gravity is also acting on them, providing a **weight force**. **Friction** between the tyres and the road opposes their motion, as well as air resistance. Finally, there is the force from the road pushing the bicycle upwards, which is called the **normal force**.

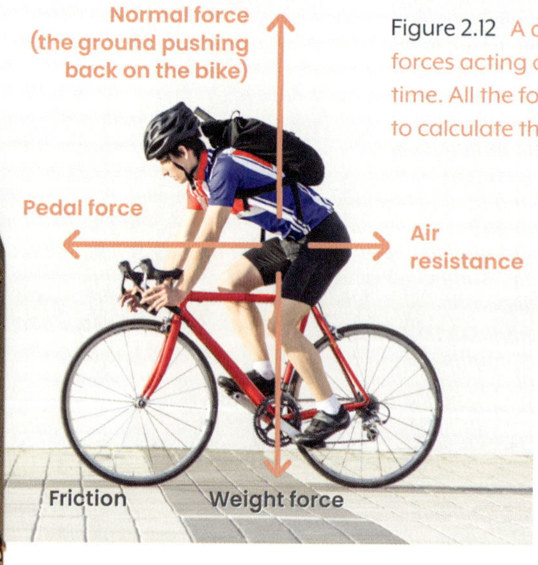

Normal force (the ground pushing back on the bike)
Pedal force
Air resistance
Friction
Weight force

Figure 2.12 A cyclist has many forces acting on them at the same time. All the forces can be added to calculate the net force.

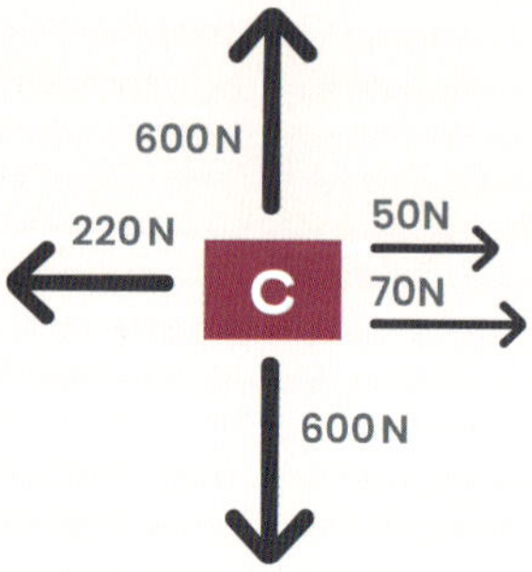

600 N
220 N
C
50N
70N
600 N

Figure 2.13 The large mass of the rocket means it requires a huge force to increase its speed.

To work out the net force acting on the cyclist, add together all the forces acting in the same direction and subtract forces acting in the opposite direction.

In the vertical direction, the weight force (600 newtons) is acting in the opposite direction to the normal force (600 N), so subtract these: 600 N – 600 N = 0 N. In the vertical direction, the sum of the forces is zero, so you can ignore them. Air resistance (50 N) and friction (70 N) act in the same direction, so add these together. The total force acting to the right is 50 N + 70 N = 120 N. The pedal force is to the left, which is the opposite direction to the friction and air resistance, so subtract them. Therefore, the net force is 220 N – 120 N = 100 N to the left.

What is a 'net force'?

2 Acceleration is any change in speed or direction over time

People often mistakenly think of acceleration as meaning 'speeding up'. However, acceleration is any change in speed or direction over time – speeding up, slowing down or turning.

If acceleration is in the same direction as an object's motion, the object will speed up. If acceleration is in the opposite direction to an object's motion, the object will slow down. If acceleration is not parallel to the object's motion, the object will change direction. Acceleration is usually measured in metres per second squared (m/s^2).

Does acceleration result in a faster speed?

3 Acceleration and net force always act in the same direction

When many forces are acting on an object, the acceleration depends on the net force and is always in the same direction as the net force.

If there is no net force, there will be no acceleration. If you are sitting in a chair reading this right now, you are not accelerating and so you have no net force acting on you. You have a weight force acting downwards and an equal-sized normal force acting upwards, but these two forces are in opposite directions so they cancel out.

What is the net force acting on a stationary object?

Figure 2.14 Net force and acceleration always act in the same direction.

INVESTIGATION 2.4
Balloon rockets

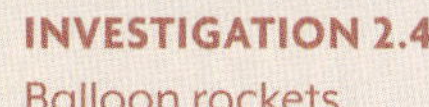

CHECKPOINT 2.4

1 Define *net force*.
2 An object is accelerating downwards. In which direction must the net force on it be acting? Give evidence to support your answer.
3 Draw a diagram and calculate the net force when an object experiences:
   a forces of 20 N to the left and 30 N to the right
   b a driving force of 15 N to the left and forces of friction of 10 N and air resistance of 5 N to the right
   c a driving force of 20 N to the left, a normal force of 8 N upwards, a weight force of 8 N downwards, and forces of friction of 10 N and air resistance of 5 N to the right.
4 A jet plane is travelling at a constant speed of 800 km/h.
   a Is the plane accelerating?
   b Given your answer to part a, what must the net force acting on the plane be?

CHALLENGE

5 Astronauts experience a very large acceleration when a rocket takes off. Use the internet to find out what can happen to humans experiencing large accelerations, and how astronauts are trained to withstand these.

SKILLS CHECK

- I can describe how an object will accelerate, given the direction of its net force.

2.5 Newton's laws

At the end of this lesson I will be able to:

- **analyse** qualitatively everyday situations involving motion in terms of Newton's laws.

KEY TERMS

action–reaction pair
two equal and opposite forces exerted by two objects on each other

balanced force
a force acting on an object that is cancelled out by another force, so the net force is zero

inertia
a property of matter that causes it to resist change in speed or direction (to remain at rest or in a state of uniform motion)

stationary
not moving, still

unbalanced force
a force acting on an object that is not cancelled out by another force, so a net force acts on the object

LITERACY LINK 

Design a poster that summarises Newton's three laws of motion.

NUMERACY LINK 

a Calculate the force required by the propeller of a small 520 kg plane if it accelerates at 7 m/s^2 during take off.

b Calculate the acceleration of a 0.2 kg golf ball if it is hit with a force of 25 N.

Figure 2.15 Newton's third law explains that while Earth pulls a skydiver down, the skydiver also pulls Earth up!

Isaac Newton was an English scientist who lived in the 1600s. He is so well regarded that the unit of force is named after him. Among his many achievements are his three laws of motion, which describe how objects move in relation to the forces applied to them.

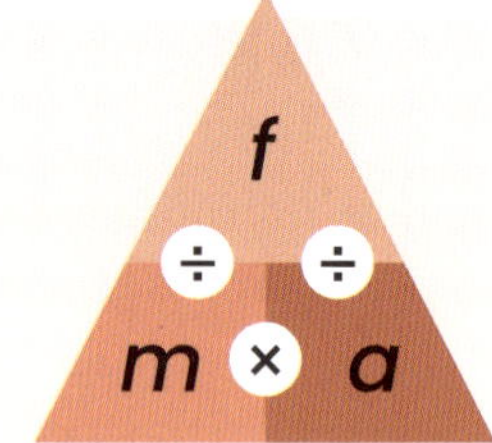

Figure 2.16 A formula triangle for force (f), mass (m) and acceleration (a)

Understanding Newton's laws helps you predict what will happen to objects experiencing forces in any situation.

1 Newton's first law: an object stays at rest or at the same speed unless an unbalanced force acts on it

Newton's first law states that an object will continue doing what it is doing unless acted on by an **unbalanced force** – a force that causes a change in motion.

This means that an object that is **stationary** (not moving) will stay stationary, and a moving object will keep moving unless a force acts on it. You rarely see this happening – if you roll a ball across the floor, it doesn't keep rolling forever. Rather, it slows down and eventually stops. This is because there is an unbalanced force acting on the ball – friction. The ball is being slowed down by a combination of friction from the ground and air resistance. If you rolled the ball where there is no friction (such as in space), it would not stop until it was acted upon by another force. The property of an object to keep doing what it is doing is called **inertia**.

A car travelling at a constant speed also experiences friction. In this case, the friction from the road is a **balanced force** – it is exactly equal to the thrust from the engine – so the car does not slow down. The backwards force on the car is exactly the same size as the forwards force, and so the forces are balanced.

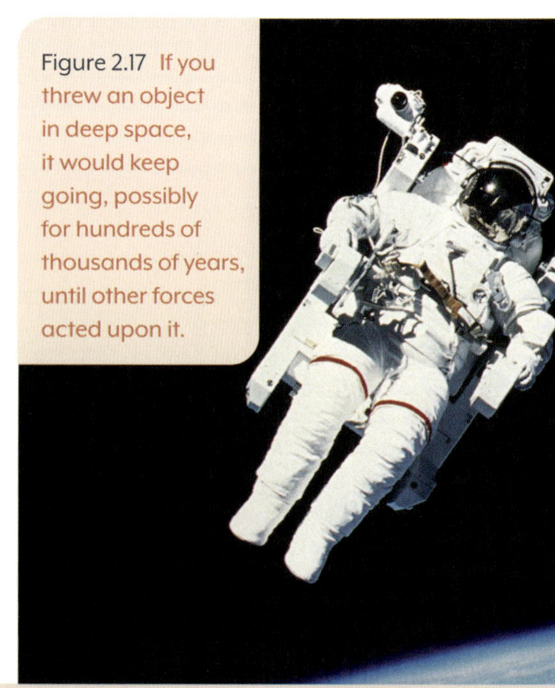

Figure 2.17 If you threw an object in deep space, it would keep going, possibly for hundreds of thousands of years, until other forces acted upon it.

What is Newton's first law?

2 Newton's second law: force = mass × acceleration

Newton's second law states that the size of the acceleration of an object is proportional to the force applied and inversely proportional to the mass of the object. In other words, the heavier an object is, the more force is required to accelerate it.

Figure 2.18 The heavier an object is, the more force is required to accelerate it.

Imagine you are standing on a skateboard and you use your foot to propel yourself. The force of your foot against the ground causes your acceleration. If someone else is standing on the skateboard with you, the total mass increases, so if you apply the same force with your foot, you will not accelerate as much. If you wanted to accelerate at the same rate as before, you would need to apply much more force.

What is Newton's second law?

3 Newton's third law: for every action, there is an equal and opposite reaction

Newton's third law is possibly the most famous. It says that for every action, there is an equal and opposite reaction. This means that if object A exerts a force on object B, object B exerts the same-sized force on object A, but in the opposite direction.

When you jump, you push the ground down with your legs, but the ground also pushes you up – that's why you can jump. This is an example of Newton's third law – the action (you pushing the ground down) has an equal and opposite reaction (the ground pushing you up).

Every force has an equal and opposite reaction force. The tyres on a car push the road backwards, and are in turn pushed forwards by the road. You are pushing down on your seat right now, and your seat is pushing you up, preventing you from falling onto the ground. When you jump, Earth's gravity pulls you back down, but you also pull Earth up very slightly. The equal and opposite forces in these examples are known as **action–reaction pairs**.

What is Newton's third law?

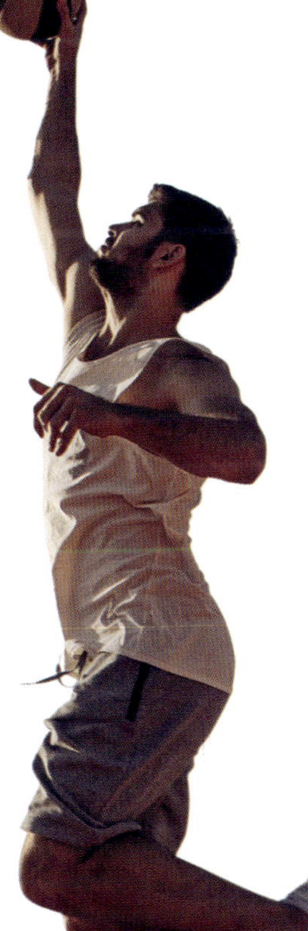

Figure 2.19 As you push down on Earth, Earth pushes up on you.

INVESTIGATION 2.5
Car safety

CHECKPOINT 2.5

1 Identify whether these definitions apply to Newton's first, second or third law.
   a The acceleration of an object depends upon the force applied and the mass of the object.
   b For every action, there is an equal and opposite reaction.
   c An object keeps doing what it's doing unless acted on by an unbalanced force.
2 Newton's first law states that an object keeps doing what it's doing unless acted on by an unbalanced force. So, why must you continue to pedal a bicycle if you want to remain at the same speed?
3 What is the equal and opposite force acting when a:
   a cricket bat pushes a ball forwards?
   b person pulls a suitcase?
4 If you push against a wall with 45 N of force, and neither you nor the wall moves, how much force is the wall pushing you with?

CHALLENGE

5 Isaac Newton was famous for more than his laws of motion. Use the internet to research more of Newton's scientific achievements.

SKILLS CHECK

- I can state each of Newton's three laws.
- I can use Newton's three laws to explain the motion of objects.

CHAPTER SUMMARY

Sir Isaac Newton ▶

Newton's first law:
An object will continue in the same state of motion unless an unbalanced force acts upon it.
- If an object is not moving, it will stay stationary.
- If an object is moving at a constant speed, it will continue to move at that same speed.

A stationary object will stay stationary as long as no unbalanced forces act on it.

A moving object will keep moving as long as no unbalanced forces act on it.

Newton's second law:
The acceleration of an object increases with the force applied, and decreases with the mass of the object.

Force = mass × acceleration

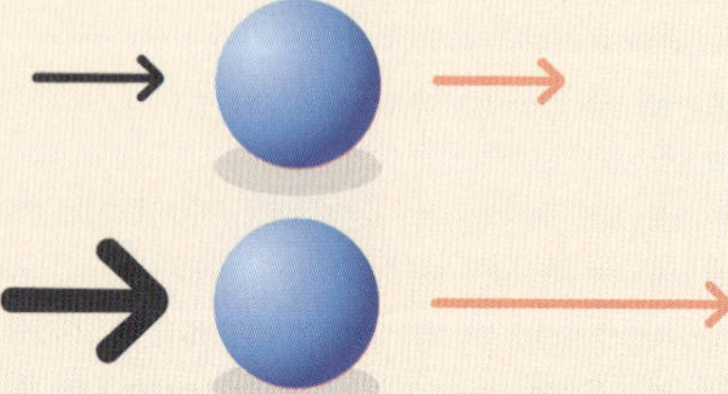

More force = more acceleration

Newton's third law:
Every action has an equal and opposite reaction.

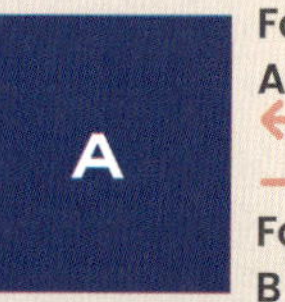

Force on A by B

Force on B by A

✓ **Distance** is how far an object actually travels, while **displacement** is how far the object is from its initial position.

Speed is how much distance changes in a certain time, while **velocity** is how much displacement changes in a certain time.

For an object to accelerate, there must be an unbalanced force. The object will always accelerate in the direction of the unbalanced force.

Car speeds up

Direction of motion

Force

Car slows down

Direction of motion

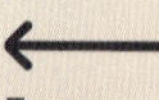

Force

Car turns

Force

Direction of motion

The speed of an object is equal to the distance travelled divided by the time taken.

Speed = distance ÷ time

To calculate distance, speed or time, you can use a formula triangle.

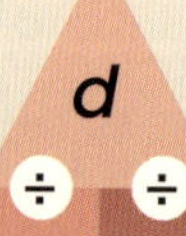

★ FINAL CHALLENGE ★

1. Copy and complete these sentences:
   a. An object will continue doing what it's doing unless it is acted on by an unbalanced ______________.
   b. The acceleration of an object increases with a larger ______________ and decreases with a larger ______________.
   c. For every ______________, there is an equal and opposite ______________.

LEVEL 1 ★☆☆ ☆☆☆ 50XP — LEVEL UP!

2. Suggest how you can increase the acceleration of an object.
3. What is the formula to calculate each of these? (Hint: Use your formula triangle.)
   a. speed
   b. time
   c. distance

4. What is the net force acting on objects A, B and C?
5. For each object, A, B and C, specify the direction of acceleration.

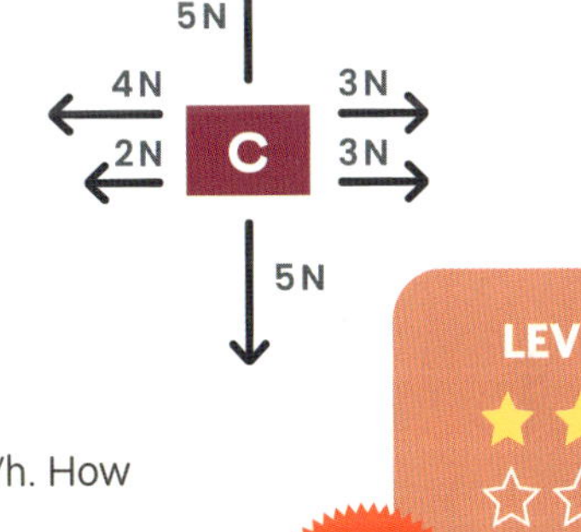

6. a. A train completes a trip of 240 km in 2.5 hours. What was the average speed of the train?
   b. The train driver thinks that he can increase the speed of the train to 140 km/h. How long would the journey take if the train travelled at this speed?

LEVEL 3 ★★★ ☆☆☆ 150XP — LEVEL UP!

7. Match each of Newton's laws with the scenario that demonstrates it.

Newton's first law	A fully loaded car takes longer to get up to top speed.
Newton's second law	As your feet push down on a trampoline, you jump higher.
Newton's third law	If you quickly pull a sheet of paper out from under a coin, the coin will remain where it is.

8. Using your understanding of Newton's first law, explain why a rocket requires much more fuel to lift off than it does to change direction once it is in space.
9. Suggest a way in which each of Newton's three laws has affected something you have done today.

3 Electricity

Electricity is a form of energy that can flow from one place to another. When you think about electricity you might automatically think of wires, light bulbs and how your television is powered, but electricity existed before all those things. Electricity, the flow of electrons, occurs in nature – in lightning, in the synapses in our bodies and even in the shock of an electrostatic 'zap'. In the 19th century, there were huge leaps forward in our understanding and use of electricity. Famous scientists such as Edison, Einstein and Tesla all joined the race to power the world. Power plants and wires spread out across Earth at an incredible rate, powering lights, and heating and cooling homes. The time before household electricity soon seemed like a distant memory, and the breakthroughs and innovations kept coming.

1 LEARNING LINKS

Brainstorm as much as you can remember about these topics.

Electrical circuits consist of different components.

Electrical symbols are used to represent the different components of electrical circuits.

The power to operate electrical devices comes from various energy sources.

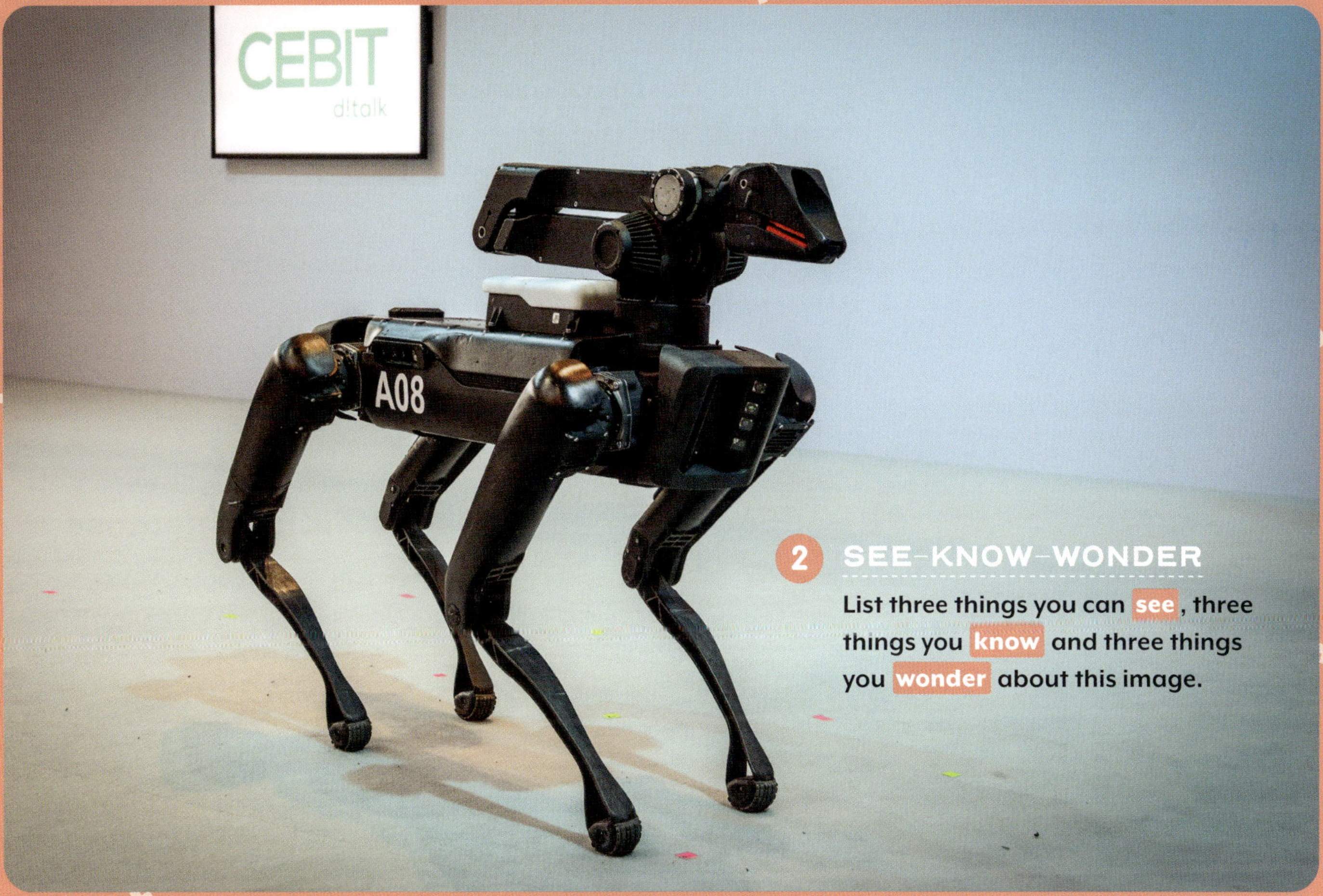

2 SEE–KNOW–WONDER

List three things you can **see**, three things you **know** and three things you **wonder** about this image.

3 CRITICAL + CREATIVE THINKING

Commonalities: How many points of commonality between an electrical circuit and a race track can you think of?

Alternatives: List ways in which you could power a light bulb without plugging it directly into a power point.

The reverse: List things that cannot be powered by electricity.

4 THE LUCKIEST UNLUCKY GUY!

Roy Sullivan was a park ranger in the United States when he was first struck by lightning in 1942. Roy went on to be struck six more times, miraculously surviving every time. He currently holds the Guinness World Record for being struck by lightning the most times. The probability of being struck by lightning more than once is *extremely* low, however, due to the nature and location of Roy's job he was more exposed to storms than the average person.

3.1 Energy in circuits

At the end of this lesson I will be able to:

- **describe** voltage, current and resistance in a circuit in terms of energy.

KEY TERMS 

circuit
a closed path containing a collection of components connected to a power source that allows charge to flow

conductor
a material that allows the movement of charge

current
a measure of how fast charge moves in a circuit

insulator
a material that resists the movement of charge

voltage
the difference in potential energy between two points in a circuit

LITERACY LINK 

Summarise the information from this section into a small postcard addressed to your teacher.

NUMERACY LINK 

Plot a graph of the following data:

I (amps)	0.12	0.29	0.51	0.67
V (volts)	0.65	1.41	2.55	3.28

Draw a line of best fit and calculate the resistance of the circuit by determining the gradient of the line.

Electricity is created by the flow of electrons and is a form of energy that can be transported from one place to another. For example, for your television to work, electrical energy must be transported from the power plant where it is generated, into your home, and finally into your television.

There are three important quantities involved in the transport of electricity. These are current (how quickly electric charges move through a circuit), voltage (how much energy each charge contains) and resistance (how much energy is used to get through a certain point in the circuit).

1 Voltage is a measure of potential difference

An electrical **circuit** is a closed path that current flows through. **Voltage**, which is also called potential difference, is the difference in potential energy between two points in the circuit. Voltage has the symbol *V* and the unit of voltage is the volt (symbol V).

The bigger the potential difference between two points in an electrical circuit, the bigger the voltage. This is like the difference in height between two points on a rollercoaster (Figure 3.1). Higher voltages have more energy and can do more work, just as higher points on a rollercoaster have more energy or potential than lower points. The difference between these points on the circuit is the voltage or potential difference. The battery causes an increase in voltage because it is an energy source. Each of the components in the circuit causes a drop in voltage when they use energy in the circuit.

If a higher voltage battery is used in a circuit, light bulbs glow more brightly. However, there is a limit to how much voltage a bulb can cope with. A bulb will fail or 'blow' if there is too much voltage in the circuit.

What is voltage?

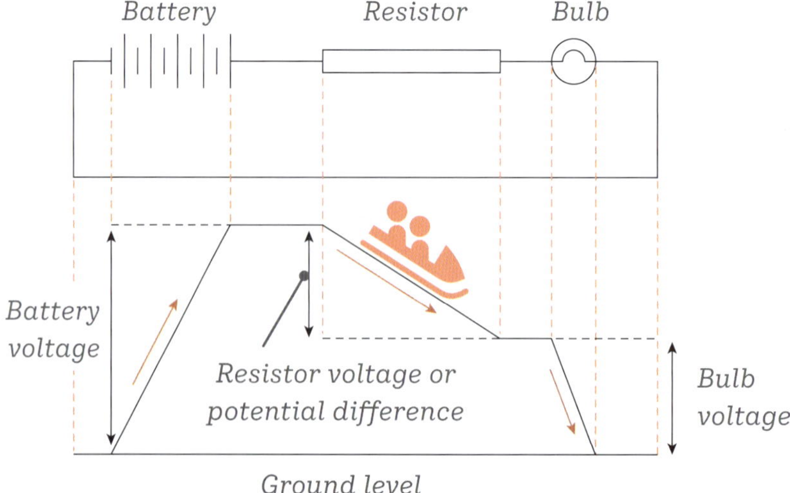

Figure 3.1 Voltage or potential difference is like the difference in height between two points on a rollercoaster.

❷ Current is the rate of movement of charge

Current is the rate of movement of charge in a circuit. Current has the symbol *I* and the unit of current is the ampere (or amp) (symbol A).

In a circuit, charge can move in either direction, depending on which way the power source 'pushes' the charge. If you changed the direction of the battery in a simple circuit, you would change the direction of current.

You can visualise current as a pipe filled with small balls. When a ball is pushed into the end of the pipe, a ball at the other end is pushed out. So, a small movement of one ball can cause an action a long distance away. In this model, the balls are the charges, you are the battery pushing the charges and the current is how fast you push the balls.

Some circuit components are sensitive to current direction while others are not. Complicated or delicate components can be damaged by charge flowing in the wrong direction.

What unit is current measured in?

Figure 3.2 Current can be visualised as a pipe filled with small balls.

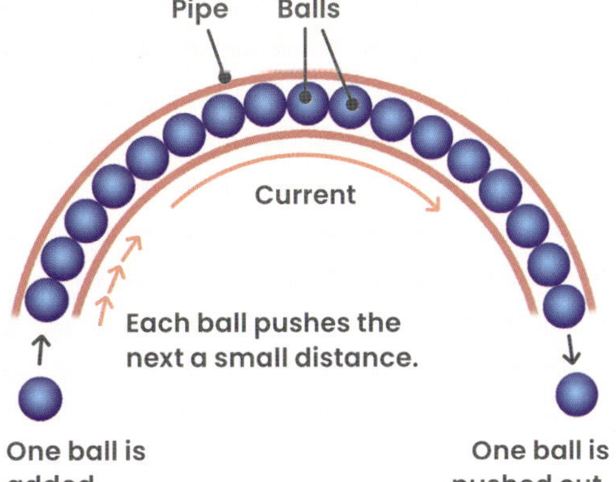

❸ Resistance opposes voltage

Circuits are made up of wires and components that conduct electrical energy. Some materials, such as copper, allow charge to flow more easily and are known as **conductors**. Other materials, such as rubber, resist the flow of charge and are called **insulators**. Resistance has the symbol *R*. The unit of resistance is the ohm (symbol Ω).

If voltage is the amount of 'push' given to a charge, then resistance is the opposing force. A component with resistance consumes electrical energy and converts it into other forms of energy, such as heat.

While each component in a circuit has resistance that 'slows' the flow of charge in a circuit, special components called resistors let you introduce exact amounts of resistance into circuits.

What is a resistor?

Figure 3.3 The coloured bands on these resistors indicate the value of the resistance in ohms.

INVESTIGATION 3.1
Modelling a simple circuit

CHECKPOINT 3.1

1 Outline the difference between voltage, current and resistance.
2 Identify these statements as true or false.
  a Voltage is supplied by a resistor.
  b Too much voltage will cause a bulb to fail.
  c Current always flows clockwise.
  d Current gets used up as it goes around a circuit.
  e Resistance is measured in ohms.
  f Conductors have zero resistance.
3 What are the units for current, voltage and resistance?
4 Explain what conductors and insulators are and provide an example of each.

CHALLENGE

5 The units that current, voltage and resistance are measured in are all named after famous scientists. Research one of these scientists to explain their contribution to modern electricity.

SKILLS CHECK

- I can explain what is meant by voltage, current and resistance.
- I can describe how voltage, current and resistance affect energy in a circuit.

3.2 Ohm's law

At the end of this lesson I will be able to:

- **describe** the relationship between voltage, resistance and current.

KEY TERMS

ammeter
a device that measures electric current

Ohm's law
a law that states that the current through a conductor between two points is directly proportional to the voltage across the two points

voltmeter
a device that measures potential difference (voltage)

LITERACY LINK 

Create a mind map using the key terms and at least two additional terms of your choice.

NUMERACY LINK

What is the reading on the voltmeter above?

The German physicist Georg Ohm (1789–1854) discovered the relationship between voltage, current and resistance in an electric circuit. This relationship became known as Ohm's law.

Ohm's law states that the current flowing between two points in a circuit is directly proportional to the voltage difference between the two points. Ohm's law forms the basis for understanding electrical circuits, and helped pave the way for the electrical devices we enjoy using today.

1 Electrical current is proportional to voltage

Ohm discovered that the relationship between current (I), voltage (V) and resistance (R) can be summarised in the formula:

$$\text{current } (I) \text{ in amperes} = \frac{\text{voltage } (V) \text{ in volts}}{\text{resistance } (R) \text{ in ohms}}$$

This means that if you have a fixed resistor (one whose resistance doesn't change), then increasing the current through the resistor will increase the voltage across the resistor. Likewise, reducing the voltage across the resistor will reduce the current flowing through it.

If you know any two of the values for voltage, current or resistance, then you can use **Ohm's law** to calculate the third one. A useful way to remember Ohm's law is to use an Ohm's law triangle (Figure 3.4). To use this triangle, place a finger over the value being calculated and the remaining two values will complete the calculation.

Conductors that obey this law are known as ohmic resistors. Some conductors, such as light-emitting diodes, don't obey Ohm's law; they are referred to as non-ohmic resistors.

What is Ohm's law used to calculate?

To find voltage:

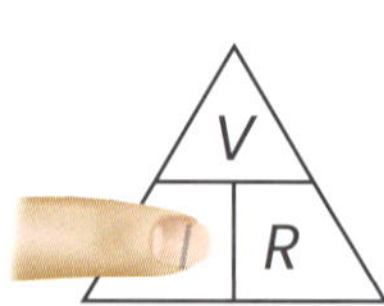

$V = I \times R$

To find current:

$I = \frac{V}{R}$

To find resistance:

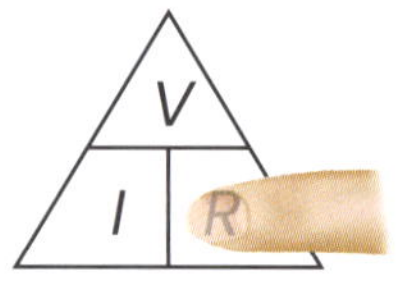

$R = \frac{V}{I}$

Figure 3.4 An Ohm's law triangle allows you to calculate voltage, current or resistance. Cover the value you need to know and use the other two values to complete the calculation.

2 Voltmeters measure potential difference

Ohm's law is more than just an equation. The tools we use to check and repair electrical circuits are based on the relationship between current and resistance.

A **voltmeter** is a device that measures the potential difference or voltage across two points in a circuit. Before you can measure voltage, the circuit must be completed. Then, add the voltmeter so that the two probes are at the two points that the voltage will be measured across.

What does a voltmeter measure?

(a)

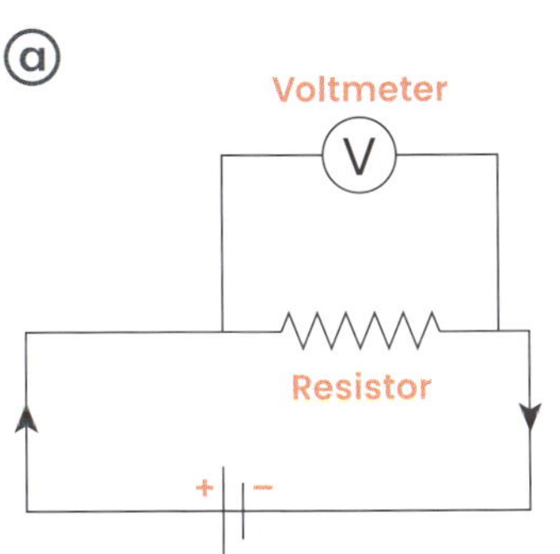

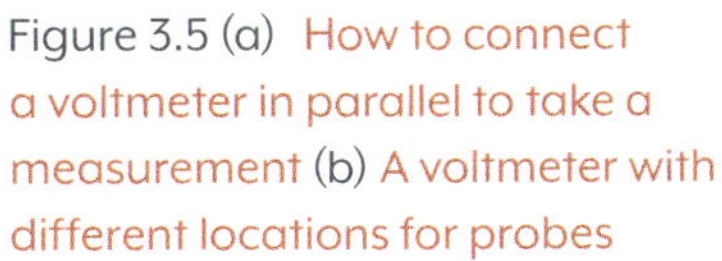

(b)

Figure 3.5 (a) How to connect a voltmeter in parallel to take a measurement (b) A voltmeter with different locations for probes

3 Ammeters measure current

An **ammeter** is a device that measures current. To correctly measure the current at a certain point in a circuit, the ammeter needs to be connected as a part of the circuit. That is, the current must have no other path but to go through the ammeter.

What does an ammeter measure?

(a)

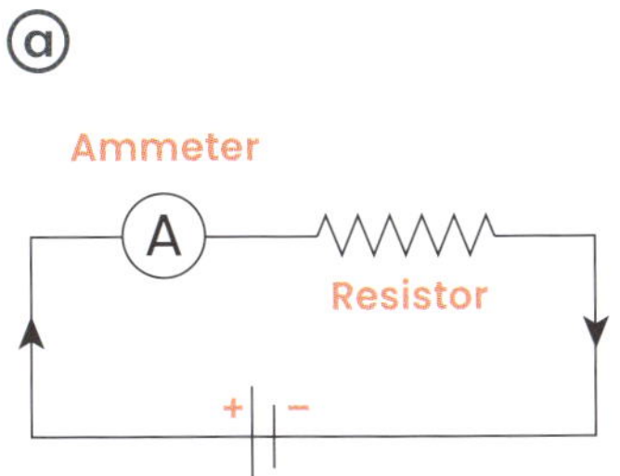

(b)

Figure 3.6 (a) How to connect an ammeter in series to take a measurement (b) An ammeter with different locations for probes

INVESTIGATION 3.2
Exploring Ohm's law

CHECKPOINT 3.2

1 Explain Ohm's law.
2 Use the Ohm's law triangle to give the equations for measuring:
   a current
   b voltage
   c resistance.
3 Outline the difference between an ammeter and a voltmeter.
4 Explain what is meant by a non-ohmic resistor.
5 Given a voltage of 110 V and a current of 7 A, calculate the resistance.
6 If the current is 8 A and the resistance is 2 Ω, calculate the voltage.
7 If a battery in a circuit is 24 V and the resistance is 14 Ω, calculate the current.

CHALLENGE

8 Carry out research to make a list of substances that have low resistance. Suggest why these substances are not used in circuits in homes.

SKILLS CHECK

- I can describe the relationship between voltage, resistance and current.
- I can calculate voltage, current and resistance using Ohm's law.

3.3 Series and parallel circuits

At the end of this lesson I will be able to:

- **compare** the features of series and parallel electrical circuits.

KEY TERMS 

parallel circuit
a circuit in which all components are connected between the same points, so the current has more than one path to take

reciprocal (of a number)
1 divided by the number

series circuit
a circuit in which components are arranged in a chain, so the current has only one path to take

LITERACY LINK 

Create a cheat sheet to summarise what you have learnt in sections 3.1–3.3. Look back at previous pages and the investigations to find key points to include. Include formulas, memory techniques and anything that will help you to remember and apply the concepts.

NUMERACY LINK 

Three resistors (3 Ω, 4 Ω and 6 Ω) are connected in parallel and then connected to a 4 V battery.

a What is the resistance of the parallel combination?

b What is the current through each resistor?

Have you ever plugged in a string of Christmas lights, only to find they won't turn on because a single bulb is out? You have to test every single globe in the string to find the faulty one. This is because the lights are part of a series circuit, and a single bulb will stop the flow of electricity through the whole circuit.

A parallel circuit allows components to work independently of each other. A string of Christmas lights that are connected in parallel would still work if there is a broken bulb.

1 In series circuits, components are connected one after another

When components in a circuit are arranged one after another, this is called a **series circuit**. If you trace your finger over the circuit or diagram, you will be able to pass through every component back to the start without having to retrace your path. There is only one path for the charge to flow through.

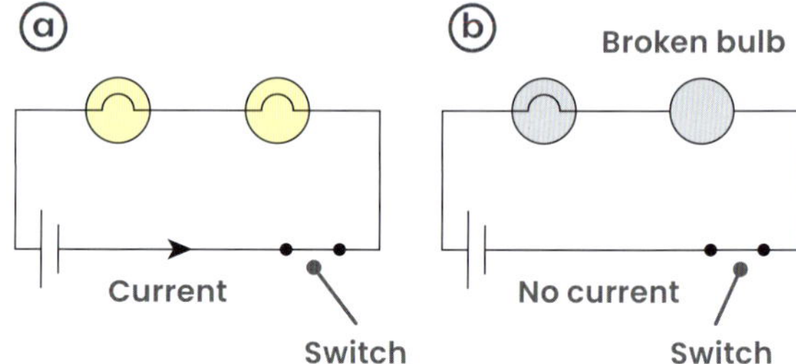

Figure 3.7 (a) In a complete series circuit, the bulbs light up. (b) A broken bulb causes an incomplete circuit and no bulbs light up.

The circuit needs to be complete for charge to flow. If one of the components breaks or is removed, then the circuit is incomplete and no charge can flow (Figure 3.7b).

What is a series circuit?

2 In parallel circuits, components are on different branches

In a **parallel circuit**, each component is connected on a different branch of the circuit. To check if components are connected in parallel, trace around the circuit. If you come to a point where you must choose between two or more branches, then this circuit is parallel.

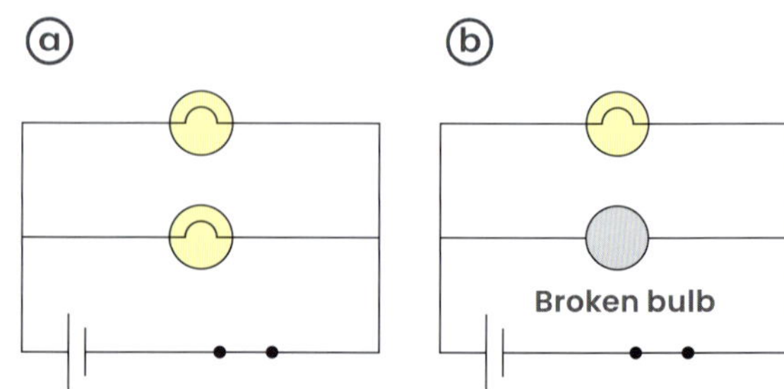

Figure 3.8(a) In the complete parallel circuit, the bulbs light up. (b) A broken bulb does not cause an incomplete circuit and the other bulbs remain lit.

The branches in a parallel circuit mean that if a component breaks, the circuit can still be complete through the connections in the other branches. A bulb in this part of the circuit stays on even when a bulb in the other branch breaks.

House wiring is an example of parallel circuits. Parallel circuits require more wiring, but if one light bulb breaks or is removed, then the rest of the house lights can still work even if they are controlled by the same switch.

INVESTIGATION 3.3
Series and parallel circuits

Table 3.1 Summary of voltage, current and resistance in series and parallel circuits

	Series circuit	Parallel circuit
Voltage	6 V 6 V Bulb 1 Bulb 2 12 V 6 V + 6 V = 12 V Bulb 1 + bulb 2 = total The voltages across components add up to the total voltage supplied by the battery.	9 V Bulb 1 9 V Bulb 2 9 V = 9 V = 9 V Bulb 1 = bulb 2 = total The voltage across each component is the same.
Current	0.3 A 0.3 A Bulb 1 Bulb 2 0.3 A = 0.3 A = 0.3 A Bulb 1 = bulb 2 = total The current is the same everywhere in the circuit.	3 A Bulb 1 3 A Bulb 2 3 A + 3 A = 6 A Bulb 1 + bulb 2 = total The current in each component adds up to the total current in the main branch containing the battery.
Resistance	10 Ω 5 Ω Resistor 1 Resistor 2 15 Ω 10 Ω + 5 Ω = 15 Ω Resistor 1 + resistor 2 = total The resistances of components add up to the total resistance of the circuit. The total resistance in the circuit is *more* than the resistance of each component.	5 Ω Resistor 2 10 Ω Resistor 1 $\frac{1}{10\ \Omega} + \frac{1}{5\ \Omega} = \frac{1}{\text{TOTAL}}$ $0.1\ \Omega + 0.2\ \Omega = \frac{1}{\text{TOTAL}}$ $0.3\ \Omega = \frac{1}{\text{TOTAL}}$ TOTAL = 3.33 Ω The **reciprocal** of the total resistance is found by adding the reciprocal of the resistance of each component. The total resistance of the circuit is less than the resistance of each component.

What is a parallel circuit?

CHECKPOINT 3.3

1 Explain the difference between series and parallel circuits.

2 Draw a simple labelled diagram of a series and parallel circuit.

3 Describe the placement of components in series and parallel circuits.

4 Give an example of a product that may contain a:
a series circuit
b parallel circuit.

5 Calculate the total resistance in a circuit in which two 10 Ω resistors are connected in:
a series
b parallel.

CHALLENGE

6 Design a circuit that contains two light bulbs that can operate independently. If possible, set up your circuit and test it in class.

SKILLS CHECK

- I can compare the features of series and parallel circuits.
- I can calculate voltage, current and resistance in series and parallel circuits.
- I can give examples of how series and parallel circuits are used.

3.4 Modern electronics

At the end of this lesson I will be able to:

- **outline** recent examples where different branches of science, engineering and technology have produced scientific developments.

KEY TERMS 

amplifier
an electronic component that boosts electrical current

transistor
an electronic component that can act as a switch or an amplifier

LITERACY LINK 

Imagine the technology used 50 years from now. Describe what daily life might be like. What devices and technology might be used?

NUMERACY LINK 

A carbon atom has a diameter of about 0.33 nanometres (0.33×10^{-9} m). Calculate the number of graphene layers that exist in 1 mm of graphite.

One of the earliest ways of storing electric charge was a Leyden jar. This was a glass jar coated inside and out in foil, with a metal bar touching the inside surface. This allowed scientists to store electric charge to create sparks.

Our understanding of electricity has come a very long way since then, and today it is hard to imagine life without electricity. How did we come to this point, and what is likely to be the future of electric energy?

1 Advances in digital circuits opened up many possibilities in electronics

Transistor

The transistor was invented in 1947 at Bell Labs in the USA. A **transistor** is a small electronic device that can be used as a switch (can be turned on or off by current) or an **amplifier** (can boost current). The transistor is the basis of all modern digital circuitry and opened up many new possibilities in electronics.

Figure 3.9 The transistor has three terminals and is made of semiconducting materials such as silicon and germanium.

Integrated circuit

The integrated circuit is also known as the microchip. It is a small chip that can contain hundreds or even millions of tiny transistors. There are many different types of microchip that can do a variety of jobs. The invention of the microchip is responsible for the digital explosion in the late half of the 20th century, giving people access to more compact personal electronic devices for the first time.

Figure 3.10 The microchip has many terminals that can be connected in a circuit to perform many different tasks.

Figure 3.11 Almost every modern electronic device contains a printed circuit board.

Printed circuit board

The printed circuit board is a mass-producible base of fibreglass or reinforced plastic with thin strips of conductor to connect components such as microchips. The invention of the printed circuit board contributed to advances in robotics, automation and industry.

What is another name for an integrated circuit?

❷ Graphene is used to make small 2D circuits

In recent years, scientists have begun developing incredibly flat circuits known as 2D circuits. These are so flat that you could stack 400 000 of them on top of each other, and the stack would still be thinner than a sheet of paper.

2D circuits are usually made of graphene, which is a sheet of carbon that is only a single atom thick. Graphene has incredible properties in terms of strength and flexibility, and is an excellent conductor of electricity.

Graphene may have started the 2D revolution in electronics, but that's just the start. Silicene, phosphorene and stanene (atom-thick forms of silicon, phosphorus and tin, respectively) have a similar honeycomb structure with different properties, allowing for different applications. All four materials could change electronics to allow for miniaturisation, higher performance and lower costs. Several companies, including Samsung and Apple, are developing applications based on graphene.

What is the benefit of a 2D circuit?

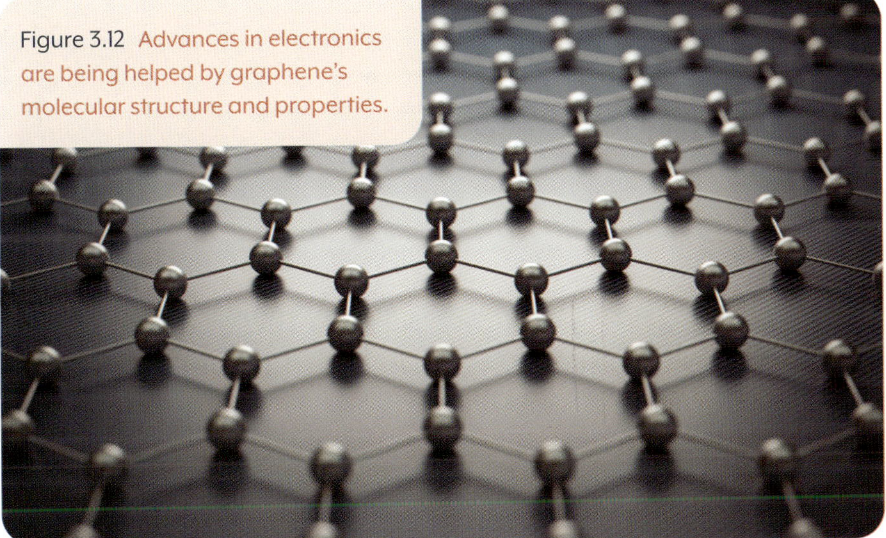

Figure 3.12 Advances in electronics are being helped by graphene's molecular structure and properties.

❸ Circuits could be built out of single molecules

Conventional electrical circuits are made from bulk materials. The goal of molecular electronics is miniaturisation. Molecular electronics uses single molecules or collections of single molecules as electronic building blocks.

In single-molecule electronics, the bulk material is replaced by single molecules. The smaller electronic components consume less power and make the device more sensitive (and sometimes perform better). Some molecular systems tend to self-assemble into functional blocks – the components of a system spontaneously come together to form a larger functional unit.

Molecular electronics is still in the early research phase, and no devices have been commercialised.

What is molecular electronics?

CHECKPOINT 3.4 

1 Describe the function of a transistor.
2 What properties of graphene make it so useful for emerging technology?
3 Explain why microchips were a huge technological breakthrough.
4 Describe some of the advantages of molecular electronics.

CHALLENGE

5 ENIAC was the first general-use computer and was made before microchips were invented. Use the internet to find out how big ENIAC was and how much it weighed, and compare this to a modern-day computer.

SKILLS CHECK

- I can outline at least one recent scientific development in the field of electricity.

CHAPTER SUMMARY

Electricity is a form of energy that can flow from one place to another.

Qualities involved in transporting electricity

- Current
- Voltage
- Resistance

Ohm's law states that current is proportional to voltage.

$$\text{current } (I) \text{ in amperes} = \frac{\text{voltage } (V) \text{ in volts}}{\text{resistance } (R) \text{ in ohms}}$$

To find current:

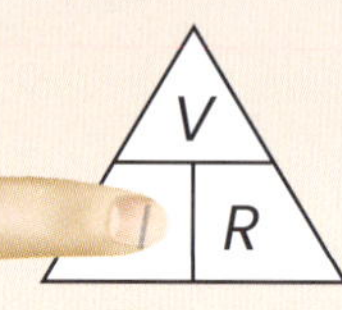

$I = \frac{V}{R}$

To find voltage:

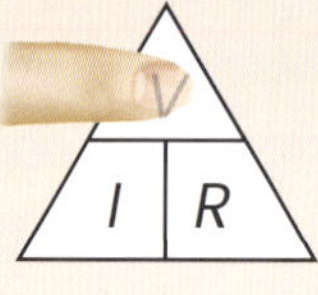

$V = I \times R$

To find resistance:

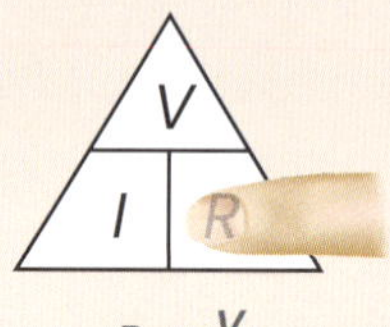

$R = \frac{V}{I}$

Ammeters measure current.

Circuits have different forms.

Current
Switch
Broken bulb
No current
Switch
Broken bulb

Series: Components are arranged one after the other.

Parallel: Each component is connected on a separate branch of the circuit.

Advances in digital circuits

The **transistor**, the basis of modern digital circuitry.

The **integrated circuit** gave people access to compact digital devices.

The **printed circuit board** contributed to advances in robotics, automation and industry.

★ FINAL CHALLENGE ★

1. Match these terms to their definitions.

conductor	the difference in potential energy between two points in a circuit
current	an electronic component that can act as a switch or an amplifier
amplifier	a material that resists the movement of charge
transistor	a measure of how fast charge moves in a circuit
insulator	a material that allows the movement of charge
voltage	an electronic component that boosts electrical current

LEVEL 1

50XP

LEVEL UP!

2. Give the symbols for current, voltage and resistance.
3. Explain what Ohm's law is in your own words.
4. Explain the difference between an amplifier and a transistor.

LEVEL 2

100XP

LEVEL UP!

5. Give an example of a new development or technology in relation to the use of electricity.
6. If you adjust the current but keep the resistance the same, will this affect the voltage? Give evidence for your answer.
7. Describe what an ammeter is, how it is used and what it measures.
8. Explain the difference between current and voltage in your own words.

LEVEL 3

150XP

LEVEL UP!

9. Calculate the:
   a. resistance of the resistor in a circuit that has a voltage of 6 V and a current of 0.7 A
   b. current in an electrical component that has a resistance of 12 Ω and a voltage of 16 V
   c. voltage in an electrical component that has a resistance of 12 Ω and a current of 24 A.

10. Draw and label a diagram of a series circuit and a parallel circuit. Annotate the diagrams to highlight how they are different (in red pen) and how they are similar (in blue pen).

4 Energy usage

When the universe formed, so too did all the energy there is and ever will be. No new energy is ever created or destroyed, it is just transferred or transformed, cycling around the universe, forever.

Most of the energy to power your house comes from burning fossil fuels such as coal. But mining and burning fossil fuels has serious impacts on the environment, such as climate change, habitat loss and air pollution.

1 LEARNING LINKS

Brainstorm as much as you can remember about these topics.

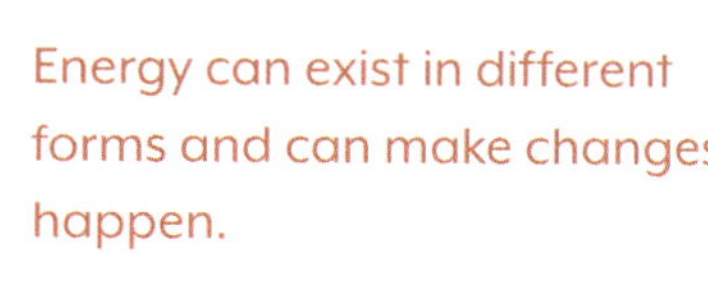

Energy can exist in different forms and can make changes happen.

Science can help us solve current local and global problems.

Advances in science may require ethical considerations.

2 SEE–KNOW–WONDER

List three things you can **see**, three things you **know** and three things you **wonder** about this image.

3 CRITICAL + CREATIVE THINKING

Mismatch: How could you stop a speeding arrow with a sheet of paper, a dog collar and a cowboy boot?

The question: Write five questions that have the answer 'energy'.

The ridiculous! Attempt to support this statement: 'We should burn forests instead of coal to generate electricity.'

4 THE BEST INTENTIONS!

In the 1920s, Thomas Midgley discovered that adding lead to petrol improved engine performance, which led to the popularity of leaded petrol. However, we now know that leaded petrol is harmful to the environment and people, so leaded petrol is banned. Midgley also helped develop CFCs, chemicals used in refrigerators, which caused a giant hole in the ozone layer. So that's two of Midgley's inventions that had devastating worldwide impacts!

4.1 Conservation of energy

At the end of this lesson I will be able to:

- **apply** the law of conservation of energy to account for the total energy involved in energy transfers and transformations.

KEY TERMS 

energy
a measure of the ability to do work

energy transfer
the movement of energy from one place to another without changing form

energy transformation
a change from one type of energy to another

LITERACY LINK 

Describe five types of energy mentioned in this section in ways that a primary school student could understand.

NUMERACY LINK 

Sketch a line graph of energy versus time, showing how the kinetic energy and gravitational potential energy of a ball change over time as it bounces on a hard surface.

Many of the devices you use every day require energy. However, this energy may need to be transferred from one place to another and transformed from one type of energy into another.

For example, a toaster will heat up to toast bread. For this to happen, the electrical energy needs to be transferred from the power station that produces it to the toaster, where it then needs to be transformed into heat energy. These types of transfers and transformations of energy are crucial for a countless number of systems to be able to operate effectively.

1 Energy cannot be created or destroyed

The conservation of energy is a fundamental law of physics. The law states that energy can never be created or destroyed; it can only change into different forms. This means that all the energy we use now existed in some form billions of years ago and has cycled around continuously. Although energy cannot be created or destroyed, it can exist in different forms. Table 4.1 lists some types of energy.

Table 4.1 Types of energy

Type of energy	Description
Chemical	Energy stored in chemical bonds
Elastic potential	Energy stored in an object that has been stretched or compressed
Electrical	Energy that travels through electrical circuits
Gravitational potential	Energy stored in an object that has been raised above the ground
Heat	Energy that can raise the temperature of an object
Kinetic	Energy that any moving object has
Light	Energy that you can see, emitted from glowing objects
Nuclear	Energy stored in the nucleus of atoms
Sound	Energy that you can hear, caused by vibrating air particles

Figure 4.1 Elastic potential energy is stored in a drawn bowstring.

What is the law of conservation of energy?

❷ Energy can be transferred from one place to another

Energy can be transferred from one location to another. A rolling marble will have kinetic energy, and if it strikes a stationary marble, both marbles will move away from the collision. There has been an **energy transfer** from the first marble to the second marble.

Another example is electrical energy, which is transferred from a power station, through power lines to your house where it comes out of a power point and into appliances. Heat energy can also be transferred from a hot stove into your finger, but that would be pretty painful!

How is electrical energy transferred from place to place?

❸ Energy can change from one form to another

Energy can transform, or change, from one form to another. This can be done using technology, such as a toaster turning electricity into heat, but it also happens naturally. When you throw a ball up in the air, the ball starts off moving quickly, slows down as it gets higher, and then accelerates back to the ground. The ball starts off with a lot of kinetic (movement) energy. As it gains height, the ball loses speed. An **energy transformation** takes place as kinetic energy is converted to gravitational potential energy.

Figure 4.2 A toaster converts electrical energy into heat energy.

When thrown straight up into the air, at its peak, the ball briefly stops – it has no kinetic energy, but its gravitational potential energy is at a maximum. As the ball comes back down, the gravitational potential energy is transformed to kinetic energy, making the ball travel faster and faster as it loses height.

As a ball is thrown upwards, what type of energy is being converted?

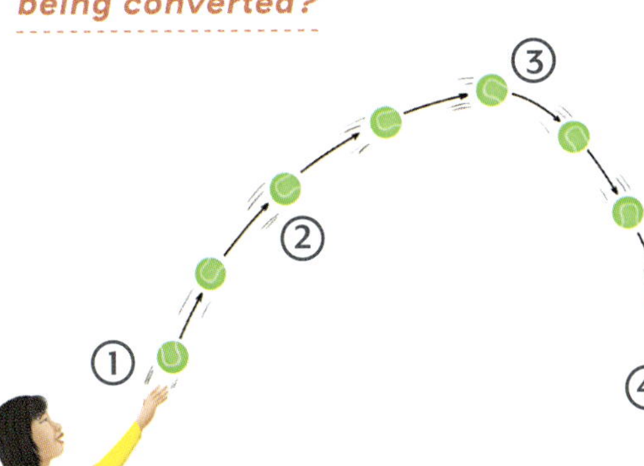

1 The ball starts with a lot of kinetic energy and moves fast.
2 It slows down and its kinetic energy is converted to gravitational potential energy.
3 The ball stops for an instant, it has no kinetic energy.
4 It speeds up and its potential energy is converted into kinetic energy.
5 The kinetic energy is transferred to the ground and is converted to sound and heat energy.

Figure 4.3 When a ball is thrown upwards, its energy changes.

INVESTIGATION 4.1
Galileo's pendulum

CHECKPOINT 4.1

1 Describe the law of conservation of energy.
2 Explain the difference between energy transfer and energy transformation.
3 What type of energy transformation is happening when a:
  a power station burns coal to provide household power?
  b pot of water is placed on a gas stove to boil?
  c slingshot is stretched back and then released, firing a marble?
4 What kind of device could transform:
  a electrical energy into light and sound energy?
  b chemical energy into heat energy?
  c light energy into chemical energy?

CHALLENGE

5 Nuclear power plants generate electricity from the nuclear energy stored within atoms. Use the internet to find out how this is done and write a summary paragraph about it.

SKILLS CHECK

- I can explain the law of conservation of energy.
- I can describe what is meant by energy transfers and transformations.

4.2 Wasted energy

At the end of this lesson I will be able to:

- **describe** how, in energy transfers and transformations, a variety of processes can occur so that usable energy is reduced and the system is not 100% efficient.

KEY TERMS 

efficient
not wasteful

energy efficiency
how much usable energy is produced compared to how much energy has been supplied

friction
a contact force that opposes motion, caused by objects rubbing against each other

joule
the unit of energy

usable energy
the type of energy that a device is designed to produce

LITERACY LINK 

Create a print advertisement encouraging people in your neighbourhood to buy energy-efficient smartphones.

NUMERACY LINK 

A TV that is 86% efficient produces 5300 J of useful energy.

a How much energy does the TV consume?

b Convert your answer to kJ.

Figure 4.4 When a light bulb produces light, it also produces some heat.

Energy can exist in many different forms, and many mechanical devices simply convert one type of energy to another. A device is usually designed to produce a particular type of energy, known as usable energy.

However, devices inevitably produce other types of energy that is not used. This energy is wasted. For example, a light bulb is designed to produce light energy, but it also produces some heat.

1 Energy is lost in energy transfers and transformations

Consider a tennis ball falling from a height. The ball starts with gravitational potential energy. As the ball falls, the gravitational potential energy is converted to kinetic energy. When the ball strikes the ground, it compresses and stores elastic potential energy. The elastic potential energy is then transformed into kinetic energy as the ball bounces back up.

The ball never bounces back to its original height even though energy can't be created or destroyed. This is because the ball is losing energy in a few places.

- As the ball is falling, it experiences air resistance, a form of **friction**, with the air around it. This causes the ball to heat up very slightly and lose some energy as heat.
- As the ball strikes the ground, it makes a sound. Some of its kinetic energy is converted to sound energy.
- As the ball strikes the ground, a small amount of heat energy is transferred to the ground.

These factors all contribute to a loss of energy, so the ball finishes with less energy than it started with.

What types of energy are lost as a ball bounces?

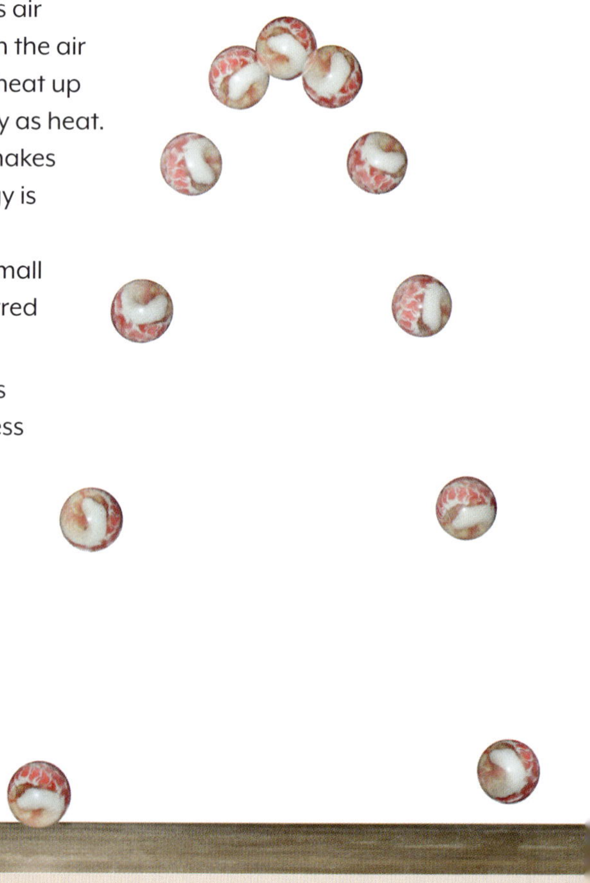

Figure 4.5 Rubber balls bounce higher than tennis balls because they lose less energy as heat and sound.

2 Most devices lose energy

Devices are designed to produce **usable energy**. Almost all devices lose energy. After riding your bicycle, your tyres will have warmed up. This is because of the friction between the tyres and the road. This 'lost' energy has been converted from usable kinetic energy into wasted heat energy.

Moving objects often lose some energy as heat, because of friction. A ball won't keep rolling forever because its kinetic energy is slowly converted to heat as a result of friction with the floor and the air. Friction can be helpful in some situations. For example, brakes in vehicles work by using friction to convert kinetic energy into heat – otherwise we'd never be able to stop!

Energy in devices can also be lost in other ways. A car loses energy through engine noise and heat. Most electrical devices in your home lose energy as heat and sound (even if the sound is just a quiet buzzing).

What causes bicycle tyres to heat up during a ride?

Figure 4.6 A spacecraft re-entering the atmosphere experiences friction with the air.

3 All devices are less than 100% energy efficient

Devices are often described as energy **efficient**. This means that they don't waste very much energy. The **energy efficiency** of a device is expressed as a percentage, calculated by dividing the amount of usable energy produced by the amount of energy put into the device. The unit of energy is the **joule** (J).

$$\text{Energy efficiency (\%)} = \frac{\text{energy output}}{\text{energy input}} \times 100$$

When a smartphone is fully charged, the energy in the battery should be converted to light energy (to show things on the screen) and sound energy (for phone calls or listening to music). If the phone heats up a lot while using it, some of the stored energy is lost as heat energy, leaving less energy to power the phone, resulting in the battery going flat earlier.

If a smartphone were 40% efficient, then for every 100 J of energy in the battery, only 40 J would be turned into light and sound, and 60 J would be lost as heat. If a smartphone were 80% efficient, it would only waste 20 J as heat and so the battery would last much longer.

However, all devices have an efficiency of less than 100%.

How is energy efficiency calculated?

INVESTIGATION 4.2
Energy efficiency of bouncing balls

CHECKPOINT 4.2

1 What does it mean if a device is 'energy efficient'?
2 Wasted energy is usually what type of energy?
3 Friction is often the reason that kinetic energy turns into heat energy. Give an example where this is a problem, and an example where this is useful.
4 What types of usable energy do smartphones produce?
5 Explain, in terms of energy, why a cannonball fired from a cannon slows down as it travels.
6 For every 100 J of electrical energy put into a kettle, 30 J is lost as unwanted sound and heat energy. What is the percentage efficiency of the kettle?

CHALLENGE

7 Research which devices in your home are the most and least energy efficient. Can you find any ways to help reduce the energy loss of the least efficient device?

SKILLS CHECK

- I can describe how energy is lost in energy transfers and transformations.
- I can explain what is meant by energy efficiency.

4.3 The importance of energy efficiency

At the end of this lesson I will be able to:

- **discuss**, using examples, how society's values and needs influence scientific research into increasing the efficiency of electricity use by individuals and society.

KEY TERMS 

finite
limited, won't last forever

fossil fuel
a fuel that is formed from the decomposition of dead animals and plants over millions of years, e.g. oil and coal

generate
produce or make something

LITERACY LINK 

Write a letter to the editor of a newspaper asking for the banning of all non-LED light bulbs. Explain your reasoning.

NUMERACY LINK 

Construct a column graph showing the relative difference in energy consumption for the different light bulbs mentioned in the text.

Imagine you lived in a city where everyone was constantly getting sick. You would probably want the scientists in your city to investigate why people were getting sick and how to cure the sickness.

Scientists are often employed to solve problems that a particular society is facing. In recent times, people have begun to realise that fossil fuels won't last forever, and that using them produces pollution and is contributing to climate change. People have looked to scientists to determine how to make fossil fuels last as long as possible, and to research potential replacements.

1 Most of our energy is generated by burning fossil fuels

Most of Australia's power and electricity is **generated** (produced) by burning fossil fuels. **Fossil fuels** include coal and oil, which take millions of years to form. In particular, Australia's electricity is mostly generated by burning coal.

Unfortunately, these resources are **finite** and will run out one day. Also, burning fossil fuels emits harmful greenhouse gases into the atmosphere. This means that eventually we will need to find other ways to generate electricity. In the meantime, it is vitally important to minimise the amount of energy we waste. For example, imagine you have the choice between two light bulbs that both produce 8 J of light energy every second. One of them is 80% efficient and the other is only 50% efficient. The first one uses 10 J of energy, but the second needs 16 J of energy for the same light output. Which will you choose?

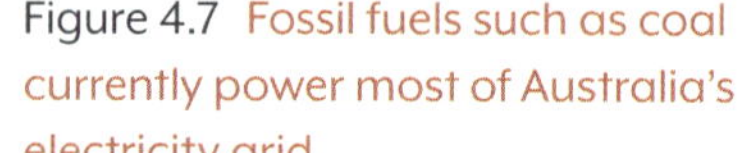

Where does most of Australia's electricity come from?

Figure 4.7 Fossil fuels such as coal currently power most of Australia's electricity grid.

❷ Energy-efficient devices are becoming more popular

As energy prices increase and we understand more about the problems arising from using fossil fuels, more people are interested in buying energy-efficient devices. Energy efficiency is becoming a popular feature of appliances for a number of reasons, including lower running costs and the benefit to the environment.

In addition, many governments around the world are interested in energy efficiency, with some enacting laws requiring minimum standards. This has meant that a lot of scientific research has been directed at improving energy efficiency of appliances and systems.

What are some benefits of energy-efficient devices?

❸ Light bulbs are an example of increasing energy efficiency

One very obvious example of society driving a change to more energy-efficient products has been with light bulbs. Old incandescent light bulbs are very inefficient, wasting 90% of their energy as heat. Halogen lamps are about 10–20% more efficient than incandescent bulbs, so governments around the world started phasing out the use of the less efficient bulbs in the mid-2000s. Compact fluorescent bulbs and LED lights use even less energy, and the Australian government began phasing out halogen bulbs in 2020.

By phasing out these older bulb types, people will pay less to light their homes and will use less energy, reducing the amount of fossil fuel we need to burn.

What is the least efficient type of light bulb?

Figure 4.8 Light bulbs have changed over time.

CHECKPOINT 4.3 

1 Describe some disadvantages of using fossil fuels to generate electricity.

2 Suggest why energy-efficient devices are better for the environment.

3 Give two reasons why society is moving towards energy-efficient devices.

4 Suppose a company that makes toasters refused to develop more energy-efficient models. What do you think would happen to this company?

5 **a** List these bulb types in order from most to least efficient: compact fluroescent, incandescent, LED, halogen.
b Which of the bulb types in part **a** would you expect to get the hottest as it runs? Explain your answer.

6 Why do you think the government started phasing out inefficient light bulbs?

CHALLENGE

7 Australia uses a star rating system to classify the energy efficiency of appliances. Find out what these stars mean and try to find three appliances with star ratings in your house.

SKILLS CHECK

- I can explain why energy-efficient devices are important.
- I can describe how the needs of society can drive scientific research.

4.4 Non-renewable energy sources

At the end of this lesson I will be able to:

- **discuss** viewpoints and choices that need to be considered when deciding about the use of non-renewable energy resources.

KEY TERMS 

abundant
present in large amounts

economy
the system of how a country makes and spends money and provides goods and services

ecosystem
a community of living things and their environment

pollution
a substance that enters the environment and has harmful or poisonous effects

LITERACY LINK 

Write an essay explaining whether Australia should continue to use fossil fuels to generate power.

NUMERACY LINK 

Carbon dioxide concentrations in the atmosphere have increased from approximately 280 ppm (parts per million) in pre-industrial times to 413 ppm in 2019. Calculate the % increase.

Most of the energy used in Australia comes from fossil fuels. Cars use petrol and oil, we heat our homes and cook with natural gas, and generate electricity by burning coal. Fossil fuels are formed when dead organisms decompose over millions of years.

Eventually fossil fuels will run out, and we will need to find other sources of energy. However, in the meantime, there are still many factors to consider when discussing the ongoing use of these fossil fuels.

Figure 4.9 Although there are benefits to burning fossil fuels, pollution is a major problem.

1 Fossil fuels contribute to Australia's economy

Historically, the fossil fuel industry has been important for Australia's **economy** – the country's wealth. Fossil fuels have been **abundant** (plentiful) in Australia, which has made them relatively cheap. Approximately 85% of all of Australia's electricity supply currently comes from coal and gas power. Australia's gas and coal industries employ more than 200 000 people, and selling coal and gas overseas has contributed to Australia's economic wealth.

Australia could eventually stop relying on fossil fuels, but this would require much more start-up investment in renewable sources such as solar, wind and water. People who work in the coal and gas industries would also need to find other jobs, although a switch to renewable energy could create a lot of new employment opportunities.

Why does Australia use fossil fuels as an energy source?

2 Burning fossil fuels damages ecosystems

One of the problems with the continued use of fossil fuels is **pollution**. When petrol, coal or gas is burned, dangerous chemicals are released into the environment. These can cause huge environmental problems, such as acid rain. Acid rain can devastate **ecosystems**, causing massive deaths of plants and animals. Acid rain is caused by pollutants such as the oxides of sulfur and nitrogen emitted by burning fossil fuels. The pollutants enter the atmosphere where they dissolve in water to produce acid rain.

Pollutants also get into the air and cause lung diseases and other medical conditions when breathed in. Burning coal also releases heavy metals such as lead and mercury, which enter water sources and contaminate fish and marine life.

How is acid rain formed?

3 Burning fossil fuels increases climate change

Carbon dioxide is a 'greenhouse gas'. Greenhouse gases in the atmosphere trap a proportion of the heat energy reflected from Earth. Trapping some heat is important for maintaining Earth's temperature so that it can support life. But trapping extra heat because of increased greenhouse gases is known as global warming, which is one symptom of climate change.

If Earth continues to heat up, there will be significant long-term damage, such as:

- sea levels will rise as polar ice melts, causing coastal cities to flood
- animals that live in frozen regions, such as penguins and polar bears, will lose their habitats
- the living things we rely on for food, like crops and fish, will not survive
- more extreme weather events such as heat waves, hurricanes, droughts and floods.

Other effects include more coral bleaching and an increase in the incidence of certain diseases.

How does carbon dioxide affect climate change?

Figure 4.10 Pollution from burning fossil fuels can cause massive damage to ecosystems.

INVESTIGATION 4.4
Effects of acid rain

CHECKPOINT 4.4

1 Describe what is meant by the term non-renewable resource.
2 What are some advantages and disadvantages of using 'non renewable' energy sources?
3 How are non-renewable resources formed?
4 What are the main benefits of using fossil fuels?
5 a What is acid rain?
  b What are the environmental effects of acid rain?
6 a What is global warming?
  b What are the negative effects of climate change?
7 Why do you think Australia hasn't changed to using only renewable energy sources? Explain your answer.

CHALLENGE

8 Iceland, Costa Rica and Uruguay get nearly 100% of their energy from renewable sources. Undertake some research to identify the sources of energy in these countries.

SKILLS CHECK

- I can describe some reasons for and against the use of non-renewable energy sources.

CHAPTER SUMMARY

Energy exists in many forms, such as kinetic, gravitational potential, elastic potential, heat, sound and light.

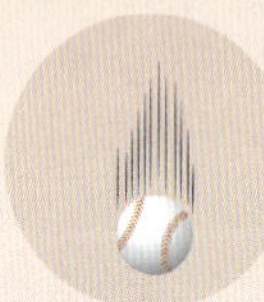
Gravitational potential energy

Electrical energy

Kinetic energy

Nuclear energy

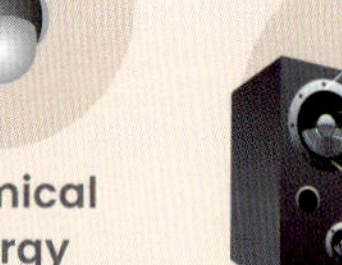
Chemical energy

Sound energy

Heat energy

Light energy

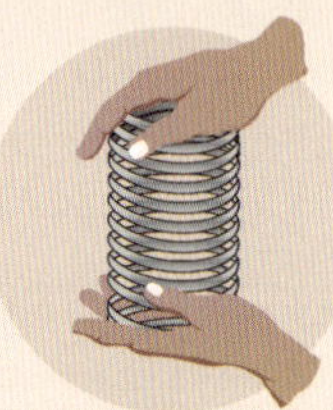
Elastic potential energy

The law of conservation of energy states that energy cannot be created or destroyed; it can only change form.

Energy can change form. Plants can convert heat and light energy into chemical energy, while a stereo converts electrical energy to sound energy.

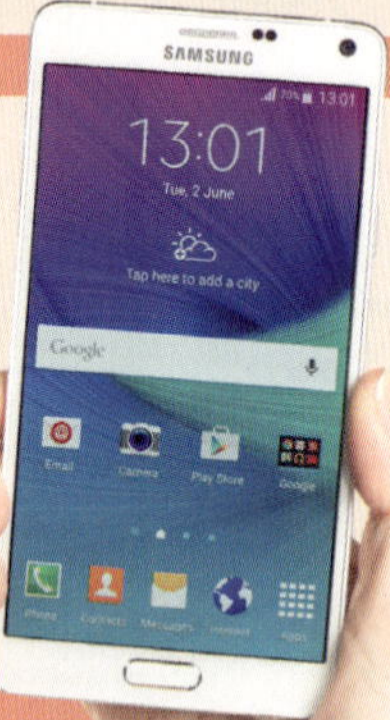

All devices are less than 100% energy efficient.

Friction produces heat and is the most common cause of energy inefficiency:

$$\text{Energy efficiency (\%)} = \frac{\text{energy output}}{\text{energy input}} \times 100$$

Most of Australia's electricity currently comes from burning non-renewable fossil fuels.

Burning coal has negative environmental effects, such as pollution and global warming. We only have a limited amount of coal left.

The benefits of burning coal are mainly economic.

★ FINAL CHALLENGE ★

1. State the law of conservation of energy.
2. Name five different types of energy, and give examples of where they can be found.

LEVEL 1
50XP
LEVEL UP!

3. Old-fashioned incandescent light bulbs are inefficient because they mostly give out what type of energy?
4. What is the source of most of Australia's electricity?
5. Into what types of energy do the following devices convert electrical energy?
   a stereo
   b television
   c kettle

LEVEL 2
100XP

6. Give three examples of situations where one type of energy is converted into another.
7. Explain why car tyres heat up after driving.
8. Describe why energy-efficient devices are better for the environment.

LEVEL 3
150XP
LEVEL UP!

9. If energy is always conserved, explain why a ball never returns to its full height after bouncing.
10. What is meant by 'energy efficiency'?
11. If light bulb A has an efficiency of 25%, and light bulb B has an efficiency of 40%, which one would be more expensive to run? Explain why.

LEVEL 4
200XP

12. Should we continue to use fossil fuels to generate electricity? Explain your point of view.
13. Imagine you are tasked with battling climate change in Australia. Create a plan that lists five recommendations in order of importance.
14. Create a flow chart that shows the loss of energy from transfers and transformations involved in running a vacuum cleaner.

LEVEL 5
300XP

5 The universe

Humans have gazed up at the night sky for millennia. Our ancestors observed that the sky changed with the seasons, used the stars to help them navigate, and linked their observations to stories of creation and mythology. We still look to the stars to help understand our place in the universe, to understand how it began and evolved, and how the Sun and solar system formed. As technology improves, scientists have been able to unlock and understand more of the mysterious universe.

1 LEARNING LINKS

Brainstorm as much as you can remember about these topics.

Earth, the Sun and the Moon interact to cause seasons and eclipses.

Technology is used to explore the solar system.

The force of gravity attracts two objects together.

2 SEE–KNOW–WONDER

List three things you can **see**, three things you **know** and three things you **wonder** about this image.

3 CRITICAL + CREATIVE THINKING

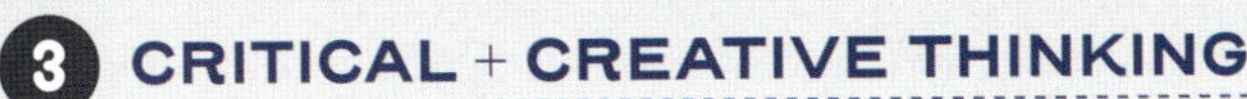

What if ... gravity had a repulsive effect instead of an attractive effect?

Variations: List as many ways as possible that we could learn about the universe from Earth.

The brick wall: Think of ways to deal with the following situation: How to continue life on Earth without the Sun.

4 BLACK HOLES!

You can imagine a black hole to be like a really heavy metal weight in the middle of a piece of fabric – the fabric is time and space and the metal weight is unimaginably dense matter left over from something like the death of a massive star. A black hole affects time and space, so things that are near a black hole start to slow down, lengthen and break apart. We are safe from black holes because there are none near Earth and they don't move around the universe gobbling things up like Pac-Man.

5.1 Stars

At the end of this lesson I will be able to:

- **outline** some of the major features of stars
- **use** appropriate scales to describe differences in the sizes of stars
- **identify** that all objects exert a force of gravity on all other objects in the universe.

KEY TERMS 

nebula
a vast region of gas and dust

neutron star
an extremely dense star left over after a supernova

red giant
a star that has stopped fusing hydrogen in its core

supernova
an explosion of a massive star at the end of its life

white dwarf
a small, very dense star formed at the end of a small star's lifetime

LITERACY LINK 

Write an interview with a white dwarf star about what they have experienced in their life.

NUMERACY LINK 

Convert the following numbers to scientific notation:

a 4200
b 0.0000065

A star is a colossal mass of gas. Inside a star, elements are made through the process of nuclear fusion. Small stars live for billions of years before expanding into **red giants**, then lose most of their mass to form a planetary nebula. Massive stars have shorter lives before expanding into red supergiants and ending as supernovas, which leave behind a neutron star or a black hole.

1 Stars are mostly hydrogen

Hydrogen (H) is the most common element in the universe, and stars are mostly made of hydrogen. In the centre of stars, hydrogen atoms fuse together to form helium atoms. This reaction releases a lot of energy, including visible light.

This release of energy pushing outwards from the core counteracts the force of gravity, and so a star fusing hydrogen in its core is balanced. When these forces are not balanced, the star changes and moves through different stages of its life cycle.

What causes stars to make so much energy?

2 Nebulae are stellar nurseries

Nebulae are interstellar regions filled with gas and dust, and are where stars are born.

A star begins to form when a region in the nebula starts to contract, and gravity brings the gas together. As more gas comes together, the mass starts spinning, forming a hot dense core called a protostar. The protostar continues to contract until the gas in the centre becomes dense enough and hot enough. This begins the fusion of hydrogen into helium, resulting in a star being born.

What is a nebula made of?

Figure 5.1 The Orion nebula is located near the tip of the sword in the constellation of Orion.

3 The size of a star determines its life cycle

The size of a star determines how long it will live for, as well as the stages in its life cycle, because the size of the star affects how quickly all the hydrogen in its core fuses.

A star such as the Sun (1 solar mass) has a lifetime of about 10 billion years. Much larger stars (more than 7 solar masses) have much shorter life spans because they fuse hydrogen quickly. Smaller stars of less than 1 solar mass live a lot longer than the Sun because they fuse hydrogen more slowly.

A large star will have more gravity than a smaller star like the Sun. Anything in the universe that has mass has gravity – the more mass, the more gravity.

How long will the Sun live for?

4 Stars eventually die

Once the amount of hydrogen within a star declines, the star enters its final phase and dies. This can happen in different ways depending on the star's mass.

A small star such as the Sun releases one last burst of energy, causing the outer layers of the star to puff outwards and form a planetary nebula. The star's core remains as an extremely hot, dense and small **white dwarf**.

In a large star, the atoms inside the core eventually fuse to become iron. More energy is required to produce iron than is given off, so gravity causes the star to rapidly contract and then explode. This is known as a **supernova**, and leaves the core of the star as an extremely dense **neutron star**.

In supermassive stars, the force of gravity is so strong that the core collapses. This forms a black hole, a body with such a strong gravitational field that it attracts all nearby matter, and even bends light around it.

How does a large star end its life?

Figure 5.2 The black hole in the centre of galaxy M87 is 6.5 billion solar masses. This image shows the silhouette of the black hole against superheated gas falling into it.

INVESTIGATION 5.1
Stargazing

CHECKPOINT 5.1

1 Identify the element that is produced in the core of the Sun.
2 Explain what might cause a nebula to start to contract.
3 What main factor determines the stages a star goes through?
4 Compare and contrast the lives of a sun-like star and a massive star.
5 Explain why solar mass is a useful measurement for comparing and contrasting stars.
6 *Planetary nebula* is the term used to describe what happens when the outer layers of a red giant are lost as it transforms into a white dwarf. Explain how this term is misleading.
7 Use diagrams to show how gravity acts on the matter inside a star at each point in its life cycle.

CHALLENGE

8 Use the information in this section, and further information from the internet, to explain the saying: 'You are made of stardust'.

SKILLS CHECK

- I can describe how stars form and some of the key features of stars.
- I can use solar mass to discuss the difference in the size of stars.

5.2 Galaxies

At the end of this lesson I will be able to:

- **outline** some of the major features of galaxies
- **use** appropriate scales to describe sizes and distances between galaxies.

KEY TERMS 

astronomer
a scientist who studies space, stars and celestial objects

astronomical unit
the average distance between Earth and the Sun (about 150 000 000 km)

galaxy
a system of millions or billions of stars

light-year
the distance that light travels in one Earth year

parsec
3.26 light-years

LITERACY LINK 

Identify some adjectives that can be used to describe galaxies.

NUMERACY LINK 

Calculate these distances in parsecs, given the distances in light-years.

- **a** Proxima Centauri to Earth: 4.3 light-years
- **b** The centre of the Milky Way to Earth: 26 000 light-years
- **c** Andromeda to Earth: 2.5 million light-years
- **d** Galaxy GN-z11 to Earth: 13.4 billion light-years

Figure 5.3 Looking towards the centre of the Milky Way galaxy over Kata Tjuta

Galaxies are massive groups of stars, gas, dust and other matter bound together by forces of gravity. The existence of galaxies beyond our own Milky Way was first proven by Edward Hubble in the 1920s. Astronomers classify galaxies by their shape, and use the observations of many galaxies both close and far away to our own Milky Way to discover how galaxies form, evolve and change.

1 The light-year is useful for measuring large distances

In the universe, distances are so great that the kilometre is too small to be a useful measurement. So, astronomers use the light-year to measure distances between stars and galaxies. A **light-year** (ly) is the distance light travels in one Earth year (365.25 days): 1 light-year is about 9.46 trillion kilometres (9.46×10^{12} km).

Our solar system is in the Milky Way galaxy and the closest galaxy to the Milky Way is Andromeda, which is 2.5 million light-years away When **astronomers** look at the light coming from the Andromeda galaxy, it's as though they are looking back in time, because it has taken that light at least 2.5 million years to reach Earth.

Astronomers also use two special units to measure distances in space: **parsecs** and **astronomical units**. A parsec is equal to 3.26 light-years. An astronomical unit (AU) is equivalent to the average distance between Earth and the Sun, about 150 million kilometres.

What is a light-year?

2 Galaxies are classified by their shape

Astronomers classify galaxies into three major groups according to their shape: elliptical, spiral and irregular (Figure 5.4).

We cannot look at our galaxy, the Milky Way, from outside it. However, by observing our night sky and comparing it with images of other galaxies, astronomers have determined that the Milky Way is a spiral galaxy. The Milky Way is part of a cluster of galaxies known as the Local Group. This includes the nearby Andromeda galaxy, the Triangulum galaxy and the Large and Small Magellanic Clouds.

What type of galaxy is the Milky Way?

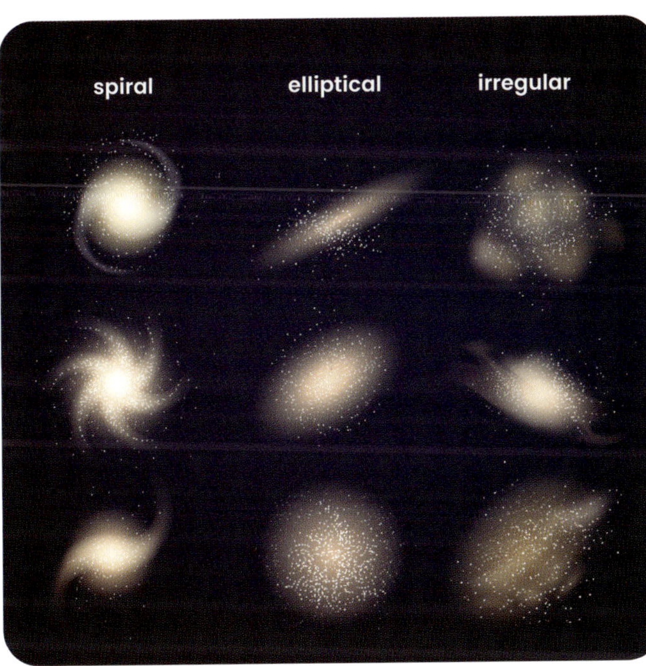

Figure 5.4 Galaxies can be classified into three main categories, spiral, elliptical and irregular.

3 Scientists are still studying how galaxies form

Astronomers do not yet have a good understanding of how the different types of galaxies form, although they know that the forces of gravity cause stars, gas and dust to clump together. One popular theory is that irregular and elliptical galaxies form when galaxies collide and interact with one another, and that spiral galaxies form on their own, with their spiral shape resulting from the spinning motion of the galaxy.

Using telescopes, astronomers have observed that galaxies that are further away appear different from closer galaxies. This is because galaxies further away are much older, but we are observing them at a much younger stage in their life.

Evidence shows that supermassive black holes are in the centres of most galaxies. Their gravitational fields influence the stars, dust and gas in the rest of the galaxy.

What is at the centre of most galaxies?

CHECKPOINT 5.2 

1 Explain how gravity is important for the formation of galaxies.
2 What evidence do astronomers use to work out how galaxies might have evolved?
3 Explain why looking at galaxies further away from us is like looking back in time.
4 What evidence have astronomers used to determine that the Milky Way is a spiral galaxy?
5 Outline some of the major features of galaxies.

CHALLENGE

6 The Milky Way is moving towards the Andromeda galaxy and they are predicted to start to collide in 3.75 billion years. Predict what the shape of the new galaxy could be. Carry out research to see if your predictions match those of astronomers.

SKILLS CHECK

- I can describe what a galaxy is.
- I can describe some major features of galaxies.
- I can identify the three main types of galaxies.
- I can describe what a light-year is, and how it is used to determine distances within and between galaxies.

5.3 The solar system

At the end of this lesson I will be able to:

- **outline** some of the major features of the solar system
- **use** appropriate scales to describe sizes and distances between planets in our solar system

KEY TERMS 

accretion
the process of matter collecting together into a bigger mass

exoplanet
a planet outside our solar system

frost line
a boundary just inside Jupiter's orbit

LITERACY LINK 

The word *planet* comes from a Greek word meaning 'wanderer'. Consider what observations were made by the Ancient Greeks to give planets this name.

NUMERACY LINK 

An astronomical unit is equal to 150 000 000 km. Use the data in Figure 5.5 to calculate the distances in kilometres between Earth and the other planets in the solar system.

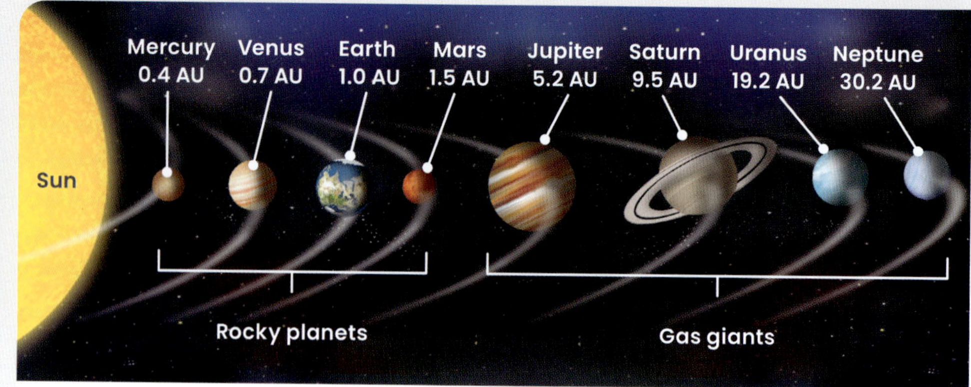

Figure 5.5 The astronomical unit is the average distance between Earth and the Sun. 1 AU = 150 000 000 km. It is useful for describing distances in the solar system.

Our solar system formed around 4.6 billion years ago. The planets formed from the dust and gas around the Sun, with the rocky planets close to the Sun, and the gas and ice giants further out. Astronomers have detected planets around other stars, and continue to search for these to learn more about our solar system.

❶ The rocky planets formed from heavy elements

The nebula that our solar system formed from was mostly made up of lighter elements such as hydrogen and helium. Heavier elements such as iron, nickel, silicon and aluminium were much rarer. This is why the inner, rocky planets, such as Venus and Earth, are relatively small, because they are mostly made up of these heavier elements.

The inner solar system was too warm for substances with low boiling points (such as water and methane) to condense and exist as liquids. So, only compounds with high melting points formed, such as metals. It was here that the rocky planets formed. These high melting compounds are quite rare in the universe so the rocky planets are relatively small.

Why are the rocky planets small?

❷ The gas giants mostly formed from hydrogen and helium

The **frost line** is found just inside Jupiter's orbit. Outside this boundary, temperatures are so low that hydrogen, helium and other compounds that are gases on Earth (such as methane and carbon dioxide) are able to condense and exist as liquids.

Gas giants have cores made up of rock and ice, but are otherwise mostly gases. Jupiter and Saturn, the largest planets in the solar system, are made up of large amounts of hydrogen and helium. Their large cores allowed the planets to attract lots of hydrogen and helium before the solar wind cleared the solar system. Uranus and Neptune are further out, with larger orbits.

What is the frost line?

3 Comets come from the Kuiper Belt and Oort Cloud

Comets are like dirty snowballs, chunks of rock and ice moving through space. The 'tails' that we observe are formed when they melt as they move closer to the Sun.

Short-period comets, which orbit the Sun fairly frequently, originate in the Kuiper Belt. The Kuiper Belt is an area of the outer solar system that extends from the orbit of Neptune to about 55 AU from the Sun. It is a region of icy bodies left over from the formation of the solar system. The Kuiper Belt is home to dwarf planets such as Pluto.

Long-period comets have very slow orbits; the Hale–Bopp comet orbits the Sun once every 2500 years. These comets come from the Oort Cloud, a spherical cloud of icy bodies also left over from the formation of the solar system. The Oort Cloud lies in the outermost parts of the solar system, 5000–100 000 AU from the Sun.

Where do long-period comets come from?

4 Many other solar systems have been detected

Astronomers have found many different types of solar systems. Some are similar to ours, while others are very different. Planets that orbit stars other than the Sun are called **exoplanets**. The first exoplanets were observed in 1995 and since then thousands of exoplanets have been discovered. Small rocky planets like Earth, Mars, Venus and Mercury are very common in other solar systems.

Studying other solar systems helps astronomers build a better picture of how our solar system formed. The main way that astronomers search for exoplanets is to look for a dip in the light coming from the star when the planet passes across it.

What is an exoplanet?

Figure 5.6 'Hot Jupiters' are gas giants in other solar systems that are orbiting their stars much closer than Earth orbits the Sun.

INVESTIGATION 5.3
Investigating orbits

CHECKPOINT 5.3

1 Outline some of the major features of our solar system.
2 What are the most common two elements in the solar system?
3 Explain the significance of the frost line in the formation of the gas giants.
4 Identify one reason for the rocky planets being much smaller than the gas giants.
5 Construct a table that compares the rocky planets with the gas giants.
6 Compare and contrast the Kuiper Belt with the Oort Cloud.
7 Describe how astronomers find exoplanets.

CHALLENGE

8 Create a scale that depicts the distances between the Sun and the planets in the solar system.

SKILLS CHECK

- I can describe some of the major features of the solar system.
- I can use appropriate units to describe the size of planets in our solar system and the distances between them.
- I can describe how our solar system formed.

5.4 Telescopes

At the end of this lesson I will be able to:

- **describe**, using examples, some technological developments that have advanced scientific understanding about the universe.

KEY TERMS 

resolution
the ability to tell two separate objects apart

LITERACY LINK 

In a table, summarise the information on this spread about telescope types, what they can detect and how they are useful to astronomers.

NUMERACY LINK 

To compare the light-gathering powers of two telescopes, you divide the area of one mirror by the area of the other. For example, a 40 cm diameter mirror on one telescope has four times the light-gathering power of a telescope with a 20 cm diameter mirror.

Using the information from the text, compare the light-gathering power of the James Webb telescope to the Hubble Space Telescope.

Figure 5.8 The Hubble Space Telescope

Telescopes have allowed us to improve our understanding about the universe: what it contains, how it formed and what our place is in it. Technologies have improved greatly from Galileo's simple telescope that allowed him to observe the Moon and planets. Telescopes can now be as large as buildings – located on remote mountain tops, and even in space. Today's telescopes can even detect more than just visible light.

1 Optical telescopes collect and amplify light

An optical telescope works by collecting and amplifying light. Galileo was the first person credited with using a telescope to study stars and the solar system. He observed four of the moons orbiting Jupiter, the rings around Saturn, and the mountains on the Moon. His refracting telescope used two glass lenses – one large lens to collect light, and a smaller eyepiece lens to magnify the image.

Today, large telescopes have mirrors to collect light, because large mirrors are easier to make than large lenses. The larger the objective lens or mirror, the more light that can be collected, and the more fine detail can be observed. This is called **resolution**. The further away an object is, the larger the telescope must be to have good resolution.

Most optical telescopes are located away from populated areas that cause light pollution. They also tend to be located at high altitudes to limit atmospheric disturbance.

Where are most optical telescopes located?

Figure 5.7 The Anglo-Australian Telescope near Coonabarabran, New South Wales, is the largest optical telescope in Australia.

2 Radio telescopes detect radio waves from excited hydrogen atoms

Hydrogen is the most common element in the universe, and when hydrogen atoms are excited by energy, they emit radio waves. Radio telescopes detect these radio waves from space, allowing astronomers to map the shape of galaxies.

Unlike light, radio waves can also travel through dust clouds, so astronomers can discover and map out objects that cannot be seen with an optical telescope. Radio astronomy has enabled the discovery and study of pulsars (rapidly spinning neutron stars), quasars (primordial galaxies with supermassive black holes), supernova remnants, and black holes in the centres of galaxies.

Radio telescopes need to be situated away from large populations where there are few radio signals from radio, television and mobile phones. Radio telescopes can be built in arrays of much smaller antennas that work together to detect faint radio signals.

What can radio telescopes be used to observe?

3 Space telescopes are located in outer space

Space telescopes allow astronomers to gather clearer images because they don't experience interference by the atmosphere or light pollution.

The Hubble Space Telescope was launched in 1990. The telescope orbits 547 km above Earth. Its main mirror has a diameter of 2.4 m, and it was able to produce images about 50% sharper than an optical telescope on Earth. The discoveries made by the Hubble Space Telescope have significantly advanced our knowledge of the universe.

The James Webb Space Telescope is due to be launched in 2021. This telescope will not orbit Earth, but will sit at a point 1 500 000 km away, orbiting the Sun. This telescope will have a mirror about 6.5 m in diameter. It will gather light in the near infrared part of the spectrum to find out more about the early universe, as well as searching in dust clouds for stars and solar systems that are forming.

What is the advantage of a space telescope over a ground-based telescope?

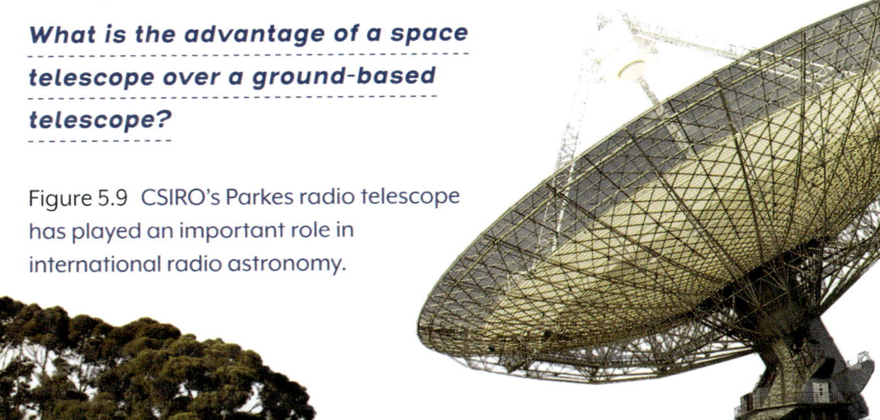

Figure 5.9 CSIRO's Parkes radio telescope has played an important role in international radio astronomy.

INVESTIGATION 5.4A
Making a telescope

INVESTIGATION 5.4B
Lens diameter and resolution

CHECKPOINT 5.4

1 Copy and complete these sentences.
Resolution is the ability to distinguish between ________ separate objects. The further away the objects, the ________ the telescope needs to be.

2 Explain the advantage of using a mirror over a lens in a large telescope.

3 A space telescope has advantages over a ground-based telescope. Explain why.

4 Why would using an array of radio telescopes be an advantage?

5 Some telescopes can focus on an object and then move as Earth rotates. This allows more energy to be gathered, like a long-exposure photograph. Explain the advantage this would have over a telescope that does not move.

CHALLENGE

6 Find out more about one telescope and present your findings to your class.

SKILLS CHECK

- I can describe optical telescopes, radio telescopes and space telescopes and list some of the discoveries they have been used for.
- I can describe the advantages of a space telescope and list some of the discoveries they have been used for.

5.5 The Big Bang

At the end of this lesson I will be able to:

- **use** scientific evidence to outline how the Big Bang theory can be used to explain the origin of the universe and its age
- **outline** how scientific thinking about the origin of the universe is refined over time through a process of review by the scientific community.

KEY TERMS 

antimatter
particles that have properties opposite to that of normal matter

astrophysicist
a scientist who studies the physics of the universe

redshift
a change in light's wavelength towards the red end of the visible spectrum

singularity
an infinitely dense point of matter that existed before the Big Bang

LITERACY LINK 

Summarise the major steps in the history of the universe since the Big Bang, including the formation of the Sun and our solar system.

NUMERACY LINK 

The three most abundant elements in the universe are as follows: 75% hydrogen, 23% helium, 1% oxygen.

How many times more abundant is hydrogen than oxygen?

As technology has improved, and more observations and calculations have been made, so has our understanding of the universe and how it formed. The Big Bang theory explains how the universe rapidly expanded from a dense singularity about 13.7 billion years ago, creating matter and forming elements that then went on to create stars and galaxies.

1 The universe began as a singularity

The Big Bang theory is the currently accepted theory for the origin of the universe. The theory states that the universe began as a small dense region called a **singularity**, which then rapidly expanded about 13.7 billion years ago. This expansion is called 'the Big Bang'.

After the Big Bang, the first events happened very quickly. As it continued to expand, the universe started to cool, allowing the formation of particles of matter and **antimatter** (particles that are the opposite of matter). At about 0.001 seconds after the Big Bang, these particles annihilated each other. The matter that we observe today is what was left over from this interaction.

After 3 minutes, the protons and neutrons then combined to form the nuclei of atoms, about 75% hydrogen, 25% helium, and a small fraction of lithium. At this stage, all the matter and energy that would ever exist was formed, and all in the time it takes to boil a kettle. Between the formation of these nuclei and 500 000 years, the universe was a plasma of hydrogen, helium and lithium nuclei and free electrons.

At about 500 000 years, the universe cooled enough for the electrons to be attracted to the nuclei and so for atoms to form. This released energy in the form of photons of light. The first stars and galaxies then began to form 1 billion years after the Big Bang.

What does the Big Bang theory explain?

2 There are three pieces of evidence for the Big Bang

The Big Bang theory is supported by three major pieces of evidence.

The universe is expanding

Edwin Hubble observed that light coming from distant galaxies was stretched into longer wavelengths – it was **redshifted**. This indicated that galaxies are moving away from us. The movements of the galaxies can be traced back to a single point, meaning that the universe must have once been contained in a small region of space.

Abundance of light elements

By mass, the universe is about 74% hydrogen and 24% helium, with the other 2% being all the other heavier elements. This abundance supports the Big Bang theory because if helium was only made by fusion in stars, there would be significantly less than 24%.

Cosmic microwave background radiation

Cosmic microwave background radiation is left-over heat released about 100 000 years after the Big Bang, when the universe had cooled enough to allow electromagnetic radiation to pass through it. Because the universe was not of uniform density when the cosmic microwave background was released, images show 'clumps' of matter. The cosmic microwave background would have originally been released as visible and UV light, but the expansion of the universe has redshifted it into the microwave band.

What is the evidence that the universe is expanding?

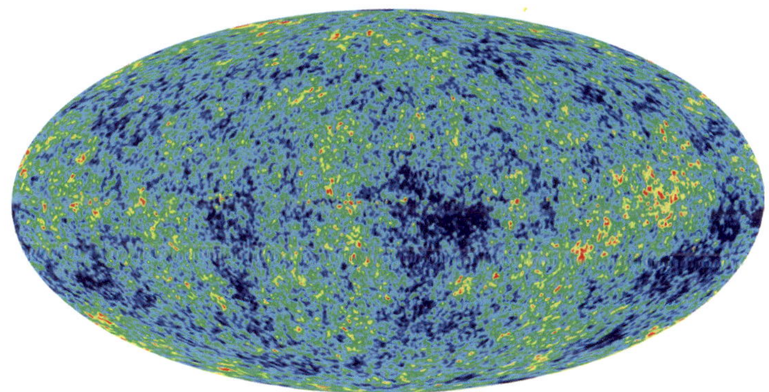

Figure 5.10 This image of cosmic microwave background radiation was taken from the Wilkinson Microwave Anisotropy Probe, a space-based microwave telescope.

3 Data from space telescopes helps scientists calculate the age of the universe

Current observations and calculations put the age of the universe at 13.7 billion years old. There are several pieces of evidence that have helped scientists to calculate this age.

The Hubble Space Telescope was able to measure how far away galaxies were from us and the speed that they were moving. This allowed **astrophysicists** to calculate how long it took galaxies to get to their current locations.

The Hubble Space Telescope was also used to calculate the ages of the oldest star clusters – the first stars that would have formed after the Big Bang. The data gathered by the Wilkinson Microwave Anisotropy Probe (Figure 5.11) also supports this age.

How can we tell how old the universe is?

Figure 5.11 The Wilkinson Microwave Anisotropy Probe (WMAP) is a space-based microwave telescope.

INVESTIGATION 5.5
Modelling the expanding universe

CHECKPOINT 5.5

1 Describe the Big Bang theory.
2 What is a singularity?
3 When did atoms begin to form?
4 When did the first stars and galaxies begin to form?
5 Identify three pieces of evidence used to calculate the age of the universe.
6 Explain why the percentage of light elements in the universe supports the Big Bang theory.

CHALLENGE

7 Create a timeline of major findings that helped us to understand how the universe formed.

SKILLS CHECK

- I can describe the Big Bang theory.
- I can outline three major pieces of evidence that support the Big Bang theory.
- I can suggest evidence that was used to calculate the age of the universe.

CHAPTER SUMMARY

Developments in technology have advanced scientific understanding of the universe.

Galaxies are groups of stars, gas and dust bound together by gravity.

Black holes form when supermassive stars die.

Stars are mostly **hydrogen**.

1
H
Hydrogen
1.0079

Hydrogen is the most common element in the universe.

Galaxies are classified by their shape.

Spiral
Elliptical
Irregular

The **Big Bang theory** is the currently accepted theory for the origin of the universe.

Evidence for the Big Bang theory includes:

- the universe is expanding
- there is an abundance of light elements
- cosmic microwave background radiation.

1 AU = 150 000 000 km = distance from the Sun to Earth

Sun
Mercury 0.4 AU
Venus 0.7 AU
Earth 1.0 AU
Mars 1.5 AU
Jupiter 5.2 AU
Saturn 9.5 AU
Uranus 19.2 AU
Neptune 30.2 AU

Rocky planets
Gas giants

FINAL CHALLENGE

1. The Sun will end its life as a:
   - **A** red dwarf
   - B white dwarf
   - C planetary nebula
   - D supernova.

2. What type of stars would you expect to end their lives as a black hole?
   - **A** stars of less than 1 solar mass
   - B stars of 1–5 solar masses
   - C stars of 10 solar masses
   - D stars of more than 20 solar masses

3. A galaxy is best defined as:
   - **A** a region where stars are forming
   - B a cloud of dust and gas
   - C a mass of stars, gas and dust bound together by gravity
   - D the gas left over after a supernova.
4. Identify the three main shapes used to classify galaxies.
5. Explain why the rocky planets are close to the Sun and the gas giants are further away.
6. Identify an appropriate situation to use distances measured in:
   - **a** astronomical units
   - b light-years.

7. Outline how gravity was important for the formation of the Sun and our solar system.
8. Identify the factors that astronomers need to consider when determining where to locate a new ground-based telescope.
9. What advantage do space telescopes have over ground-based telescopes?

10. Identify and explain three pieces of evidence that support the Big Bang theory of the origin of the universe.
11. Discuss this statement using evidence to support your response.
    'Our understanding of the universe cannot improve unless technology to observe the universe also improves.'

6 Plate tectonics

The theory of plate tectonics explains how Earth functions from a geological point of view. This one theory accounts for why mountain ranges have formed where they are, why volcanoes erupt, why earthquakes occur, why we observe certain patterns in the fossil record and why animals and plants on one continent can be related to those on the other side of the planet.

1 LEARNING LINKS

Brainstorm as much as you can remember about these topics.

Rocks change during the rock cycle.

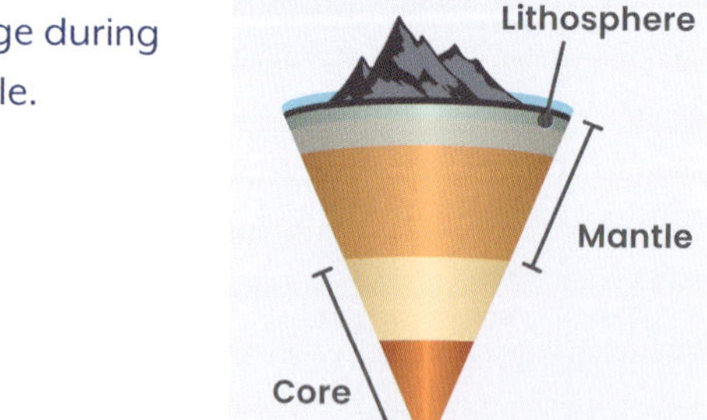

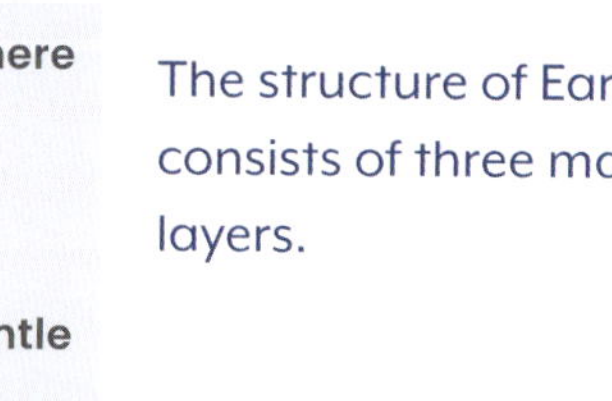

The structure of Earth consists of three main layers.

Fossils change throughout the geological time scale.

Natural geological events can change Earth's surface.

2 SEE–KNOW–WONDER

List three things you can **see**, three things you **know** and three things you **wonder** about this image.

3 CRITICAL + CREATIVE THINKING

What if ... earthquakes hit NSW every day?

Five questions: Write five questions that have the answer 'plate'.

Prediction: What will the surface of Earth look like in 250 million years?

4 THE BIGGEST!

The most powerful earthquake ever recorded was the 1960 earthquake in Valdivia, Chile, which geologists think measured 9.5 on the moment magnitude scale. The energy released was large enough to shift Earth's axis and shorten the length of a day by 1.26 microseconds. The earthquake also caused a tsunami that hit the coastline of Chile and travelled across the Pacific Ocean at more than 300 km/h. The tsunami killed people in Hawaii, Japan and the Philippines, and damaged infrastructure in many other countries.

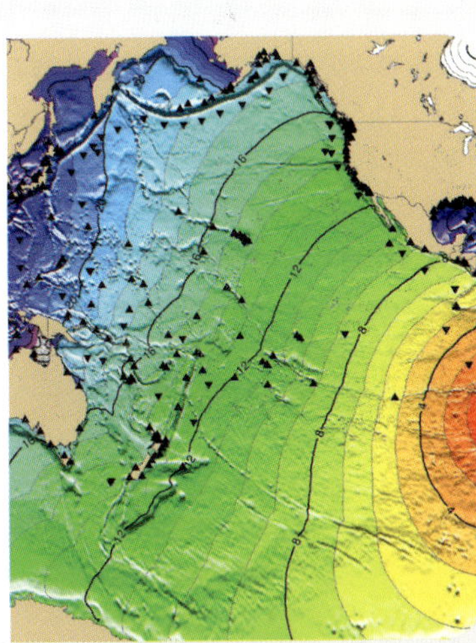

6.1 Evidence for plate tectonics

At the end of this lesson I will be able to:

- **outline** how the theory of plate tectonics changed ideas about Earth and how Earth has changed over geological time.

KEY TERMS

asthenosphere
the portion of Earth's mantle underneath the lithosphere that can flow

continental drift
the theory that the continents have moved position over time

lithosphere
Earth's rigid outer zone (crust and upper mantle), made up of tectonic plates

tectonic plate
a section of Earth's lithosphere

LITERACY LINK

Write a status update or tweet from Alfred Wegener that summarises the evidence for his theory of continental drift.

NUMERACY LINK

Earth has a radius of 6000 km. If the core has a radius of 3000 km, what is the volume of the mantle and crust, together, in m^3?

(Use the equation $V = 4/3\pi r^3$)

The theory of plate tectonics states that Earth's outer layer – the **lithosphere** – is divided into more than 15 massive pieces called **tectonic plates**. These plates move around on the **asthenosphere** (upper mantle), interacting at boundaries, shifting the continents and producing new landforms.

For a long time, people thought that the continents were in the same place as when Earth first formed. However, evidence gathered in the first half of the 20th century indicated that the continents are moving.

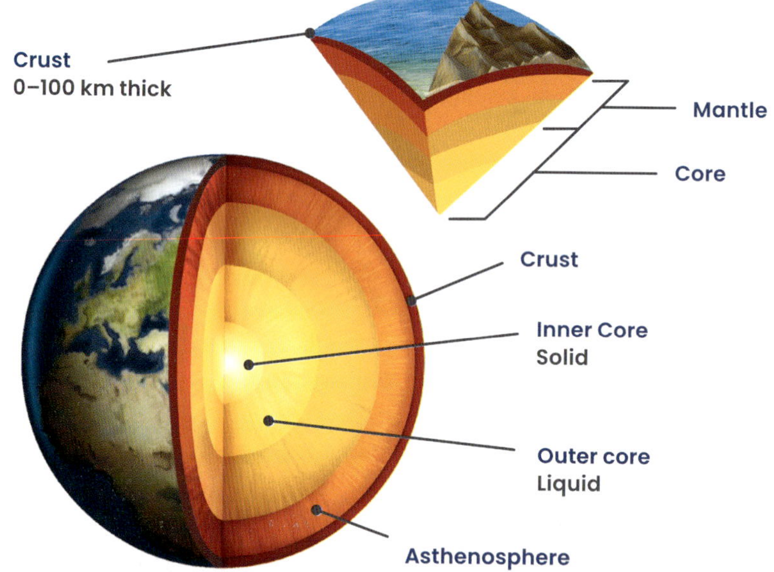

Figure 6.1 The lithosphere is made up of more than 15 tectonic plates that move around on the upper layer of the mantle.

❶ The continents move over time

In 1912, German meteorologist Alfred Wegener published his theory on **continental drift**. He proposed that the continents had once been joined in one large landmass that he called Pangaea. Over time, this landmass split apart and the continents moved to their current positions.

Wegener found evidence for past glacial climates in equatorial Africa and tropical climates in northwestern Europe. The only way to explain this was that the continents had moved.

Further evidence to support Wegener's theory includes:

- how the continental shelves of continents fit together like pieces of a jigsaw
- identical rock formations on either side of the Atlantic Ocean
- identical plant and animal fossils on different continents separated by oceans.

INVESTIGATION 6.1
Modelling seafloor spreading

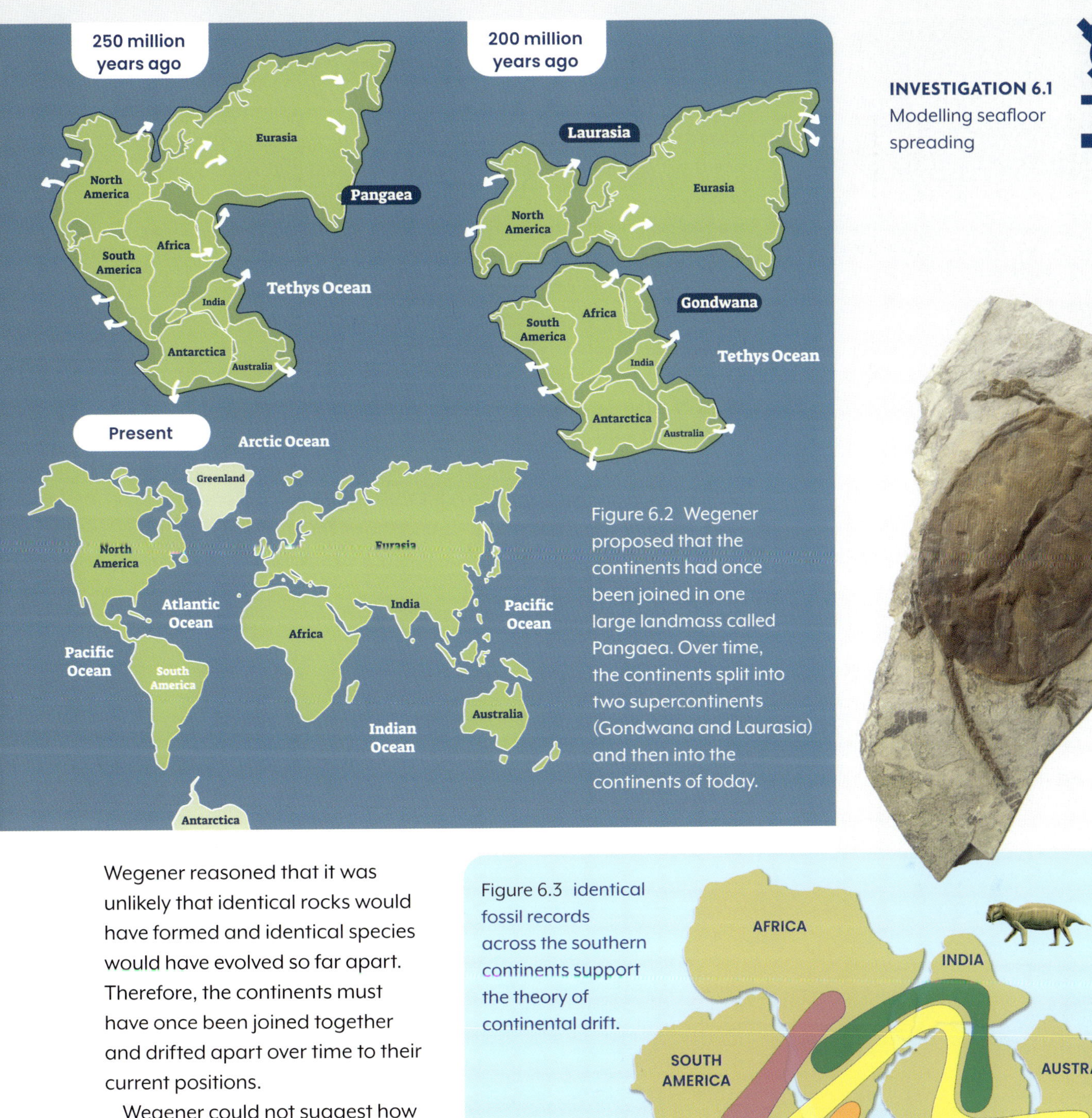

Figure 6.2 Wegener proposed that the continents had once been joined in one large landmass called Pangaea. Over time, the continents split into two supercontinents (Gondwana and Laurasia) and then into the continents of today.

Wegener reasoned that it was unlikely that identical rocks would have formed and identical species would have evolved so far apart. Therefore, the continents must have once been joined together and drifted apart over time to their current positions.

Wegener could not suggest how the continents moved, and so his theory was not supported by most of the scientific community. He continually refined and published his ideas as new evidence came to light. Modern geologists accept his theory as correct.

What three pieces of evidence support Wegener's theory of continental drift?

Figure 6.3 identical fossil records across the southern continents support the theory of continental drift.

AFRICA
INDIA
SOUTH AMERICA
AUSTRALIA
ANTARCTICA

Fossil evidence of freshwater reptile *Mesosaurus* has been found in Brazil and Africa.

Fossil evidence of the land reptile *Lystrosaurus* has been found in Africa, Antarctica and India.

Fossils of the fern *Glossopteris* have been found in all southern continents.

Fossil evidence of land reptile *Cynognathus* has been found in Argentina and southern Africa.

6.1 continued...

6.1 continued...

Evidence for plate tectonics

KEY TERMS 

mantle
Earth's middle layer, made up of two layers

mid-ocean ridge
a long chain of mountains under the ocean formed by plate tectonics

rift valley
a valley formed when a continent is being pulled apart

subduction
when one tectonic plate moves underneath another

2 The sea floor is spreading apart

Seafloor spreading happens when molten rock rises up from the **mantle** at the **mid-ocean ridges** and solidifies to form new oceanic lithosphere.

There are four major pieces of evidence to show that the sea floor is spreading.

- *Rift valleys along mid-ocean ridges*: In 1952, US geologist Marie Tharp found that there was a V-shaped valley running along the bottom of the Atlantic Ocean. This **rift valley** is where the new lithosphere is formed.
- *Magnetic striping*: As lava cools, the magnetic minerals in it align with Earth's magnetic poles, just like a compass needle does. The positions of the magnetic poles have changed over time, even reversed, and so in rocks formed at different times, the minerals are aligned differently.
- *Depth of sediments*: The depth of sediments on the oceanic crust are deeper closer to the continents. This implies that those rocks are older because there has been more time for the sediment to accumulate.
- *Age of the sea floor*: Radiometric dating shows that the oceanic crust closer to the continents is much older than the rock closer to the mid-ocean ridges.

What is a rift valley?

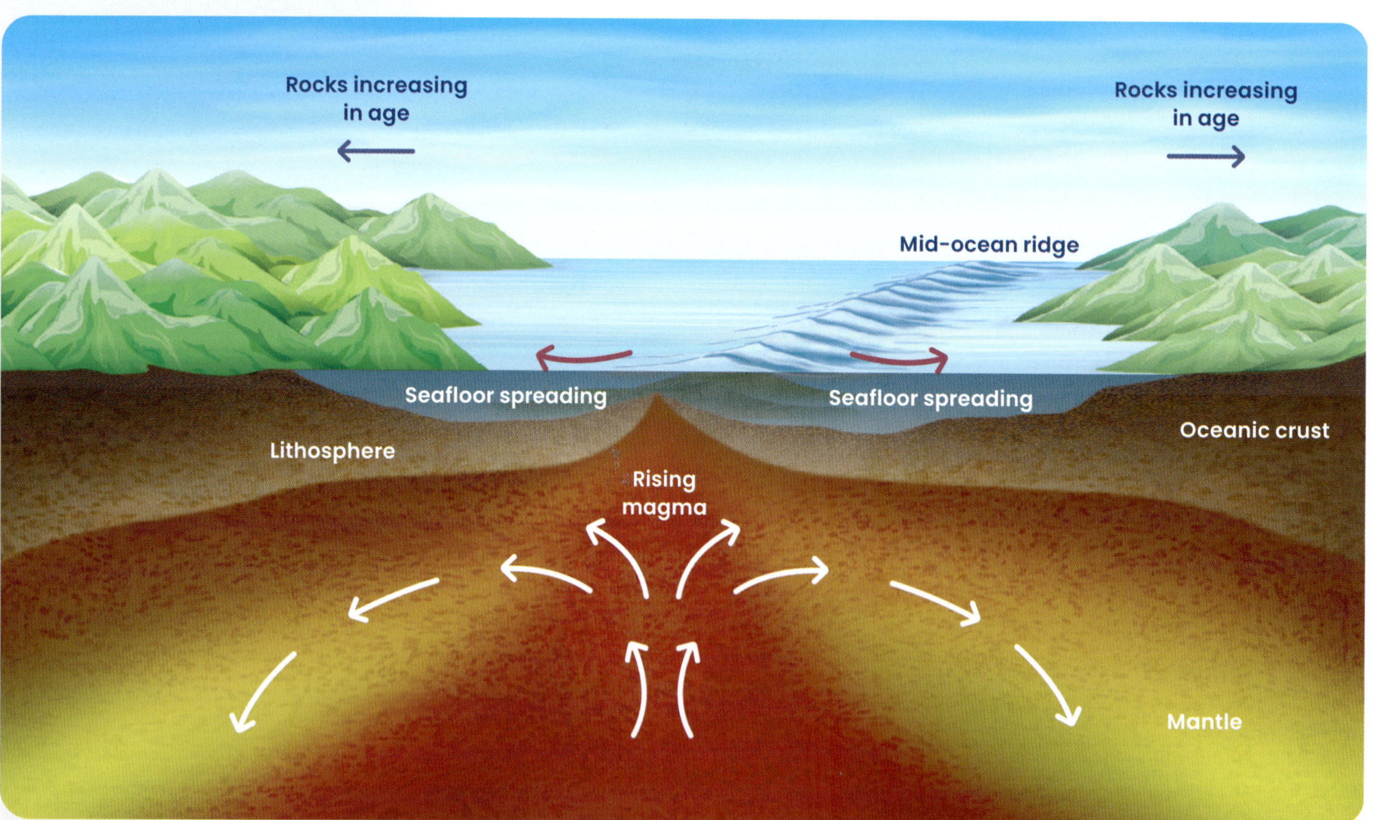

Figure 6.4 Molten rock rises up from the mantle at the mid-ocean ridges and solidifies to form new oceanic lithosphere.

❸ Old lithosphere is subducted

If new lithosphere is being formed at mid-ocean ridges, why isn't Earth getting larger? Scientists discovered that this is because of **subduction** – where the edge of one tectonic plate is pushed under the one next to it.

Geologists use three main pieces of evidence to show that the lithosphere is being subducted.

- *Ocean trenches*: Seafloor mapping shows the existence of deep ocean trenches. We now know this is where two tectonic plates are colliding.
- *Earthquake locations*: Earthquakes get deeper further away from ocean trenches. This is evidence that one plate is moving down under the other. Regions where this happens are called Wadati–Benioff zones, after the scientists who discovered them.
- *Volcano chains above such zones:* The formation of chains of volcanoes above a Wadati–Benioff zone shows that one plate is moving under another. When a plate subducts, the subducting lithosphere starts to melt and the molten rock rises to the surface to form volcanoes.

What is subduction?

Figure 6.5 Denser crust subducts underneath less-dense crust. This forms a deep ocean trench, causes deep earthquakes and forms a chain of volcanoes.

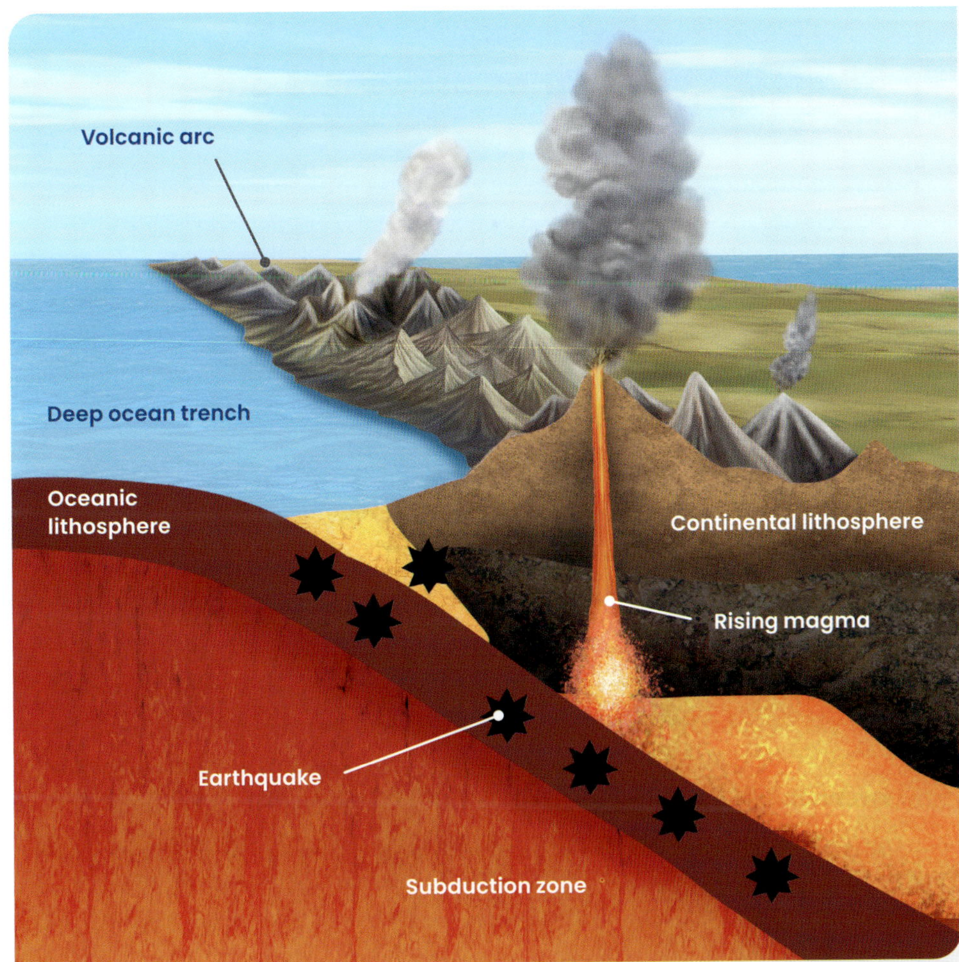

CHECKPOINT 6.1 

1 Copy and complete this sentence: The theory of ________ ________ states that Earth's continents were once ________ together and moved ________ over time.

2 Outline why the discovery of seafloor spreading was important for the development of the theory of plate tectonics.

3 The discovery of subduction was important for the development of plate tectonic theory. Suggest why.

4 Explain why Wegener's theory of continental drift was not supported by the scientific community.

5 If plate tectonics did not exist and the continents had not moved, what would Wegener have observed in the geological record instead? How would this have been different from what he actually observed?

CHALLENGE

6 Use the internet to research and learn more about Wegener, Tharp, Harry Hess or John Tuzo Wilson. Find out about their contribution, why it was important for our understanding of Earth and any challenges they faced.

SKILLS CHECK

- I can explain the theory of plate tectonics.
- I can explain some of the evidence that was used to support the theory of plate tectonics.

6.2 Plate boundaries

At the end of this lesson I will be able to:

- **describe** how tectonic plates interact with each other.

KEY TERMS

convergent boundary
where two tectonic plates are moving towards each other

divergent boundary
where two tectonic plates are moving away from each other

fault
a break in Earth's surface where blocks of rock slide past each other

fold mountain
a mountain formed by the folding of continental crust when tectonic plates collide

transform boundary
where two tectonic plates are sliding past one another

LITERACY LINK

Convergent boundaries are often called destructive, divergent are called constructive and transform are called conservative. Can you think of any other adjectives to describe how the plates are moving at the three different boundary types?

NUMERACY LINK

Plate boundaries can move up to 7 cm per year.

Calculate the speed of plate boundary movement in cm/week.

As tectonic plates move on the flowing, semi-molten asthenosphere, they slowly transform Earth's surface. Violent geological changes can occur when plates collide, move apart or grind past each other.

Figure 6.6 The major tectonic plates and their boundaries

1 Divergent boundaries are where plates move apart

Divergent boundaries exist where two plates are moving away from each other. Magma rises up in the gap between the two plates and solidifies to form new lithosphere.

When this happens between two oceanic plates, it forms a mid-ocean ridge. When this happens on a continent, it forms a rift valley and volcanoes. A divergent boundary is sometimes referred to as a constructive boundary because new lithosphere is being made.

In what direction do plates move at a divergent boundary?

2 Convergent boundaries are where plates collide

Convergent boundaries are where two plates are moving towards each other and colliding. A convergent boundary is sometimes referred to as a destructive boundary because the lithosphere is destroyed.

If two continental plates collide, the crust buckles and pushes together to form **fold mountains**.

In what direction do the plates move at a convergent boundary?

3 Transform boundaries are where two plates slide past each other

Transform boundaries are where two plates slide past each other. This causes a break in Earth's surface called a **fault**. As the plates slowly move, they cause earthquakes.

Transform boundaries are also called conservative boundaries because lithosphere is neither created nor destroyed.

The San Andreas Fault near San Francisco in the USA is the most famous transform fault. It has caused many destructive earthquakes.

In what direction do tectonic plates move at a transform boundary?

Divergent plate boundary

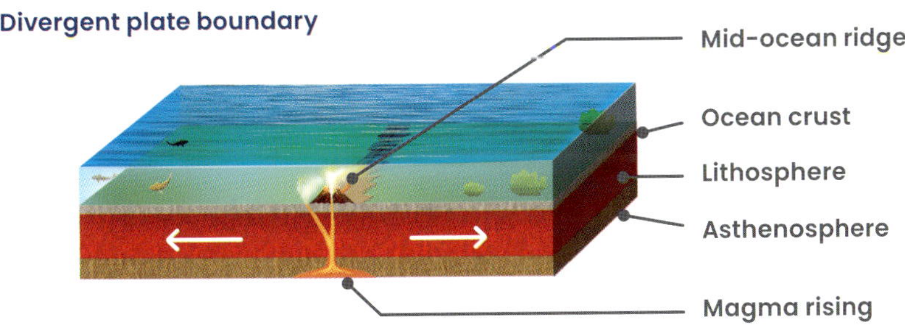

Convergent plate boundary

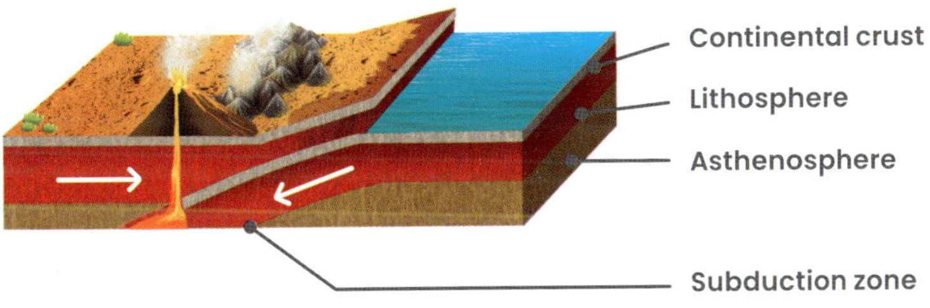

Transform plate boundary

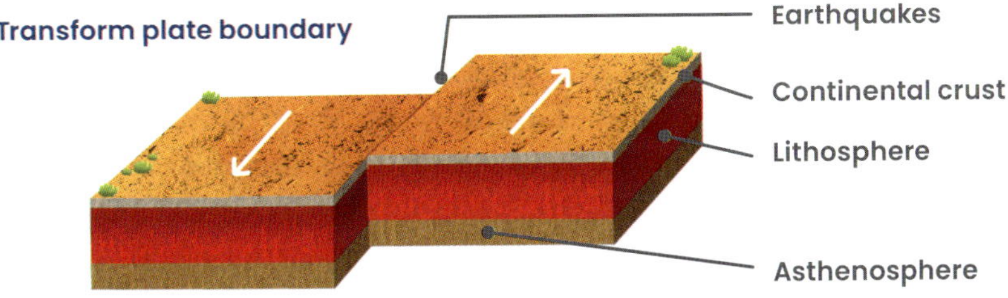

Figure 6.7 At a divergent boundary, two plates move apart. At a convergent boundary, two plates collide and the denser plate subducts. At a transform plate boundary, two plates slide past each other.

INVESTIGATION 6.2
Modelling plate boundaries

CHECKPOINT 6.2

1 Describe subduction.

2 A convergent boundary is sometimes called a destructive boundary. Explain why.

3 Draw a diagram to illustrate what you would expect to occur in these situations.
   a Plate A is moving towards plate B. Plate A is denser than plate B.
   b Plates C and D are both continental plates. They are moving towards each other.

4 The African continent is beginning to split apart along the East African Rift Valley. What type of crust would you expect to be forming in the rift? Justify your response.

5 What type of plate boundaries are needed to:
   a bring continents together?
   b break continents apart?

CHALLENGE

6 Use the internet to research how the Australian and Pacific plates interact in New Zealand. Explain why active volcanoes are found on the North Island, while a long mountain range is found on the South Island.

SKILLS CHECK

- I can describe the three types of movement that can happen at tectonic plate boundaries.
- I can describe how tectonic plates interact with each other.

6.3 What causes the plates to move?

At the end of this lesson I will be able to:

- **describe** the forces that move tectonic plates.

KEY TERMS 

asthenosphere
the portion of Earth's mantle underneath the lithosphere that can flow

convection
the transfer of heat by movement of a liquid or gas

LITERACY LINK 

Convection is the transfer of heat energy and occurs in Earth's mantle as well as many other situations. Can you think of other examples where convection is important?

NUMERACY LINK 

Density is a physical property determined by the amount of mass in a given volume. The density of metals is greater than non-metals. If the amount of iron is doubled in the crust and the oxygen is decreased, how would the density of the crust be affected?

How would the density of the crust change if silicon replaced the iron in the crust?

Most geologists currently think that three factors cause tectonic plates to move on top of the **asthenosphere**. Gravity is an important force, pulling subducting plates down into the mantle and pushing newly formed lithosphere along at mid-ocean ridges. Convection currents in the asthenosphere play a minor role as they bring hot rock up towards the crust. Most plates have a convergent boundary on one side and a divergent boundary on the other. These three factors work together to move the plates along like a conveyor belt.

1 Big plates can pull each other down

When a denser plate moves under another plate (subduction) at a convergent plate boundary, it will start to pull the rest of the plate along with it. This is known as slab pull.

Slab pull is thought to be the major cause of the movement of tectonic plates. As a subducting plate is cooler and denser than the warmer mantle, gravity causes it to sink towards Earth's core, pulling the rest of the plate along behind it. Plates with long subduction zones often move faster than plates with shorter subduction zones.

What is slab pull?

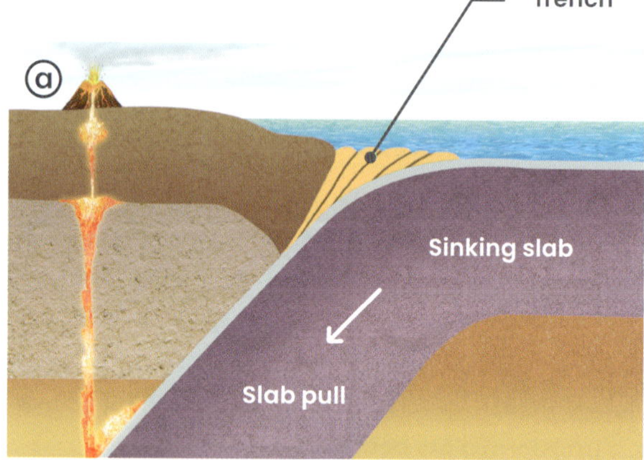

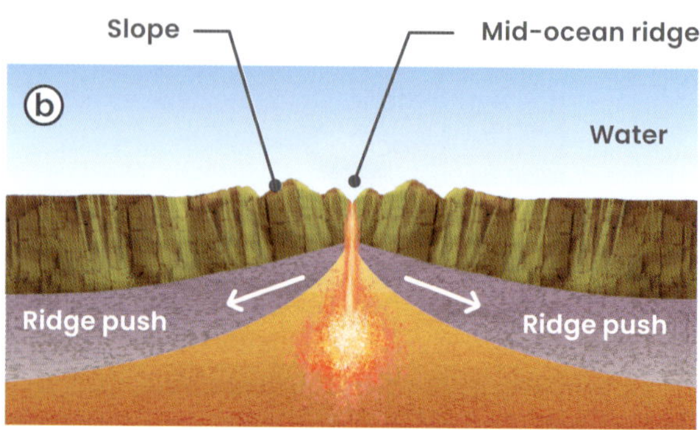

Figure 6.8 (a) Slab pull happens when the denser lithosphere slowly sinks underneath the less dense asthenosphere. This is like a rock slowly sinking in water. (b) Ridge push happens when the region of a rift is lifted up and the mass of the ridge pushes sideways. This is like a wedge of honey with a sloping surface.

2 Ridge push is caused by gravity

Ridge push happens at mid-ocean ridges. When magma rises up at a mid-ocean ridge, it forms new lithosphere. This new lithosphere sits higher than the old one and so gravity causes it to slide downhill, pushing the old lithosphere in front of it. This push helps to move the tectonic plate along, away from the mid-ocean ridge towards the subduction zone, similar to the movement of a conveyor belt.

Where does ridge push happen?

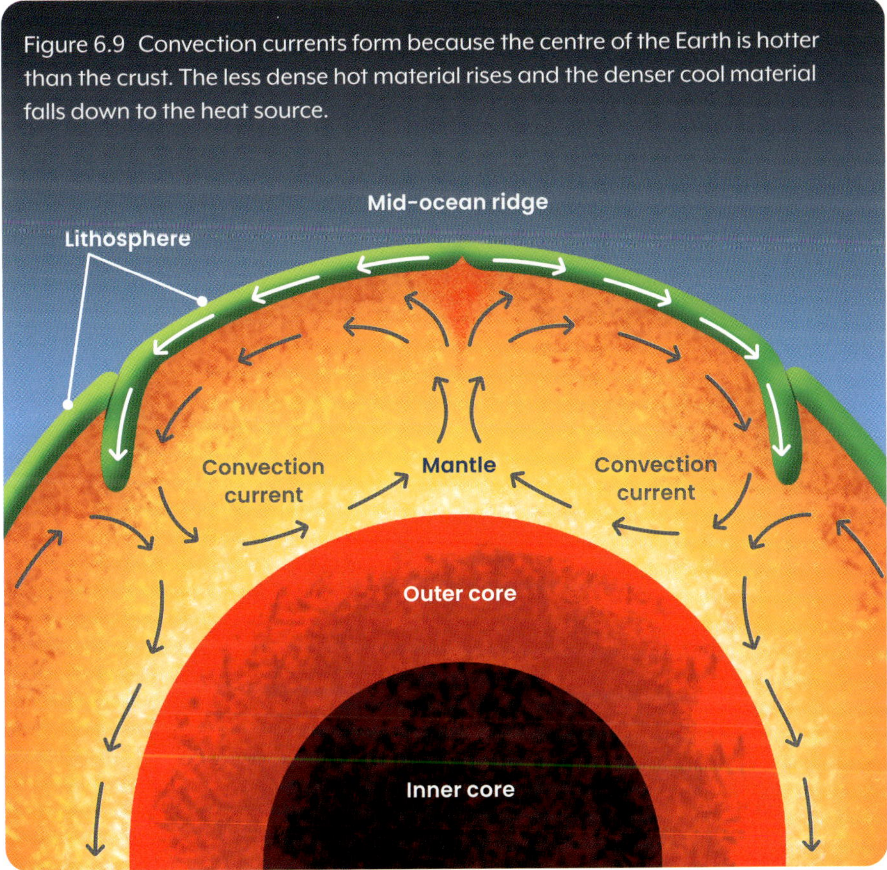

Figure 6.9 Convection currents form because the centre of the Earth is hotter than the crust. The less dense hot material rises and the denser cool material falls down to the heat source.

3 Convection causes hot rock to rise and cooler rock to sink

Convection is a way of transferring heat. A convection current is the movement of material from a hot area to a cool area and back again.

In the mantle, rock in the asthenosphere moves slowly in convection currents formed by hot rock near the core rising up, cooling and falling back down again. This brings molten material up to the mid-ocean ridges. As the rock in the asthenosphere slowly moves, it drags the plate along like a conveyor belt moving out and away from the mid-ocean ridges towards the subduction zones.

What is a convection current?

INVESTIGATION 6.3A
Modelling slab pull

INVESTIGATION 6.3B
Observing convection currents

CHECKPOINT 6.3

1 Copy and complete this sentence: Tectonic plates are thought to be able to move around on the ________ due to three ________, ________ pull, ridge ________ and mantle ________.

2 Describe the forces that move plate tectonics.

3 Outline the difference between ridge push and mantle convection.

4 Explain how density is important for the movement of the tectonic plates.

CHALLENGE

5 Use the map of the tectonic plates in Figure 6.6 on page 74 to put these plates in order from longest subduction zone to smallest subduction zone.
- Australian Plate
- Nazca Plate
- Pacific Plate
- Philippine Plate
- Arabian Plate

Use the internet to find out how fast these plates are moving. Is there a relationship between speed and length of subduction zone? Why do you think that this is important evidence for slab pull being the major factor that moves tectonic plates?

SKILLS CHECK

- I can describe how slab pull, ridge push and mantle convection move tectonic plates.

6.4 Earthquakes

At the end of this lesson I will be able to:

- **outline** how earthquakes can be explained by the theory of plate tectonics.

KEY TERMS 

epicentre
the point on Earth's surface directly above the focus of an earthquake

focus
the origin of an earthquake

seismic wave
a wave of energy that passes through Earth's layers and is caused by an earthquake

seismometer
a scientific instrument that detects seismic waves

LITERACY LINK 

Write a short story about a scientist who finds out a category XII earthquake is about to hit Sydney.

NUMERACY LINK 

How long would it take a P wave to travel from the epicentre to a location 6000 km away? How long for an S wave?

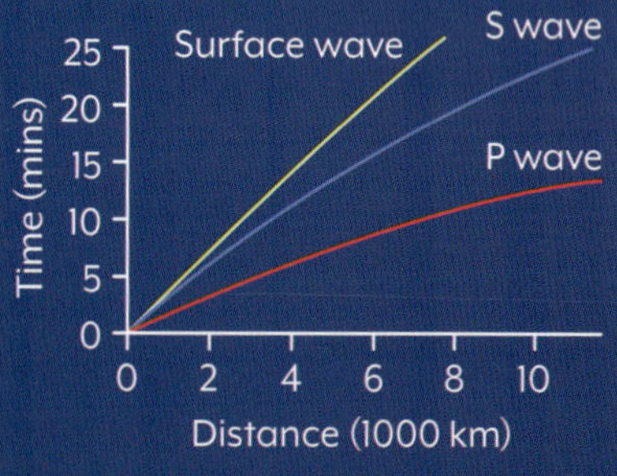

Earthquakes are caused by a build-up of pressure and the release of energy in Earth's crust. Most earthquakes happen along the boundaries of the tectonic plates, but they can also happen within a plate. A fault is the name given to a part of the Earth's surface where blocks of rock slide past each other. Faults can be as large as a tectonic plate boundary or much smaller.

1 Earthquakes happen when tectonic plates move

Blocks of rocks don't slide smoothly past each other. They catch and lock together, almost like Velcro. As they catch, pressure builds up until the rock is forced to move, releasing energy. The energy passes through Earth as **seismic waves** that move and shake the crust. This is an earthquake.

The point where an earthquake starts is called the **focus**. This is where the pressure has built up and been released, often causing rocks to rupture (break) and move along the fault. The point on the surface directly above the focus is called the **epicentre**.

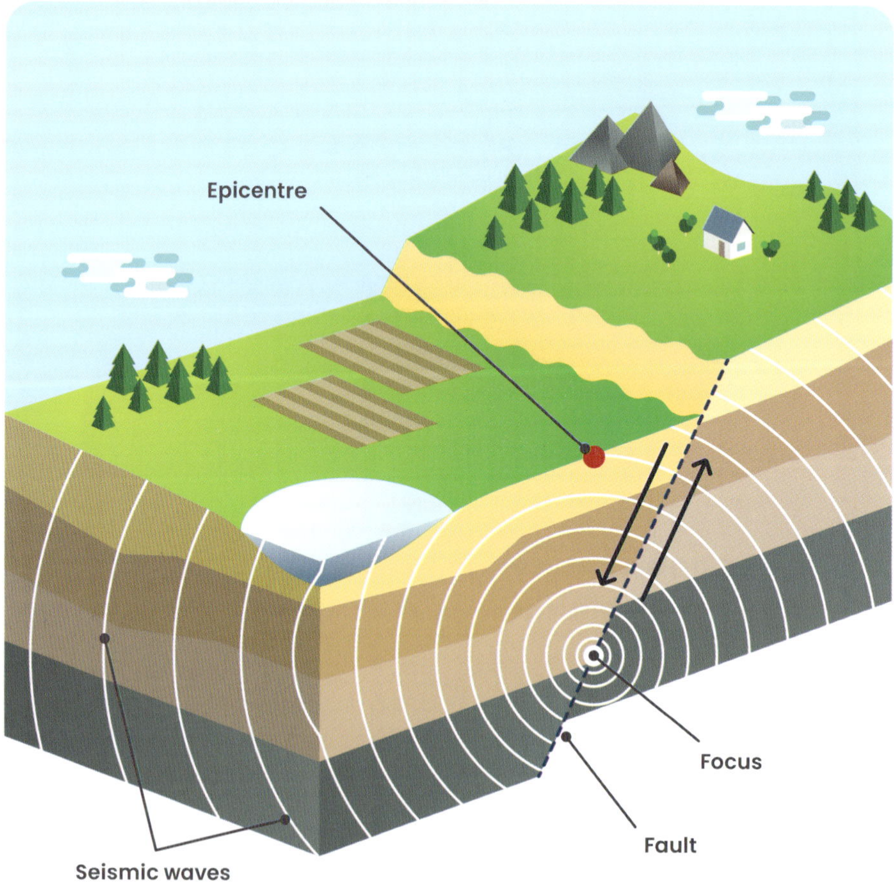

Figure 6.10 The focus is the point where an earthquake starts, and the epicentre is the point on the surface above it.

INVESTIGATION 6.4
Slinky waves

Epicentres are usually located along plate boundaries. This is because this is where most movement happens. Shallow earthquakes happen at divergent boundaries as the plates move apart. Shallow earthquakes also happen at transform boundaries because the plates are sliding past each other. The deepest earthquakes happen at subduction zones because the subducting plate is moving down into the asthenosphere. Earthquakes that take place within plates are called intraplate earthquakes. Intraplate earthquakes happen along fault lines and are caused by the build-up of pressure within the plate.

What causes earthquakes?

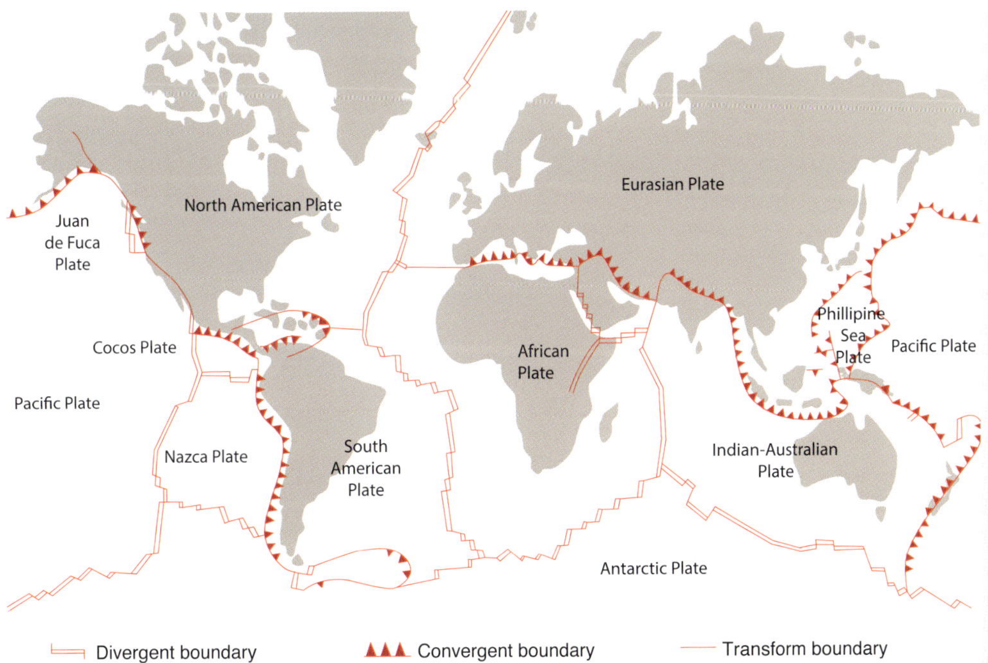

Figure 6.11 Most earthquake epicentres are located on plate boundaries and are caused by the movement of the plates

2 Earthquakes produce seismic waves

Seismologists are scientists who study earthquakes. They use sensitive equipment called **seismometers** to measure how much and how often Earth's outer layers move as a result of a seismic wave.

There are two main types of seismic wave. Body waves travel through Earth and surface waves travel around Earth's surface.

There are two main types of body wave – primary (P) waves and secondary (S) waves. Primary waves travel faster and are longitudinal waves. Secondary waves are transverse waves.

There are also two types of surface waves, known as Rayleigh and Love waves. They can move across the surface in all directions, like a rolling ocean wave. These waves travel more slowly than P and S waves, but their surface motion causes more destruction.

What are the two main types of seismic wave?

6.4 continued...

6.4 continued... Earthquakes

KEY TERMS 

intensity
a measure of the amount of destruction caused by an earthquake

magnitude
a measure of the energy released by an earthquake

moment magnitude scale
a logarithmic scale used to compare the amount of energy released by earthquakes

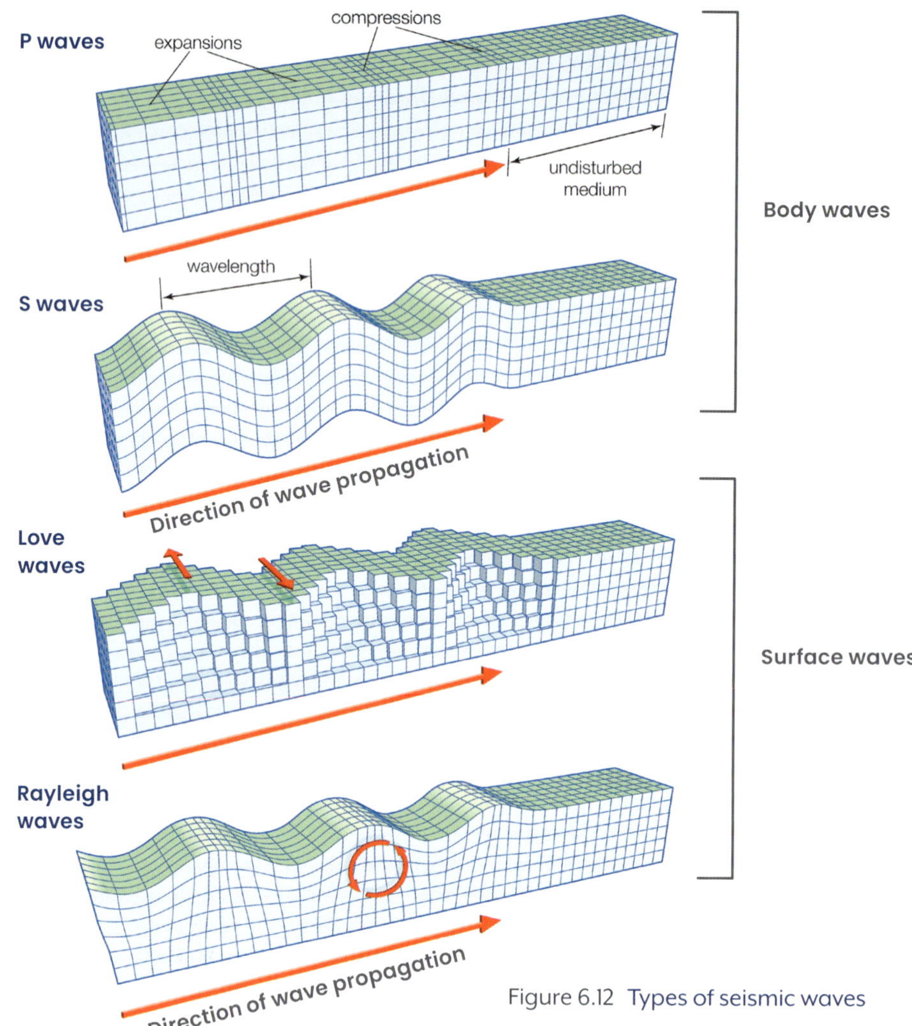

Figure 6.12 Types of seismic waves

❸ Earthquakes are measured by magnitude and intensity

Scientists describe the size of the earthquake in terms of its magnitude and intensity.

The **magnitude** of an earthquake depends on how much energy it releases, and is related to the area of the fault that ruptured and how far it moved.

The **moment magnitude scale** is used to compare earthquakes. On the moment magnitude scale, a magnitude of 10 is equivalent to the energy released by 100 000 atomic bombs. Seismologists now use the moment magnitude scale instead of the Richter scale because it allows them to more accurately compare the size of earthquakes all over the world.

An earthquake's **intensity** refers to the amount of damage it causes and is measured by the modified Mercalli intensity scale (Table 6.1). The intensity of an earthquake is influenced by the magnitude of the earthquake, the distance from the epicentre, the local geology, and any building and other structures in the area.

How is the size of an earthquake usually described?

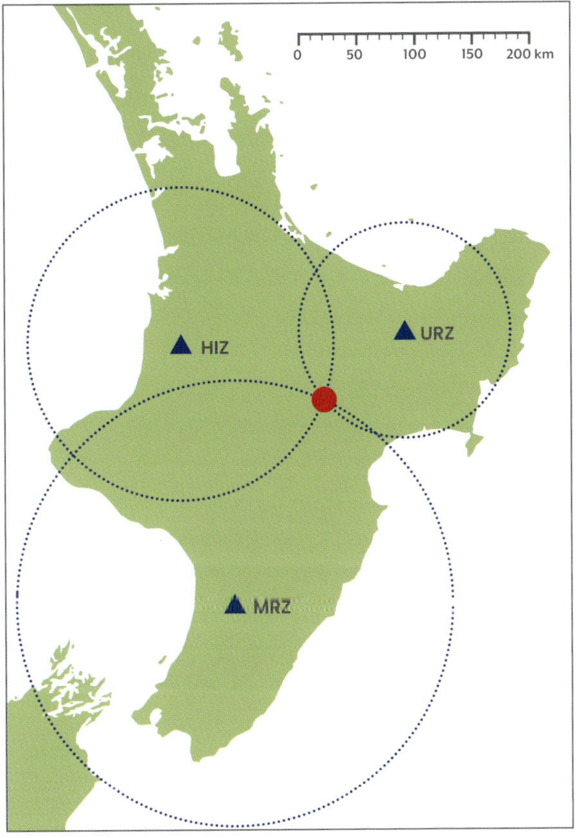

Figure 6.13
Triangulation is used to locate the epicentre of an earthquake in New Zealand. If you draw a circle around each seismic monitoring station (HIZ, URZ and MRZ) with the radius being the distance to the epicentre, then the epicentre is where the three circles overlap.

Table 6.1 **The modified Mercalli intensity scale refers to the amount of damage caused by an earthquake.**

Intensity	Shaking	Common observations
I	Not felt	Detected only by instruments
II	Weak	Noticed only by people at rest
III	Weak	Noticed by people indoors. Vibrations similar to a passing truck
IV	Light	Felt by people indoors and some people outdoors. Loose objects disturbed
V	Moderate	Felt by most people. Unstable objects overturned. Pendulum clocks stopped
VI	Strong	Felt by everyone. Slight structural damage
VII	Very strong	Felt by people in vehicles. Damage to poorly designed structures
VIII	Severe	Slight damage to well-designed structures. Much damage to other buildings
IX	Violent	Much damage to substantial structures
X	Extreme	Many buildings destroyed
XI	Very disastrous	Few structures left standing
XII	Catastrophic	Total destruction

CHECKPOINT 6.4 

1 Describe what causes earthquakes.
2 Describe the difference between primary and secondary waves.
3 Which waves would cause more destruction to buildings – body waves or surface waves? Justify your response.
4 What information do seismologists need to know to determine how far an earthquake epicentre is from a seismic station?
5 Explain why a high-magnitude earthquake in an unpopulated area could have a lower value on the intensity scale than a lower magnitude earthquake in a populated area.
6 Explain why Australia has fewer earthquakes than New Zealand.

CHALLENGE

7 Use the internet to research the five largest earthquakes that have been recorded. Plot their locations on a map that shows tectonic plate boundaries.
What do their locations have in common? Select one earthquake and conduct further research to find out what major impacts it had on the population.

SKILLS CHECK

- I can explain what causes earthquakes and how this is linked to plate tectonics.

6.5 Volcanoes

At the end of this lesson I will be able to:

- **outline** how volcanoes can be explained by the theory of plate tectonics.

KEY TERMS 

lava
molten rock above Earth's surface

magma
molten rock below Earth's surface

strato volcano
a volcano formed at a subduction zone

volcano
a point in Earth's crust where lava erupts

LITERACY LINK 

Research to find out the origin of the word *volcano*.

NUMERACY LINK 

The VEI (volcanic explosivity index) is a logarithmic scale used to measure the intensity of an eruption. For example, an eruption classified as 3 is ten times more explosive than one at 2. Compare the explosivity of a volcano that is 5 on the VEI scale with one that is 8.

A **volcano** is where molten rock erupts at Earth's surface. Most volcanoes are found along the boundaries of tectonic plates, which helps explain how they formed and their behaviour.

Volcanoes along diverging boundaries have runny lava that spreads out over large areas. Volcanoes formed along subduction zones have thicker lava and tend to be more explosive. Hot spot volcanoes are formed in the middle of a plate, rather than at a boundary.

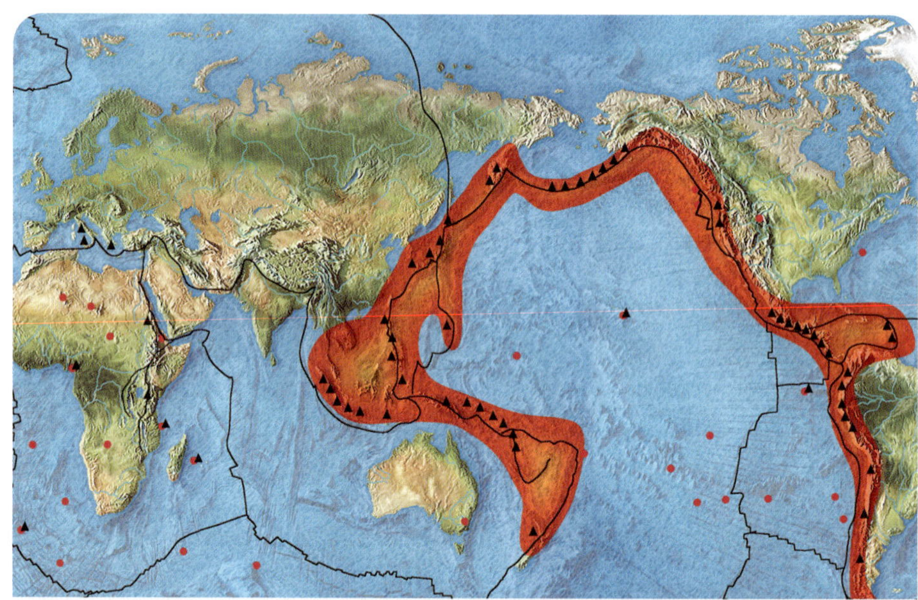

Figure 6.14 The Ring of Fire is a chain of volcanoes around the north, east and western edges of the Pacific Ocean.

1 Some volcanoes form at divergent plate boundaries

Figure 6.15 Basalt is an igneous rock that is formed from the fast cooling of lava from volcanoes at divergent boundaries. It makes up oceanic crust.

At divergent plate boundaries, volcanoes form when **magma** (molten underground rock) rises up to fill the gap between the two diverging plates.

Most volcanoes on divergent plate boundaries form as fissure volcanoes – long fractures in the crust from which the **lava** erupts – and can be many kilometres long. Because the lava has come from the asthenosphere, it contains a lot of dark minerals and is very hot and runny, so spreads out over large areas. When it cools and solidifies, the lava forms an igneous rock called basalt.

What causes volcanoes to form along divergent plates?

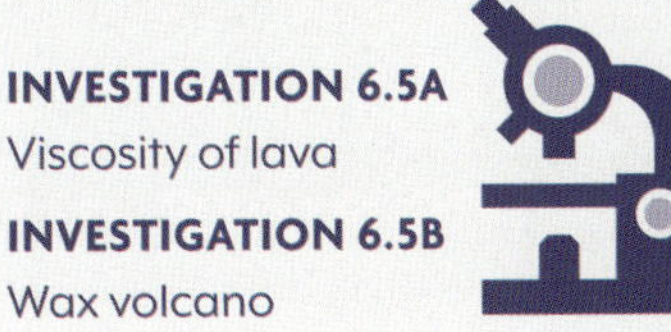

INVESTIGATION 6.5A
Viscosity of lava

INVESTIGATION 6.5B
Wax volcano

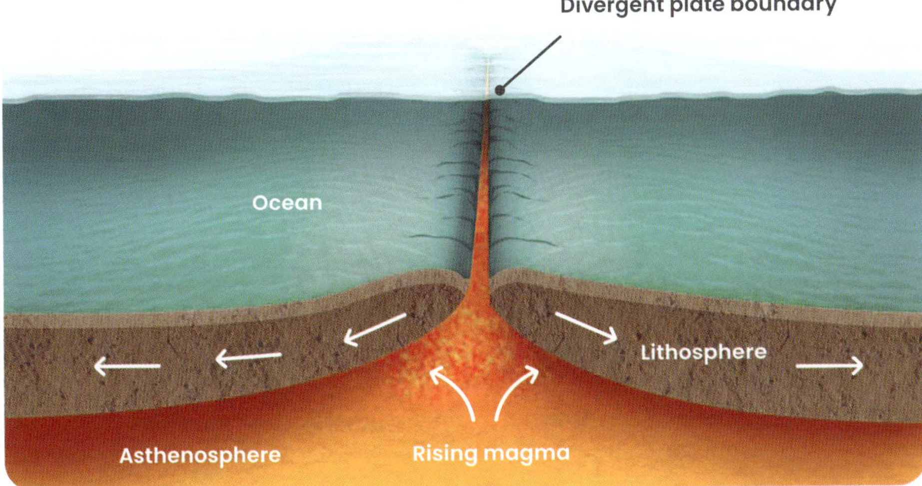

Figure 6.16 Formation of a volcano on a divergent plate boundary under the ocean. As magma rises up in the asthenosphere, it forms a fissure volcano many kilometres long.

❷ Volcano arcs form at subduction zones

Volcanoes also form in chains known as arcs along the length of subduction zones.

Subduction zone volcanoes erupt violently because the pressure of gases moving up to the surface can build up and cause explosive eruptions. Because the lava is sticky, it doesn't spread out very far. Instead, volcanoes build up with steep sides.

Volcanoes along subduction zones are known as **strato volcanoes**.

Where do strato volcanoes form?

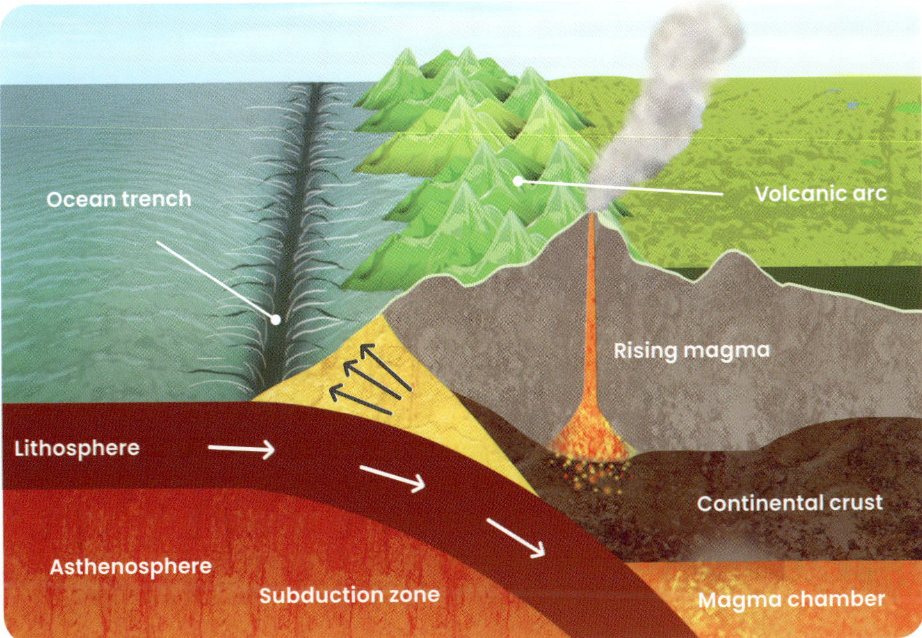

Figure 6.17 Strato volcanoes form as one plate subducts under the other. This causes magma to form and move up towards the surface.

Figure 6.18 Rhyolite is an igneous rock formed from the rapid cooling of lava from a strato volcano.

6.5 continued...

6.5 continued...

Volcanoes

KEY TERMS 

hot spot volcano
a volcano formed by magma upwelling underneath a tectonic plate

❸ Some volcanoes form in the middle of plates

Some volcanoes are not on plate boundaries, but are in the middle of the tectonic plates. These volcanoes are formed when there is an upwelling of magma (a hot spot) in a single location, so they are known as **hot spot volcanoes** or shield volcanoes. The lava that erupts out of hot spot volcanoes is similar to that of volcanoes at divergent boundaries because it comes from the asthenosphere. It is made of dark minerals, is very runny and forms the igneous rock basalt when it solidifies. The lava spreads out over large areas and over time forms volcanoes that are wide at the base compared to their height.

The hot spot in the mantle does not move, but the tectonic plates do and over time new volcanoes form in a chain (Figure 6.20). Unlike volcanoes that form along a subduction zone, only the volcano over the hot spot is active and erupts. The volcanoes that have moved away from the hot spot no longer have a magma source and so are said to be extinct.

Where on the tectonic plates do hot spot volcanoes form?

Figure 6.19 A fissure eruption at Iceland's Eyjafjallajökull volcano. This volcano has formed as the North American and European Plates move apart.

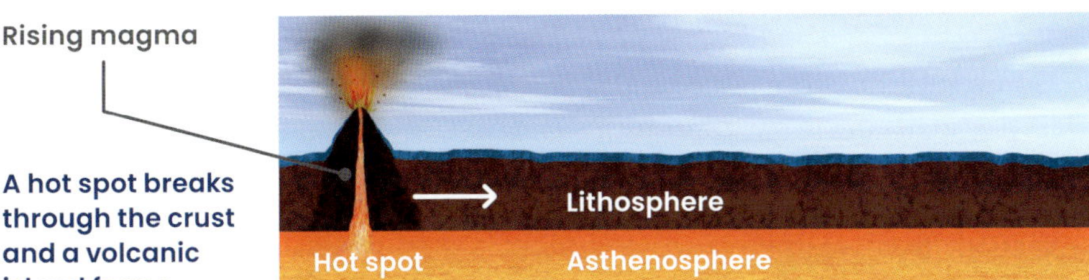

The crustal plate is constantly moving, so the island eventually moves off the hot spot. This makes room for a second volcanic island.

More islands form as the crustal plate continues to move over the hot spot.

Figure 6.20 Steps in the formation of a volcanic island chain as a result of hot spot volcanism

4 Volcanoes create new landforms

We often think of volcanoes as destructive, but they are actually the main way in which landforms, especially islands, are created.

The Hawaiian Islands and Galapagos Islands have been formed by hot spot volcanism, and the positions of the islands in the chain can provide evidence for the direction and speed at which the plate is moving. Volcanoes in eastern Australia are part of a chain of hotspot volcanoes that formed as the continent moved north away from Antarctica. The oldest volcanoes are in Queensland, and the youngest is Mount Gambier in South Australia, which was last active about 10 000 years ago.

What kind of volcanoes formed the Hawaiian Islands?

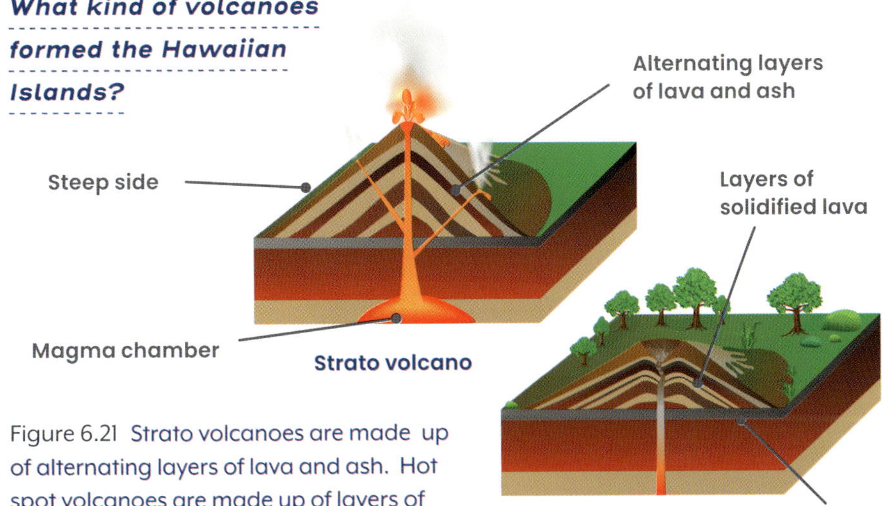

Figure 6.21 Strato volcanoes are made up of alternating layers of lava and ash. Hot spot volcanoes are made up of layers of solidified lava.

CHECKPOINT 6.5 ✓

1 State the type of volcano that is under the ocean and is many kilometres long.

2 State the type of volcano that forms chains of volcanoes in the middle of tectonic plates.

3 State the type of volcano that is made up of alternating layers of ash and lava flows.

4 Describe the relationship between the theory of plate tectonics and volcanoes.

5 How is the lava that erupts from strato volcanoes formed?

6 Compare and contrast a chain of islands formed by a hot spot with a chain of islands formed by a subduction zone.

CHALLENGE

7 You observe some volcanoes on a continent. What evidence do you need to look for to determine how they were formed? Justify your reasoning.

SKILLS CHECK

- I can explain how volcanoes are formed.
- I can name and describe at least two types of volcanoes.

6.6 Monitoring our Earth

At the end of this lesson I will be able to:

- **describe** how technology has allowed a greater scientific understanding of global patterns of geological activity.

KEY TERMS 

GPS
global positioning system

tsunami
a sea wave caused by the displacement of water as a result of an earthquake or other disturbance

LITERACY LINK 

Technology can now give some warning for earthquakes and tsunamis. Create a pamphlet that tells residents what to do in case of either an earthquake or tsunami.

NUMERACY LINK 

Use the formula below to determine the speed (in m/s) of a tsunami wave at a depth of 4,500 m:

$$\text{speed} = \sqrt{g \times d}$$

Where g = acceleration due to gravity (9.81 m/s^2) and d = water depth in metres.

Convert your answer to km/h.

As technology has improved, scientists have been able to collect more data and develop our understanding about Earth. In the 20th century, new technologies enabled scientists to discover evidence for plate tectonics. More recently, very precise instruments, satellites and internet communication technologies have allowed us to collect even more data and information about Earth.

Some countries have technology to warn citizens about geological hazards such as earthquakes and tsunamis.

1 Geologists use seismic data to predict the size of earthquakes

Computer analysis of seismic data can help geologists build better models of local geology, such as how plates move at subduction zones. As one plate subducts under another, it causes many earthquakes to occur. Scientists can plot this information to produce a three-dimensional image of the subduction zone. The data can also identify if the plate tends to move down at a steady rate, causing many smaller earthquakes, or if it tends to get stuck and move suddenly, causing fewer larger earthquakes.

Seismologists can determine patterns and predict the possible magnitudes and intensities of earthquakes in a particular region, but they can't predict exactly when an earthquake is going to occur.

How can earthquakes be predicted?

Figure 6.22 The 2004 Boxing Day tsunami devastated more than half of Sri Lanka's coastline.

2 Warning systems can detect tsunamis

Tsunamis are fast-moving waves generated from a sudden massive movement of water. Most tsunamis are caused by earthquakes, but they can also be caused by undersea volcanoes or even a meteorite impact. Tsunamis can result in mass destruction and loss of life in populated areas.

Earthquakes that cause tsunamis happen along subduction zones because when one plate moves under during an earthquake, it pushes the other plate upwards suddenly, displacing the water above it.

After the Boxing Day tsunami hit Indonesia in 2004, the Australian Government funded the Joint Australian Tsunami Warning System, which uses various technologies to detect tsunamis. The system involves a network of seismic sensors that measure earthquakes and send out warnings when an earthquake occurs that could cause a tsunami.

Deep ocean detection buoys have also been deployed near the subduction zones that surround Australia to monitor sea level changes. The buoys will send out an alert via satellites if they detect a change in sea level that could have been caused by a tsunami, providing the warning centre with further information.

How can tsunamis be detected?

Figure 6.23 A deep ocean detection buoy is deployed in the Tasman Sea to monitor changes in sea levels that may have been caused by a tsunami.

3 Australia is slowly moving

Geologists know from **GPS** (global positioning system) satellite and laser technology that Australia is moving north at about 7 cm a year.

As the continent moves, the latitude and longitude coordinates change. GPS navigation systems rely on reference sets of data to determine location. If the longitude and latitude of the continents are not kept up to date, systems that rely on GPS for navigation and positioning will be incorrect.

Geoscience Australia recently updated the data set that was being used by GPS navigation systems because it was out by about 1.5 m!

What technologies are used to measure the movement of the Australian continent?

CHECKPOINT 6.6 

1 Technology has led to greater scientific understanding of geological activity. Explain how.

2 Explain why seismic data can help us understand what is happening at subduction zones.

3 Explain why it is better for tsunami warning centres to use data from both earthquakes and sea level changes to determine if a tsunami could affect the Australian coastline.

4 If Australia is moving north at 7 cm a year, when will the continent be 1.5 m further north than it is today?

5 Explain why it is important to be able to know exactly where continents are located when using GPS and other satellite navigation technologies.

CHALLENGE

6 Research tsunami warning systems and prepare a short report about which countries have them and how they work.

SKILLS CHECK

- I can describe how technology has allowed a greater understanding geological activity.
- I can name at least two pieces of technology that have led to greater scientific understanding of geological activity.

CHAPTER SUMMARY

The theory of **continental drift** states that the continents have moved over time.

Evidence for continental drift includes:

- the continental shelves of continents fitting together like pieces of a jigsaw
- identical rock formations on either side of the Atlantic Ocean
- identical plant and animal fossils on different continents now separated by oceans.

SOUTH AMERICA
AUSTRALIA
ANTARCTICA

Fossil evidence of freshwater reptile *Mesosaurus* has been found in Brazil and Africa.	Fossil evidence of the land reptile *Lystrosaurus* has been found in Africa, Antarctica and India.	Fossils of the fern *Glossopteris* have been found in all southern continents.	Fossil evidence of land reptile *Cynognathus* has been found in Argentina and southern Africa.

▲ **GPS data** can be used to track the movement of the continents.

The theory of **plate tectonics** explains how the continents move.

Tectonic plates interact in three different ways:

- divergent – plates move away from each other
- convergent – plates move towards each other
- transform – plates slide past each other.

Divergent plate boundary

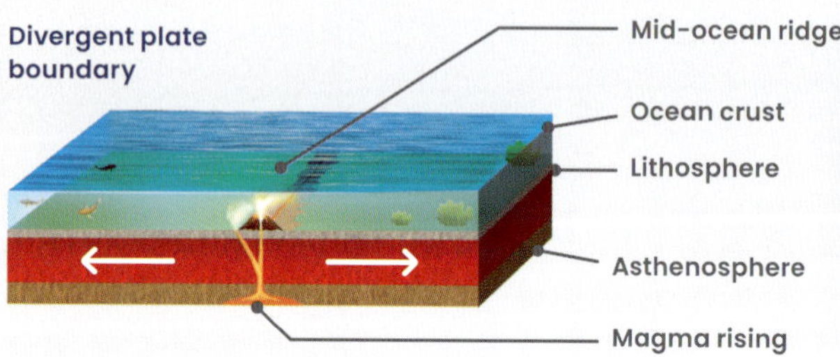

Convergent plate boundary

Transform plate boundary

Earthquakes
Continental crust
Lithosphere
Asthenosphere

Most **earthquakes** happen when tectonic plates move.

The **epicentre** is the point on the surface directly above the focus.

The **focus** is the point where the earthquake starts.

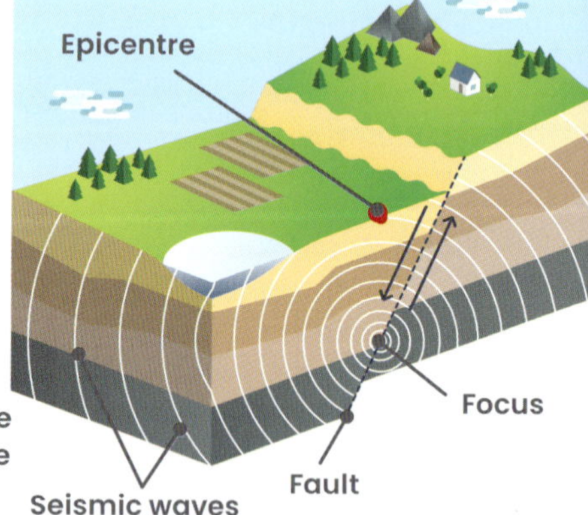

◄ Earthquake data can be used to warn about **tsunami** risks.

FINAL CHALLENGE

1. Identify this statement as true or false. The theory of continental drift is the same as the theory of plate tectonics.
2. List three observations that Alfred Wegener identified as evidence that the continents have drifted over time.

3. What layer of Earth do tectonic plates 'float' on?
4. Identify three pieces of evidence for the theory of plate tectonics.
5. Match the following plate boundaries with the types of plate movement.

Divergent	Plates slide past each other
Convergent	Plates move apart
Transform	Plates move together

6. Copy and complete these sentences.
   a Earthquakes produce _________ waves. _________ waves travel through Earth, and _________ travel along the surface of Earth.
   b There are two main types of body wave; _________ or P waves travel fastest, _________ or S waves travel _________.
   c Surface waves cause the _________ damage.

7. What is subduction? Use a labelled diagram to help you describe the process.
8. About 85 million years ago the continents of Australia and Antarctica began to separate. The continents were completely separated 30 million years ago.
   a Identify the type of plate boundary that would have formed to separate the continents.
   b Draw a series of labelled diagrams that illustrate two continents separating and the features that would be observed.

9. Identify two ways that earthquake data can be used to tell us more about patterns of geological activity.
10. Explain why more earthquakes and volcanic eruptions are experienced in Indonesia than in Australia.

7 Global systems

No matter how big or small, everything that happens on Earth has an effect on everything else. You can consider Earth to have four interacting spheres – the atmosphere (air), hydrosphere (water), lithosphere (rock) and biosphere (living things) – which help explain how the natural systems and cycles work. Knowledge of Earth's four spheres is especially important for understanding how human activity affects Earth and how we can preserve natural environments for future generations.

1 LEARNING LINKS

Brainstorm as much as you can remember about these topics.

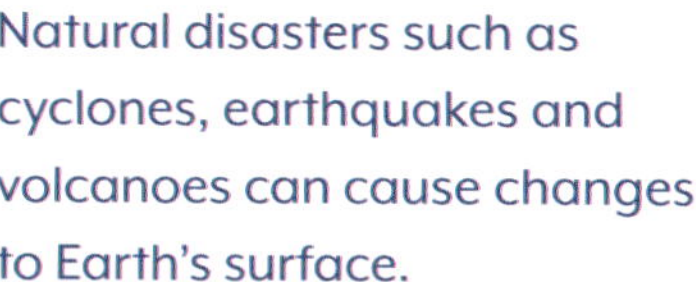

Natural disasters such as cyclones, earthquakes and volcanoes can cause changes to Earth's surface.

Science and technology can be used to plan for and manage natural disasters.

Earth is made up of layers – core, mantle and lithosphere.

2 SEE-KNOW-WONDER

List three things you can **see**, three things you **know** and three things you **wonder** about this image.

3 CRITICAL + CREATIVE THINKING

Alphabet key: Create an A–Z list of words associated with Earth's four spheres.

The reverse: What would happen if carbon dioxide was a product of photosynthesis, not a reactant?

What if: All the ice on Earth melted overnight?

4 THE NEWEST GEOLOGICAL EPOCH?

If geologists in the future were to look back on rocks that formed today, they would see big differences between these rocks and those formed in the past. Evidence such as plastics in sediments, an increase in carbon dioxide in the atmosphere and the rapid extinction of species is showing that humans have significantly altered Earth's spheres in the last 100 years. It has been proposed that these changes are significant enough to mark the start of a new geological epoch – the Anthropocene.

7.1 Earth's spheres

At the end of this lesson I will be able to:

- **describe** Earth's atmosphere, hydrosphere, lithosphere and biosphere.

KEY TERMS 

atmosphere
the layer of gas that surrounds Earth and is 600 km thick

biosphere
all of the living things on Earth

hydrosphere
all the water on Earth

lithosphere
Earth's rigid outer zone (crust and upper mantle), made up of tectonic plates

LITERACY LINK 

Etymology is the study of the origin of words. The word *sphere* comes from the Greek word *sphaira* meaning 'ball'. Propose meanings for the words *hydro, atmos, lithos* and *bio*.

NUMERACY LINK 

Create a pie chart to show the allocation of fresh water on Earth.

- Glaciers and ice caps: 69%
- Groundwater: 30%
- Available fresh water: 1%

Earth scientists consider Earth to be a system made up of four interacting spheres – the atmosphere (air), hydrosphere (water), lithosphere (rock) and biosphere (life). By studying the spheres and how they interact, we can gain a better understanding of how Earth works and how humans are affecting it.

1 The atmosphere is a layer of gas

The **atmosphere** is the 600-kilometre thick thick layer of gas that surrounds Earth. The atmosphere is made up of four major layers.

- The troposphere is the bottom layer and is where we live, jet planes fly and weather takes place.
- The next layer is the stratosphere, which contains the ozone layer.
- Above the stratosphere is the mesosphere, where meteorites burn up.
- The upper layer is the thermosphere, which is where aurora happen.

The atmosphere contains oxygen, which we breathe, and carbon dioxide, which plants need for photosynthesis. Oxygen makes up about 21% of the gases in the atmosphere. Other gases are nitrogen (78%), argon (0.93%) and carbon dioxide (0.04%). The atmosphere also contains minute quantities of neon, helium, methane, water vapour, krypton, hydrogen, xenon and ozone.

The atmosphere protects life on Earth from harmful radiation by either absorbing it or reflecting it back into space.

What is the atmosphere?

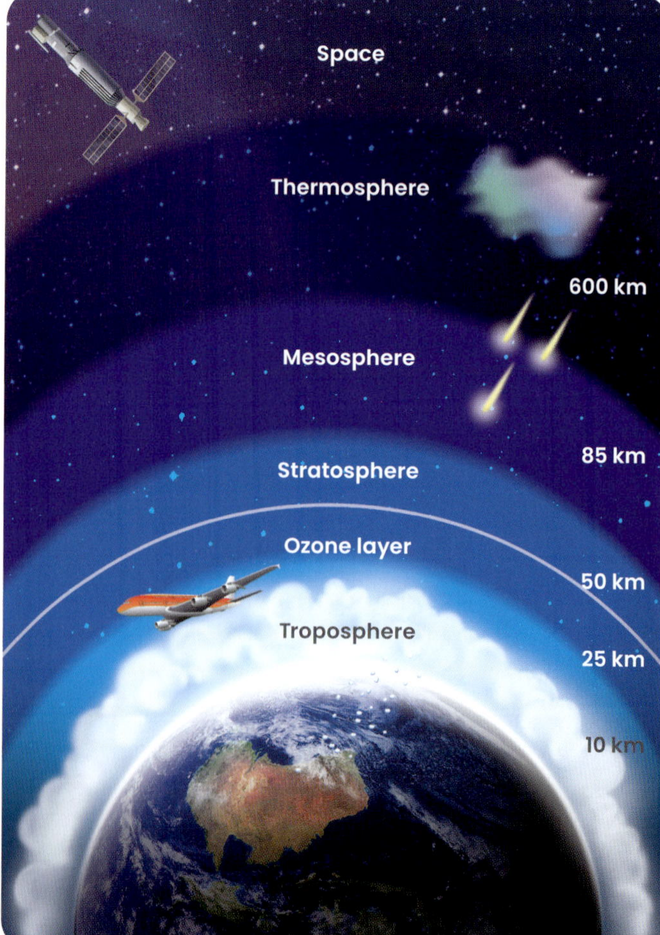

Figure 7.1 Earth's atmosphere is made up of four major layers: the troposphere, stratosphere, mesosphere and thermosphere.

❷ The hydrosphere is all the water on Earth

The **hydrosphere** is all the water on Earth, in all its forms – solid (snow/ice), liquid and gas (water vapour). The water is located on the surface (in oceans, rivers and lakes), in the atmosphere (clouds and water vapour), underground (groundwater) and in the bodies of living things (biological water).

More than 97% of the water on Earth is saline (salty). Less than 3% of the water on Earth is fresh water. Most of this fresh water is in glaciers and ice caps (69%), or as groundwater (30%). The remaining 1% is found in ground ice and permafrost, lakes, the atmosphere, living things, rivers, swamps, marshes and soil. All this water is connected through the water cycle.

What is the hydrosphere?

❸ The lithosphere is Earth's crust and upper mantle

The **lithosphere** is made up of Earth's crust and the top 100 km of the rigid upper mantle. It includes the rocks and soil, and landscape features such as mountains and valleys. The lithosphere is made up of 15 major tectonic plates that 'float' on the upper mantle. Movement of these plates results in volcanic eruptions and earthquakes.

What is the lithosphere?

❹ The biosphere includes all living things on Earth

The **biosphere** includes all the living things on Earth, from microbes to humans. When studying the biosphere, scientists consider how populations of different species interact with each other and their environment.

Biologists use food chains and food webs to represent interactions between organisms in ecosystems. Taxonomists use different levels to classify living things to show how they are related.

How do plants (in the biosphere) interact with the other spheres?

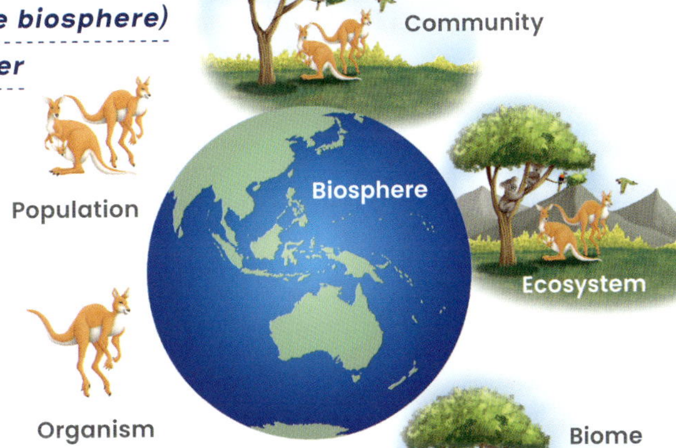

Figure 7.2 The biosphere includes all living things on Earth, how they interact with each other and their environment.

INVESTIGATION 7.1
Observing interactions between the spheres

CHECKPOINT 7.1

1 How does the atmosphere protect Earth?
2 Identify the layer of the atmosphere where:
  a we live
  b the ozone layer is
  c meteorites burn up
  d aurora occur.
3 Describe the different ways in which water is stored in the biosphere.
4 Identify features of Earth that are part of the lithosphere.
5 Identify the sphere(s) that are involved in:
  a a thunderstorm
  b a wombat digging a burrow
  c humans burning fossil fuel
  d waves eroding a beach.
6 Suggest at least three ways in which the biosphere and hydrosphere interact.

CHALLENGE

7 Use the internet to research the following problems: the melting of the ice caps, erosion, species extinction and air pollution. Explain how they relate to Earth's spheres.

SKILLS CHECK

- I can describe the atmosphere, hydrosphere, lithosphere and biosphere.
- I can discuss some ways in which the four spheres interact.

7.2 The carbon cycle

At the end of this lesson I will be able to:

- **explain** how the carbon cycle is an example of the connections between Earth's spheres.

KEY TERMS 

carbon cycle
the cycle that explains how carbon moves between Earth's spheres

cellular respiration
the process that all living things use to produce cellular energy from glucose and oxygen

photosynthesis
the process that plants use to make glucose from carbon dioxide and water

NUMERACY LINK 

Use the graph to answer the following questions.

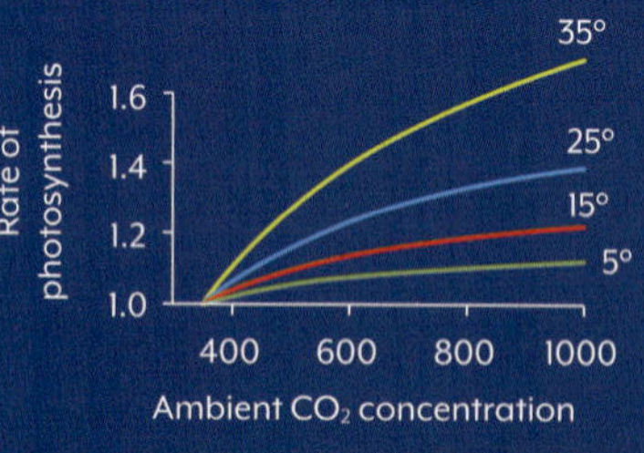

a Describe the pattern of results for this graph.

b Construct a table showing the rate of photosynthesis at different temperatures when the ambient CO_2 concentration is 600 μmol mol^{-1}.

Matter cannot be created or destroyed – instead, it is recycled. One day, an atom is part of the atmosphere; the next day, it's part of a plant. A week later, that atom is part of an animal, and eventually it moves into the soil and then the ocean. The same atoms are recycled over and over again as they move through the four spheres.

We use models called cycles to better understand the movement of matter between the spheres. The carbon cycle is a particularly important cycle that affects all of Earth's systems.

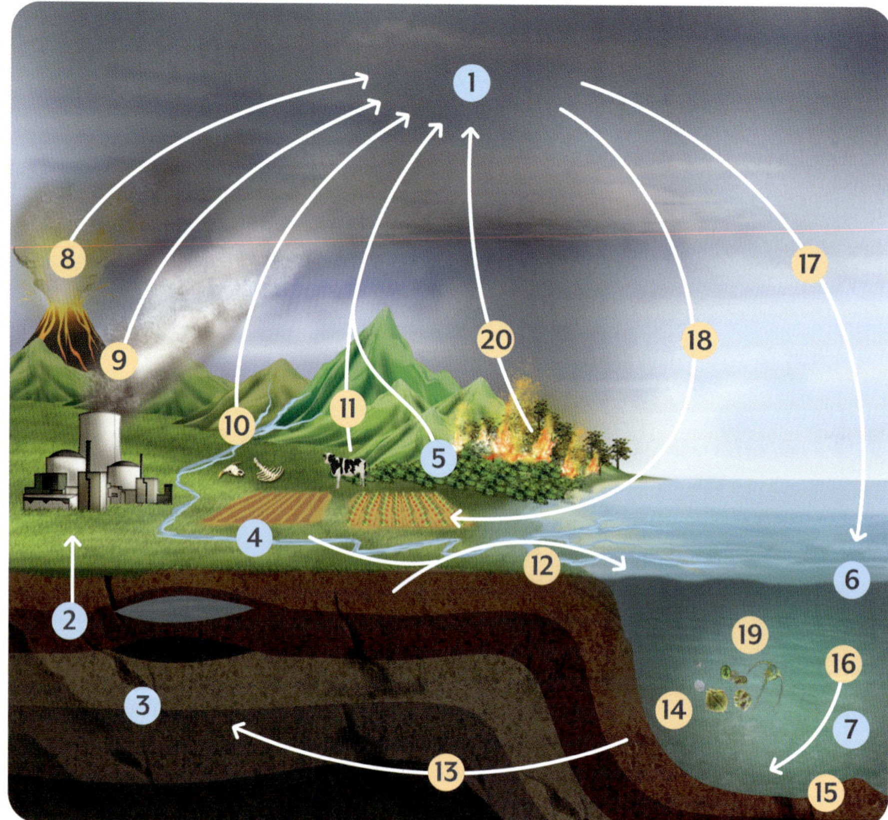

CARBON STORES
1. Atmosphere
2. Coal, oil, gas
3. Sediments and sedimentary rock
4. Soil and organic matter
5. Plants
6. Ocean surface
7. Deep ocean

PROCESSES
8. Erupting volcanoes
9. Burning fossil fuels
10. Decomposition
11. Respiration
12. Weathering, erosion and run-off
13. Rock formation
14. Shellfish and corals
15. Sinking sediment
16. Deep ocean currents
17. Carbon dioxide exchange
18. Photosynthesis
19. Phytoplankton
20. Burning

Figure 7.3 The carbon cycle shows how atoms of carbon move between the atmosphere, biosphere, hydrosphere and lithosphere.

INVESTIGATION 7.2A
Releasing dinosaur breath

INVESTIGATION 7.2B
The effect of temperature on soil respiration

❶ Carbon is a vital element

Carbon is a key element for life. It is the second most abundant element (after oxygen) in the human body and just as important for other living things. Carbon is the main building block of carbohydrates, fats and proteins – which make up pretty much everything in your body. When you eat food, you are consuming carbon. When you exhale, you are removing carbon from your body in the form of carbon dioxide.

The **carbon cycle** explains how carbon moves between Earth's spheres. In the atmosphere, carbon is found in molecules such as carbon dioxide, which helps to regulate Earth's temperature. In the lithosphere, carbon is found in soils as decomposing matter, and in rocks, often in the form of calcium carbonate ($CaCO_3$). In the hydrosphere, carbon is in the form of carbon dioxide dissolved in oceans, lakes and rivers.

Why is carbon a key element for life?

❷ Carbon cycles through Earth's spheres

Carbon moves between Earth's spheres through some of these processes:

- Photosynthesis moves carbon from the atmosphere into plants and other producer organisms. **Photosynthesis** is the process by which plants synthesise glucose, which can then be used for growth and respiration.
- Carbon moves through food webs from producers to consumers.
- **Cellular respiration** moves carbon from the biosphere to the atmosphere. In cellular respiration, living things use oxygen and glucose to produce energy, with carbon dioxide being a waste product.
- Decomposition moves carbon from living things into the soil. When living things die and start to decay, their remains become part of the soil. Decomposers in the soil consume the dead matter and release the carbon back into the atmosphere through cellular respiration. If the dead matter is not decomposed, it can form fossil fuels such as oil and coal, locking the carbon away for hundreds of millions of years.
- Carbon dioxide dissolves into bodies of water where the surface mixes with the atmosphere. Here it often forms carbonate ions (CO_3^{2-}).
- Marine organisms use the carbon dissolved in sea water (in the form of carbonate ions) to build their shells.
- Tectonic movements bring rocks containing carbon, such as limestone (calcium carbonate) or fossil fuels (e.g. coal), to the surface. The carbon can then be washed into waterways, or returned to the atmosphere.
- Burning of forests and fossil fuels releases carbon into the atmosphere.
- Volcanic eruptions release carbon from the lithosphere into the atmosphere as carbon dioxide.

How can carbon move between the biosphere and the hydrosphere?

Figure 7.4 Decomposition releases carbon back into the atmosphere through cellular respiration.

7.2 continued...

7.2 continued...

The carbon cycle

KEY TERMS 

carbon sink
a place where carbon is stored in the carbon cycle

permafrost
frozen soil in the Arctic

3 Carbon sinks are stores of carbon

Carbon sinks are where carbon is taken from the atmosphere and stored within the carbon cycle. This means that the carbon is locked up and cannot cycle for long periods of time. Examples of carbon sinks are forests and the ocean, although fossil fuel deposits and rocks that contain large amounts of calcium carbonate, such as limestone, are also significant.

Soil is one of the most important carbon sinks because more carbon is stored in soil than in living things and the atmosphere combined. This is because a lot of decomposed organisms end up as part of the soil. Once a carbon atom is locked up in the lithosphere, it can be buried for many millions of years until natural movement of the tectonic plates brings it up to the surface again.

The ocean is the world's largest carbon sink. When air from the atmosphere mixes with water in the ocean, carbon dioxide gas dissolves into the water. This carbon can then be used by marine creatures, such as molluscs and corals, to build their shells and skeletons.

What is a carbon sink?

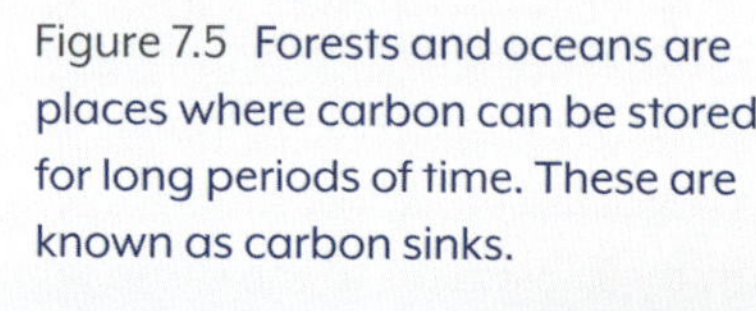

Figure 7.5 Forests and oceans are places where carbon can be stored for long periods of time. These are known as carbon sinks.

4 Carbon sources release carbon

Carbon sources release carbon into the atmosphere as part of the carbon cycle. Volcanoes, decomposition, respiration and fires are all natural carbon sources. Weathering of rocks containing calcium carbonate can also release carbon back into the atmosphere.

Decomposers, such as bacteria and fungi, play an important role in the carbon cycle. They return the carbon locked up in the bodies of dead organisms to the atmosphere as carbon dioxide, by consuming them and undergoing cellular respiration.

What is a carbon source?

5 Humans influence the carbon cycle

The processes of mining and burning fossil fuels is interrupting the natural carbon cycle. Instead of the carbon atoms being locked away in sinks, they have been brought to the surface and added back into the active carbon cycle much faster than they would naturally.

This activity has greatly increased the amount of carbon in the atmosphere, mostly as carbon dioxide, but also as other greenhouse gases such as methane (CH_4), resulting in the enhanced greenhouse effect. Natural processes of the carbon cycle that remove carbon from the atmosphere and store them in sinks, such as photosynthesis, are unable to significantly reduce this excess carbon. Deforestation has added to this problem by reducing the size of forests and limiting the amount of photosynthesis that can take place.

Scientists have found that the ocean has absorbed some of the excess carbon from the atmosphere. However, this has led to ocean water becoming more acidic, having negative impacts on marine life. If ocean temperatures increase, the ability for the ocean to act as a carbon sink will decrease, as it will be unable to absorb as much carbon dioxide from the atmosphere.

There is a real threat that warming caused by the enhanced greenhouse effect will cause **permafrost** to thaw, releasing even more carbon into the atmosphere as the trapped methane and carbon dioxide it holds escape.

How does the burning of fossil fuels change the natural carbon cycle?

CHECKPOINT 7.2 

1 Identify the natural processes illustrated in Figure 7.3.

2 Identify the human-influenced processes illustrated in Figure 7.3.

3 What is the role of the carbon cycle?

4 Describe one way that carbon is moved from one sphere into another.

5 Create a flow chart to illustrate the effects of burning fossil fuels on the natural carbon cycle.

6 Another name for a matter cycle like the carbon cycle is a biogeochemical cycle. Propose why this is a suitable name.

7 Deforestation also affects the natural carbon cycle. Use information provided in this section to explain how.

CHALLENGE

8 Use the internet to research new technology for carbon capture and storage. Provide a summary, and state and explain your opinion on whether it is the answer to climate change.

9 Explain how the oxygen cycle is linked with the carbon cycle.

SKILLS CHECK

- I can describe four major steps in the carbon cycle.
- I can describe how the carbon cycle connects the biosphere, lithosphere, hydrosphere and atmosphere.

7.3 The impact of volcanic eruptions

At the end of this lesson I will be able to:

- **describe** some impacts of a volcanic eruption on Earth's spheres.

KEY TERMS 

aerosol
tiny droplets in the atmosphere

igneous rock
rock formed when molten rock cools and becomes solid

lava
molten rock at Earth's surface

magma
molten rock under Earth's surface

pyroclastic flow
a dense and fast-moving mass of extremely hot ash and gas

LITERACY LINK 

Brainstorm and write down as many impacts on each sphere caused by the eruption of volcanoes as you can think of.

NUMERACY LINK 

A scientist measured the sulfur dioxide emissions, in Dobson units, around an active volcano over a number of days and obtained the following data: 12.2, 10.3, 11.8, 11.2 and 10.9.

Calculate the average value of the sulfur dioxide emissions.

When a volcano erupts, we observe a series of changes and impacts to the lithosphere, atmosphere, hydrosphere and biosphere. These changes occur both in the region immediately surrounding the volcano, but also on a global scale.

1 Volcanic eruptions cause ash particles to circulate in the atmosphere

When a volcano erupts it adds gases and ash particles into the atmosphere, which are quickly spread around the globe by air currents. The major gases are water vapour (H_2O), sulfur dioxide (SO_2), and carbon dioxide (CO_2).

While incredibly hot, a volcanic eruption may actually cause surface temperatures around the world to drop. This is because the ash particles in the stratosphere reflect solar energy back into space. Sulfur dioxide in the stratosphere combines with water to form tiny droplets (**aerosols**) of sulphuric acid, and these also reflect solar energy.

The greenhouse gases emitted by a volcano, such as water vapour and carbon dioxide, are usually present in very small amounts and do not make a significant contribution to global warming. However, there is evidence from the geological record that massive eruptions in the past have contributed to climate change.

How can gases and ash particles from a volcanic eruption spread around the globe?

Figure 7.6 The June 1991 eruption of Mt Pinatubo in the Philippines was the second largest eruption of the 20th century. This image was taken from the space shuttle in August 1991 and shows the dark double layers of aerosols from the eruption.

2 Volcanoes can add new rock to the lithosphere

A volcano can be considered part of the lithosphere. However, an eruption will also change the landscape. **Lava** flows can destroy old landforms and create new ones, forming new **igneous rock** as it solidifies. Some eruptions are so powerful that the whole volcano will be destroyed. Ash adds minerals and nutrients to soils, making volcanic soils highly fertile.

How can volcanic eruptions affect the lithosphere?

3 Volcanoes add water vapour to the hydrosphere

Volcanic eruptions contribute to the hydrosphere. Water vapour from volcanic eruptions condenses in the atmosphere to form clouds. The water then falls as rain and snow. This may be how Earth's oceans formed many millions of years ago.

Sulfur dioxide from volcanic eruptions reacts with water in the atmosphere to form sulfuric acid. This can produce acid rain or lower the pH of the water surrounding the volcano.

Groundwater is also affected by **magma** (molten rock) rising up through Earth's crust. The rising magma can increase the acidity of the water. Volcanologists sample water around volcanoes to look for chemical clues for magma rising and an impending eruption.

Volcanoes that are under water can heat the water around them, in turn killing life.

How can volcanoes add water to the atmosphere?

4 Volcanoes can harm the biosphere but also provide rich soils

A volcanic eruption has a direct impact on the surrounding ecosystem. Lava (molten rock), ash, gas and **pyroclastic flows** (fast-moving masses of extremely hot ash and gas) can kill plants and wildlife, and destroy habitats for those that survive.

Volcanic soils are nutrient rich, so many plants grow in areas where there have been eruptions. This is one reason why areas around some volcanoes are highly populated.

What can cause the deaths of plants and wildlife?

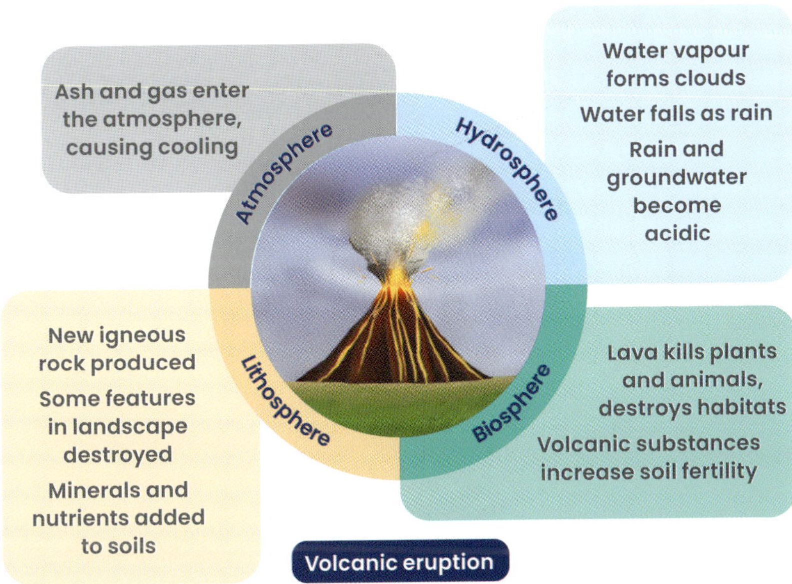

Figure 7.7 A volcanic eruption affects the atmosphere, hydrosphere, lithosphere and biosphere.

CHECKPOINT 7.3 

1 Identify the three major gases that a volcano releases into the atmosphere.
2 Why are soils around volcanoes highly fertile?
3 Does volcanic activity make water more acidic or more basic?
4 Explain how volcanic eruptions can be both good and bad for surrounding life forms.
5 Explain how volcanic ash can cool the surface of Earth.
6 Explain how a volcanic eruption in Indonesia can affect Europe.
7 Volcanic eruptions are devastating for plant life in the short term but beneficial in the long term. Explain this statement.

CHALLENGE

8 Find out how earthquakes can affect Earth's spheres. Create a diagram to illustrate the relationships.

SKILLS CHECK

- I can describe at least one impact a volcanic eruption has on the atmosphere, hydrosphere, lithosphere and biosphere.

7.4 The impact of cyclones

At the end of this lesson I will be able to:

- **describe** some impacts of cyclones on Earth's spheres.

KEY TERMS 

Coriolis force
a force caused by the rotation of Earth

cyclone
an intense tropical storm that forms over warm oceans

erosion
the natural process of wearing away by wind, water or other natural agents

LITERACY LINK 

Cyclones, hurricanes and typhoons are different names for the same phenomenon. Find out more about the origins of the names, which depend on the locations of the storms, as well as the regions that they are used in.

NUMERACY LINK 

The wind speeds for a particular cyclone were measured as follows:

Distance from cyclone centre (km)	Wind speed (knots)
100	23
50	55
30	72
25	35
0	5

Plot a graph for this data and draw a curve of best fit. Describe the pattern of results.

Cyclones are intense tropical storms that form over warm oceans and away from the equator. Cyclones are mostly an interaction between the atmosphere and the hydrosphere, although they also affect the biosphere and the lithosphere.

1 Cyclones are tropical storms over oceans

A **cyclone** is an intense tropical storm that forms over warm oceans with sea surface temperatures of at least 26.5°C and at least 5° latitude from the equator. The storm forms when warm seawater heats the air above it, causing the air to rise. Warm seas have a high rate of evaporation, adding a large amount of water vapour to the rising hot air. This water condenses into clouds when it hits cold air above.

Once the mass of water vapour is large enough, the **Coriolis force** will cause the storm to start to rotate. The Coriolis force is caused by Earth spinning; it causes air and water to rotate clockwise in the Southern Hemisphere and anticlockwise in the Northern Hemisphere.

Cyclones continue to increase in size while they are over warm water. They start to break down as they move over land.

What conditions cause cyclones?

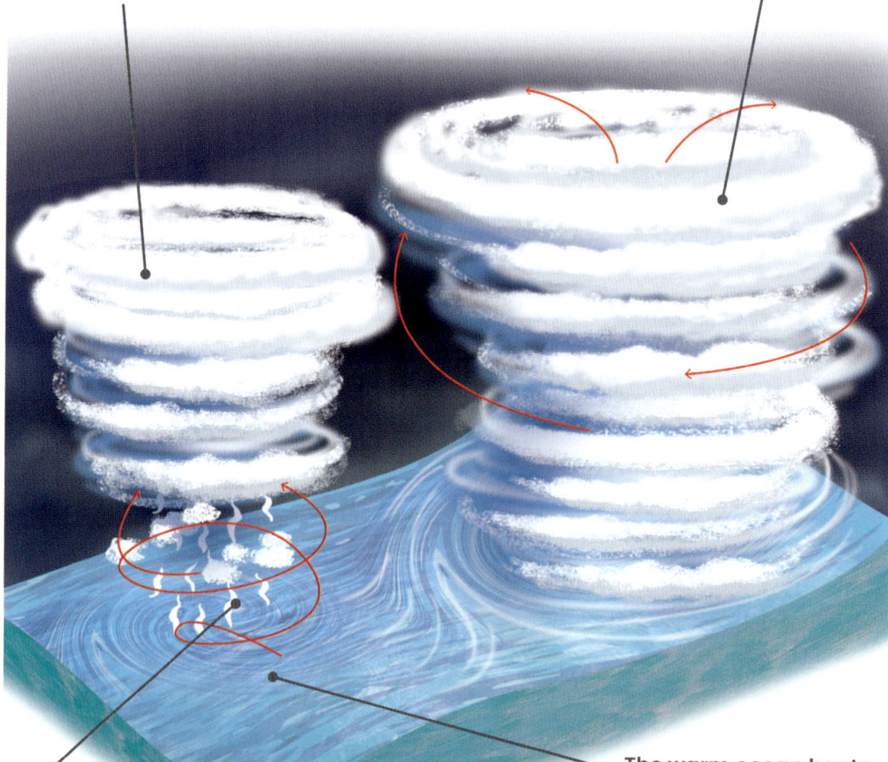

Figure 7.8 How cyclones are formed in the Southern Hemisphere

2 Cyclones involve the atmosphere and hydrosphere

Cyclones cause high-energy winds and a mass movement of water. The winds are the result of air moving from areas of high air pressure (particles are cooler and closer together) to areas of low air pressure (particles are warmer and further apart). In a cyclone, the winds spiral from the outside towards the centre (the eye) where the air rises. In the eye of the storm, the winds are a lot calmer.

As the winds move the storm over the ocean, the water that originally evaporated from the warm ocean surface falls as rain. If this happens over land, the high level of rainfall can cause flooding. Floods can also happen when strong winds cause storm surges in coastal areas, pushing waves much higher than normal tides.

How do cyclones move water?

3 Cyclones cause erosion and damage habitats

Erosion is a major impact that cyclones have on the lithosphere, especially around beaches and sand dunes, due to storm surges. Storm surges damage beaches, dunes and urban areas. Flood waters pick up topsoil and transport it into the oceans.

The intense wind and wave action of a cyclone can kill organisms and damage habitats both on land and in the ocean. Not only are corals damaged due to high energy waves, further damage occurs when sediments from the land are washed into the ocean near the reef.

In 2011, Cyclone Yasi caused significant damage to the Great Barrier Reef and the rainforest habitat of the endangered southern cassowary and mahogany glider.

What causes damage to a reef during a cyclone?

Figure 7.9 Cyclone Yasi did a lot of damage to the Great Barrier Reef.

CHECKPOINT 7.4 

1 Explain what cyclones are and how they form.
2 What is the difference between cyclones in the Southern and the Northern Hemispheres?
3 What direction is the air moving in the eye of a cyclone?
4 Identify where the water originates that falls as rain over land during a cyclone.
5 Describe how cyclones can change the lithosphere.
6 Explain why cyclones are a significant threat to coral reef communities.
7 Propose how the destruction of habitats by cyclones affects the species living there.
8 Climate change is causing sea surface temperatures to rise. Propose what impact this could have on cyclone formation.

CHALLENGE

9 Research the impacts of a natural hazard such as a bushfire, flood or drought. Identify how the hazard affects the atmosphere, hydrosphere, lithosphere and biosphere. Create a diagram to illustrate the relationships.

SKILLS CHECK

- I can describe what a cyclone is.
- I can describe at least one impact a cyclone has on the atmosphere, hydrosphere, lithosphere and biosphere.

7.5 The enhanced greenhouse effect

At the end of this lesson I will be able to:

- **evaluate** scientific evidence for the effect that human activity has on the greenhouse effect
- **discuss** some reasons different groups in society may use or weight criteria differently to evaluate claims, explanations or predictions in making decisions about the enhanced greenhouse effect and its impact on Earth's spheres.

KEY TERMS 

enhanced greenhouse effect
an increase in the greenhouse effect due to human greenhouse gas emissions

greenhouse effect
the trapping of the Sun's warmth by the atmosphere

greenhouse gas
a gas that traps heat energy in the atmosphere

LITERACY LINK 

Identify an issue or argument that is preventing action on reducing greenhouse gas emissions. Research and prepare some dot points that could be used to counter this issue or argument.

A greenhouse is a glass structure that gardeners use to trap the Sun's energy and keep their plants warm. The glass allows the energy through but prevents some of it from escaping. Earth's atmosphere acts in a similar way. Carbon dioxide is a significant greenhouse gas, and emissions from the burning of fossil fuels have increased the amount of carbon dioxide in the atmosphere, artificially enhancing the greenhouse effect and raising the average temperature of Earth.

1 The greenhouse effect keeps Earth's surface warm

Without the **greenhouse effect**, the surface of Earth would be very cold. When the Sun's radiation hits the surface of Earth, some of it is reflected back out into space and some is absorbed by Earth's surface, warming it. The warm surface emits energy as infrared radiation.

A **greenhouse gas** is a gas that stops this infrared energy from going straight back out into space. Molecules of greenhouse gases absorb and then re-emit this energy, warming the lower atmosphere and the surface of Earth. The major greenhouse gases in the atmosphere are water vapour (H_2O), carbon dioxide (CO_2), methane (CH_4), nitrous oxide (N_2O) and ozone (O_3).

Why is the greenhouse effect important for life on Earth?

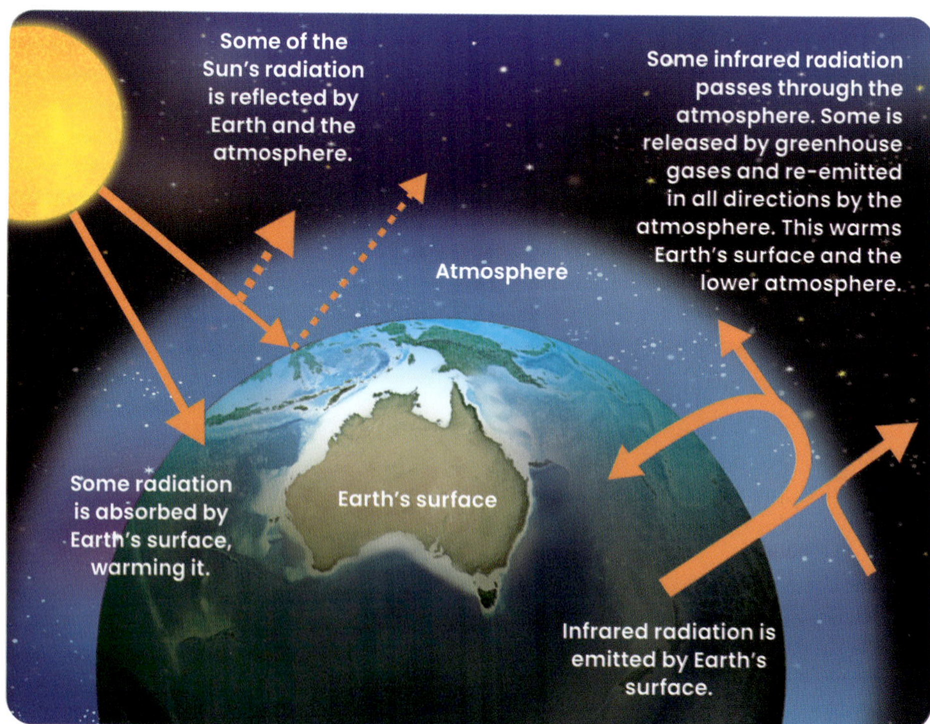

Figure 7.10 The greenhouse effect is important for maintaining the temperature on the surface of Earth.

INVESTIGATION 7.5
The greenhouse effect

2 Humans have increased the greenhouse effect

Human activity, such as the burning of fossil fuels and factory emissions, increases the levels of greenhouse gases in the atmosphere. Because there are more gas particles in the atmosphere, this means that more infrared radiation stays in Earth's atmosphere and less is radiated back into space. This has increased Earth's surface temperature. This is called the **enhanced greenhouse effect**.

How have humans enhanced the greenhouse effect?

3 Atmospheric measurements show CO_2 levels are increasing

Scientists can measure the amounts of the greenhouse gases in the atmosphere and correlate this with average temperature curves to demonstrate the relationship between the two. A global network of stations samples the atmosphere regularly. The most well-known station, Mauna Loa Observatory on the Big Island of Hawaii, has been operating since the 1950s.

The Cape Grim Baseline Air Pollution Station in north-west Tasmania was established in 1976. Its location has some of the cleanest air in the world, because it is well away from sources of pollution. Measurements taken at Cape Grim show that carbon dioxide levels in the atmosphere are increasing. In 2019, the atmospheric concentration of carbon dioxide was about 405 ppm.

Why is Cape Grim a good place to take air samples?

4 The Keeling curve plots CO_2 levels in the atmosphere

In 1958, US scientist Charles Keeling started monitoring carbon dioxide concentrations in the atmosphere at Mauna Loa, Hawaii. He was the first person to take regular measurements that showed how the concentration of carbon dioxide in the atmosphere was changing.

Keeling's measurements showed a steady increase in carbon dioxide concentration in the atmosphere and that the amount of carbon dioxide in the atmosphere changes throughout the year.

The annual cycle is because most of Earth's land mass and plant life are in the Northern Hemisphere. Over the northern spring and summer, plant growth and photosynthesis increase, which reduces carbon dioxide in the atmosphere. During autumn and winter, plant growth and photosynthesis decline and levels of carbon dioxide in the atmosphere increase. When plotted as a graph, this data is known as the Keeling curve.

Where did Keeling gather his data on atmospheric carbon dioxide concentrations?

Figure 7.11 The Cape Grim Baseline Air Pollution Station samples the air coming off the Southern Ocean.

7.5 continued...

7.5 continued...

The enhanced greenhouse effect

NUMERACY LINK

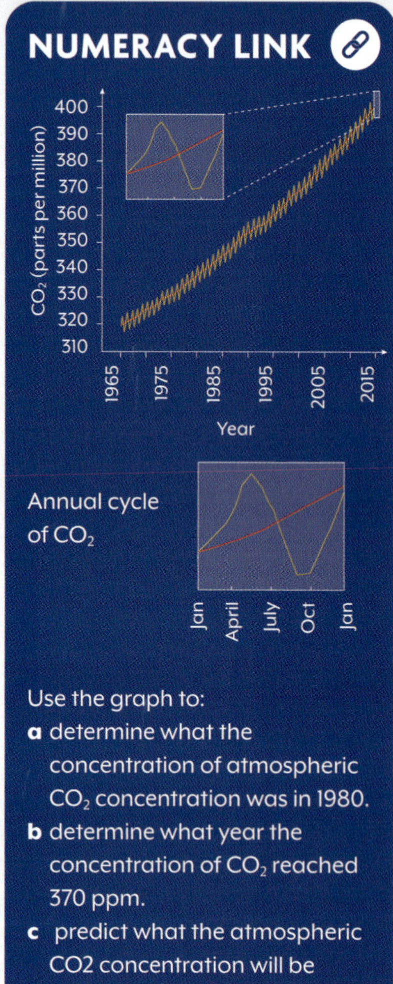

Use the graph to:

a determine what the concentration of atmospheric CO_2 concentration was in 1980.

b determine what year the concentration of CO_2 reached 370 ppm.

c predict what the atmospheric CO2 concentration will be in 2030.

5 Ice core data also shows CO_2 levels are increasing

Ice cores are cylinders of ice drilled from ice sheets. Scientists use ice cores from Greenland (up to 123 000 years old) and Antarctica (up to 800 000 years old) to find out more about Earth's past temperature and climate. Importantly, they can analyse air bubbles trapped in the ice to determine the concentration of gases in Earth's atmosphere over time.

Data from the Vostok and Law Dome ice cores in east Antarctica shows that although carbon dioxide levels have varied in the past, the recent increase in carbon dioxide levels due to human activity is unprecedented.

What do scientists analyse from ice cores?

Figure 7.12 Combined data from the Vostok and Law Dome ice cores and Mauna Loa show how the carbon dioxide concentration in the atmosphere has varied for the last 400 000 years.

6 Impacts of an enhanced greenhouse effect

As the level of carbon dioxide in the atmosphere has risen, the average temperature of Earth has increased. This has led to many changes to Earth's spheres, all of which have flow-on effects.

As the oceans get warmer, more water evaporates and atmospheric circulation patterns change. This means that cyclones and hurricanes can form further away from the equator, and that storms become larger as more water moves into the atmosphere. Rainfall patterns also change, so some areas receive more rain and other areas receive less.

Sea ice coverage in the Arctic has decreased. Solar radiation is reflected by light-coloured surfaces such as ice, and absorbed by darker surfaces. The reduction of ice has caused the Arctic Ocean to absorb more solar radiation, further warming the ocean and decreasing the sea ice.

As carbon dioxide levels in the atmosphere increase, more carbon dioxide is absorbed by the ocean. This forms carbonic acid, making the oceans more acidic.

What is one way the enhanced greenhouse effect is impacting on Earth's systems?

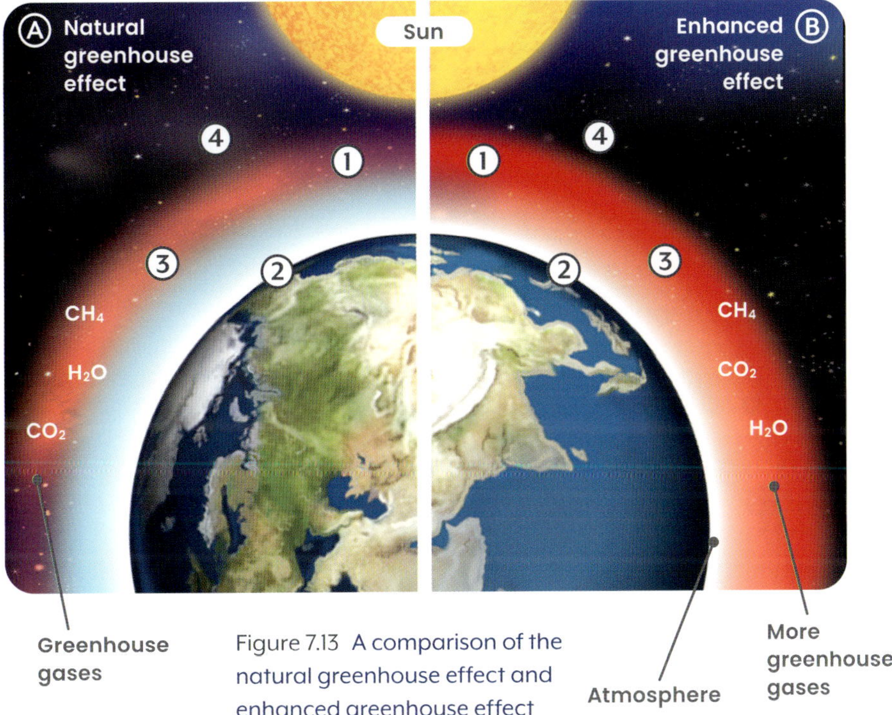

Figure 7.13 A comparison of the natural greenhouse effect and enhanced greenhouse effect

7 The Kyoto Protocol and Paris Agreement are attempts to reduce greenhouse gas emissions

The Intergovernmental Panel on Climate Change (IPCC) is a United Nations group that provides scientific information about climate change, its impacts and ways to respond. The IPCC has stated that reducing emissions of carbon dioxide and other greenhouse gases will reduce the concentration of these gases in the atmosphere, and thus reduce global temperature increases.

There have been several agreements by global leaders to address these issues and reduce the greenhouse gas emissions of their countries. The Kyoto Protocol (1997) and the Paris Agreement (2015) have been the major agreements between countries to reduce emissions to address climate change.

Barriers to reducing emissions are primarily economical. For some countries, fossil fuels are still the cheapest form of energy. Many countries, including Australia, rely on the income from the mining and sale of fossil fuels. Across the world, some governments are encouraging people to choose renewable energy options by subsidising the costs involved. Other governments encourage people to purchase electric or hybrid cars by removing taxes involved with buying those cars. Australia has pledged to reduce its emissions by 26–28% of its 2005 levels by 2030.

How can the enhanced greenhouse effect be reduced?

CHECKPOINT 7.5

1 List some common greenhouse gases.
2 Explain the difference between the enhanced greenhouse effect and the natural greenhouse effect.
3 Use your knowledge of the carbon cycle to explain how mining and burning fossil fuels affects Earth's spheres.
4 The annual fluctuations in carbon dioxide level shown on the Keeling curve have been likened to 'Earth breathing'. Explain why.
5 Explain how data from ice cores is significant when considering how carbon dioxide emissions are affecting the temperature of Earth.
6 Identify some issues that are preventing countries from reducing their greenhouse gas emissions.

CHALLENGE

7 Research one of the impacts that global warming is having on Earth (e.g. changing weather patterns, coral bleaching, decreased sea ice coverage). Present your findings to your class creatively.

SKILLS CHECK

- I can describe the greenhouse effect.
- I can describe how human activity has enhanced the greenhouse effect.
- I can discuss some of the evidence for the enhanced greenhouse effect and explain how it contributes to scientific knowledge.
- I can describe some impacts of the enhanced greenhouse effect.

7.6 The ozone layer

At the end of this lesson I will be able to:

- **evaluate** scientific evidence for the effect of human activity on the hole in the ozone layer
- **discuss** some reasons different people may use or weight criteria differently to evaluate claims, explanations or predictions when making decisions about the hole in the ozone layer and its impact on Earth's spheres.

KEY TERMS

ozone
a molecule made of three oxygen atoms bonded together

ozone-depleting substance
a chemical that reacts with ozone and breaks the molecule down as well as stopping the molecule from re-forming

ozone layer
a region of the stratosphere with a high concentration of ozone

ultraviolet radiation
harmful rays from the Sun, which can cause cancers and harm animals and plants

NUMERACY LINK

In 2000, the ozone hole above the Antarctic was approximately 24.8 million km^2 and by 2017 it had decreased to 17.4 million km^2. Calculate the percentage change in the ozone hole.

The ozone layer is a part of the stratosphere that protects life from harmful ultraviolet (UV) radiation. In 1976, it was found that chemical pollution from industry reacted with the ozone in the ozone layer, reducing ozone levels and preventing more ozone from forming. This pollution resulted in a world-wide depletion of ozone in the stratosphere. The depletion is especially high over Antarctica, where a 'hole' has formed in the ozone layer. This reduction in ozone means that more UV radiation can reach the surface, harming life forms. Once the cause of the ozone hole was determined, countries acted quickly to ban the use of ozone-depleting chemicals, which means the ozone hole has begun to close over.

1 The ozone layer absorbs UV radiation

Ozone (O_3) is a molecule made of three oxygen atoms bonded together. Ozone is found throughout the atmosphere, but is especially concentrated in the **ozone layer**.

Ozone absorbs harmful **ultraviolet (UV) radiation**, and prevents it from getting to Earth's surface. There are different wavelengths of UV radiation, which have different effects. Ozone absorbs all of the short-wave UV-C (the most harmful), and 90% of the medium-wave UV-B (which causes sunburn), but lets through 50% of the least harmful long-wave UV-A. Without ozone in the atmosphere, life on Earth could not survive.

In what layer of the atmosphere is the ozone layer found?

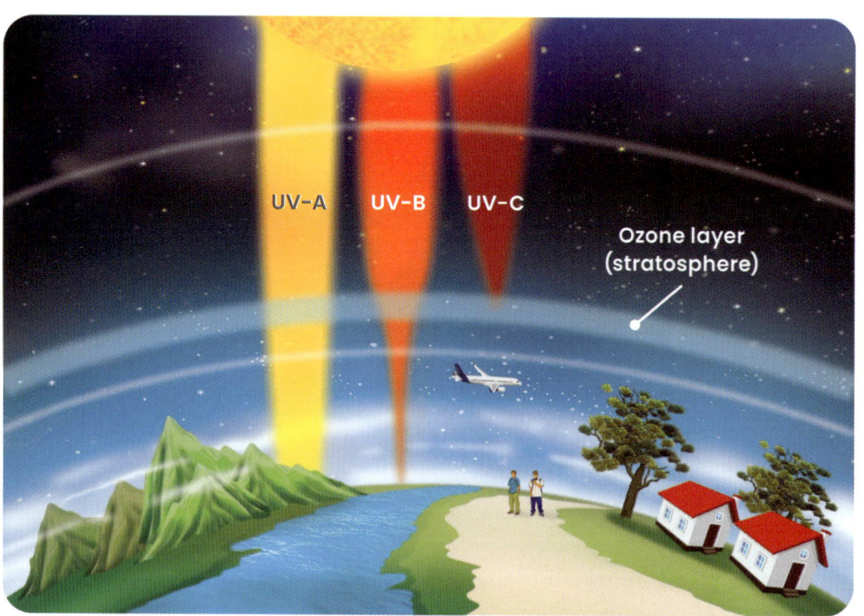

Figure 7.14 The ozone layer is a layer in the stratosphere where ozone (O_3) is abundant. The ozone layer prevents most of the harmful UV radiation from reaching Earth's surface.

2 Ozone is depleted by substances in pollution

In 1976, atmospheric researchers found that the ozone layer was becoming depleted by substances released by industry, such as chlorofluorocarbons (CFCs). CFCs were once commonly used in air conditioners and as a propellant for aerosol sprays, such as hairspray and deodorant. When CFCs are released into the atmosphere, they make their way into the stratosphere. Here they react with ozone, breaking some bonds and releasing O_2. They also prevent new ozone from forming.

Chemicals that react in this way are known as **ozone-depleting substances**. Most ozone-depleting substances stay in the stratosphere for decades. The ozone layer has reduced all over Earth by about 5%, which allows more UV radiation through.

What is the major group of chemicals that deplete the ozone layer?

3 The Montreal Protocol was designed to protect the ozone layer

Satellites and weather balloons have collected data about the amount of ozone in the stratosphere since the 1970s. In 1985, scientists published results about a hole in the ozone layer.

In 1987, as part of the Montreal Protocol on Substances that Deplete the Ozone Layer, all countries agreed to reduce and phase out their use of ozone-depleting substances. Different countries had different emissions targets depending on their economies and their current use of ozone-depleting substances.

CFCs were relatively easy to replace, and most countries banned the use of CFCs in the 1990s. However, many industries replaced CFCs with hydrofluorocarbons (HFCs), which are potent greenhouse gases. The Protocol was amended to include phasing out HFCs.

The Montreal Protocol is one of the most successful international environmental agreements. The amount of ozone-depleting chemicals in the atmosphere has decreased, and the ozone layer has shown some signs of recovery. The hole is expected to close by 2050.

What does the Montreal Protocol require countries to do?

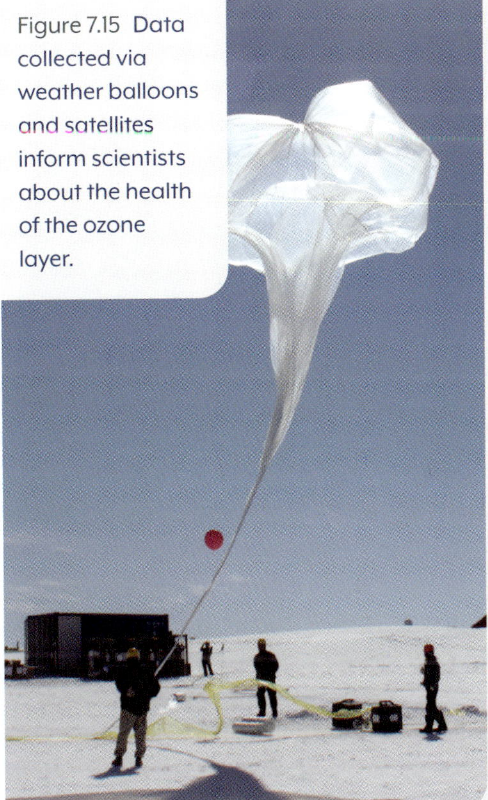

Figure 7.15 Data collected via weather balloons and satellites inform scientists about the health of the ozone layer.

INVESTIGATION 7.6
Investigating sunscreens

CHECKPOINT 7.6

1 What is the chemical formula for ozone?
2 Why is the ozone layer important?
3 Describe what an ozone-depleting substance does to the ozone layer.
4 Identify the source of the increased levels of ozone-depleting substances in the atmosphere.
5 How do scientists measure the amount of ozone in the atmosphere?
6 What is the relationship between the implementation of the Montreal Protocol and the amount of ozone-depleting substances in the atmosphere?

CHALLENGE

7 Because of the circulation of the atmosphere and the location of the continents, an ozone hole over the Arctic wasn't observed until 2010–2011. Do you think a Northern Hemisphere ozone hole would have different consequences from the ozone hole over Antarctica? Justify your response.

SKILLS CHECK

- I can describe the importance of the ozone layer.
- I can identify how human activity has depleted the ozone layer.
- I can outline the international response to the discovery of the ozone hole.

CHAPTER SUMMARY

Earth's spheres

The **atmosphere** is the 600-kilometre thick layer of gas that surrounds Earth.

The **hydrosphere** is all the water on Earth in all its forms.

The **lithosphere** is Earth's crust and upper mantle.

The **biosphere** is all of the living things on Earth.

A cyclone affects the atmosphere, hydrosphere, lithosphere and biosphere.

Strong winds

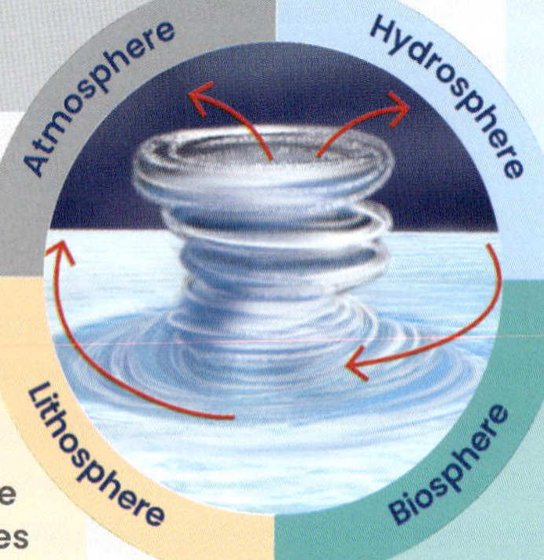

Water evaporates and forms clouds

Water falls as heavy rain, causing floods

Storm surges at the coast cause floods

Erosion occurs at beaches and sand dunes

There is damage to beaches, dunes and urban areas

Floods carry soil to oceans

Strong wind and rain kill plants and animals, damage habitats

Waves and sediments from land damage cora reefs

Cyclone

Cyclones are interactions between the atmosphere and hydrosphere.

The **carbon cycle** shows how carbon atoms are recycled as they move through the different spheres.

The **ozone layer** protects life from harmful UV radiation.

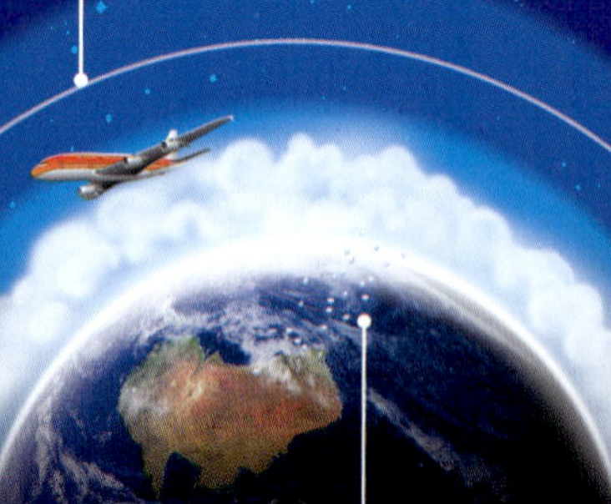

Carbon dioxide emissions are contributing to an enhanced greenhouse effect.

▼ Earth is kept warm by greenhouse gases that prevent heat being lost into space.

Some of the Sun's radiation is reflected by Earth and the atmosphere.

Some infrared radiation passes through the atmosphere. Some is released by greenhouse gases and re-emitted in all directions by the atmosphere. This warms Earth's surface and the lower atmosphere.

Atmosphere

Some radiation is absorbed by Earth's surface, warming it.

Earth's surface

Infrared radiation is emitted by Earth's surface.

FINAL CHALLENGE

1. In one sentence each, describe the hydrosphere, atmosphere, lithosphere and biosphere.
2. Identify an example of an interaction that includes the:
   a. lithosphere and biosphere
   b. atmosphere and hydrosphere
   c. atmosphere and biosphere.

3. Use a flow chart to show how carbon moves from one sphere to the next in the carbon cycle.
4. Explain how a volcanic eruption can affect Earth's spheres.
5. Explain what a carbon sink is, and give two examples.

6. Use an annotated diagram to illustrate how the greenhouse effect works.
7. Describe the relationship between the concentration of carbon dioxide in the atmosphere and the average surface temperature of Earth.
8. Explain why the Keeling curve is significant.
9. Provide these facts about the ozone layer:
   a. what it is composed of
   b. its effect
   c. how a hole was created in it
   d. what actions were taken to reduce the hole.

10. Burning fossil fuels adds carbon dioxide to the atmosphere. Identify one impact of this on each of the atmosphere, the hydrosphere, the lithosphere and the biosphere.
11. Describe how cyclones are formed and some of their impacts.

12. Countries acted swiftly when it was discovered that CFCs and other chemicals were depleting the ozone layer. Although we know the impact that burning fossil fuels is having on the atmosphere, suggest why it has taken longer for countries to reach an agreement on how to address the issue.

8 Body systems

The human body can be thought of as a balanced machine, constantly working to stay stable in an ever-changing environment. Most humans can survive only three days without water, around three weeks without food, three minutes without oxygen and just over ten days without sleep.

Keeping the body stable and meeting these needs is the function of different systems of organs. Each body system has its own specific purpose, but they all coordinate together so that the body works efficiently to ensure survival.

1 LEARNING LINKS

Brainstorm as much as you can remember about these topics.

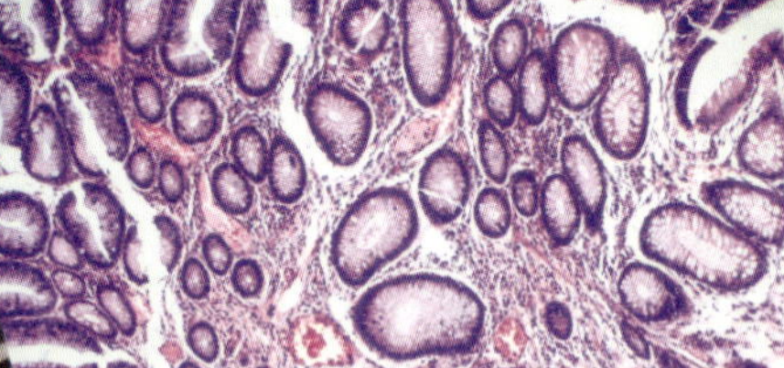

Cells are the basic units of living things.

Different types of cells make up the tissues, organs and organ systems of multicellular organisms.

Organ systems all have roles in maintaining a human as a functioning multicellular organism

2 SEE–KNOW–WONDER

List three things you can **see**, three things you **know** and three things you **wonder** about this image.

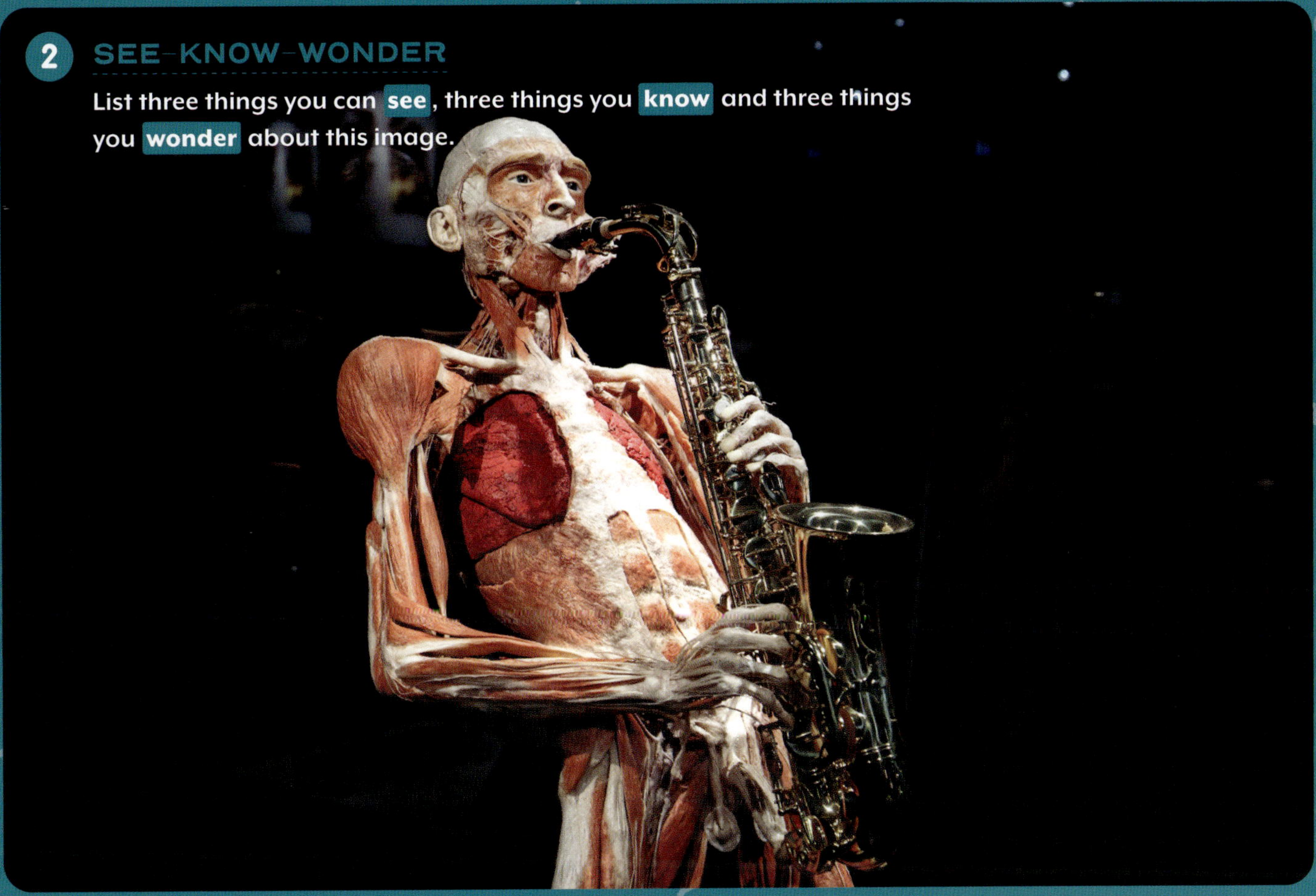

3 CRITICAL + CREATIVE THINKING

The BAR: What would you make **B**igger, **A**dd and **R**eplace in the human brain?

The alphabet: Use the topic 'Human body' to write a list of words from A–Z.

What if ... Earth's gravitational pull were to decrease by 25 per cent? Suggest what might happen to the human body.

4 THE BIGGEST!

The blue whale has the biggest heart of any living creature. It is about 180 centimetres long and weighs about 200 kilograms, or about 1 per cent of the whale's body weight. It has the capacity to pump 220 litres of blood per beat! Adult human hearts are around 12 centimetres long and make up less than 0.5 per cent of their body weight – an average adult male heart weighs between 280 and 340 grams, and an average adult female heart weighs between 230 and 280 grams.

8.1 Bodies in balance

At the end of this lesson
I will be able to:

- **describe** some examples of how multicellular organisms respond to changes in their environment.

KEY TERMS 

homeostasis
maintaining a stable, balanced state within the body

negative feedback
a response that counteracts the stimulus

receptors
specialised cells and organs that detect changes

LITERACY LINK 

Find the origin of the word 'homeostasis'. Think about how knowing its meaning can help you remember the definition of the word.

NUMERACY LINK 

A particular person sweats 0.8 litres per hour during exercise. Calculate how much sweat is produced per minute. Convert this value to millilitres.

Multicellular organisms tend to have complex bodies, consisting of multiple organs and systems that work together to maintain life and health. Whenever something changes in or around the organism, it has the potential to affect the organism's health and survival. Human bodies deal with these changes by working to keep all of the body systems in balance.

1 Many different types of changes affect the body

There are many types of changes within the environment that affect the body. Some of these are obvious, such as changes in temperature. If it gets hot, we sweat; if it gets cold, we shiver. Other changes are more subtle, but our bodies react to them all the same.

Physical changes affect the position of the body, such as balancing on a beam. As you move back and forth to keep your balance, your body reacts so that you stay in control. Movement is another physical change; not only does your body change position, but body parts such as your feet feel impacts as they push against surfaces.

Chemical changes involve your body's reaction to different chemicals. If the quality of the air you breathe changes, your body will react. Whenever you eat something, your body's systems react to it – and if it's something noxious or poisonous, that reaction could be extreme.

What are three changes in the environment that can affect the body?

2 Receptors detect changes and trigger feedback

Whenever something changes in the environment, the first step is for the body to actually detect that change. It does this thanks to different **receptors** – cells and organs that identify both internal and external changes.

Sensory receptors can be classified into four broad categories.

- *Photoreceptors* respond to changes in light. Human photoreceptors are located in the eye. Some other organisms, such as plants, have different types of photoreceptors.
- *Chemoreceptors* respond to different chemicals. The chemoreceptors in your tongue are what allow you to detect different flavours.
- *Thermoreceptors* respond to changes in temperature. Most of your thermoreceptors are on your skin; you also have a few on some organs.
- *Mechanoreceptors* respond to changes in pressure or position. Again, many of these are on your skin, but the ones in your inner ear are vital for maintaining balance.

Once a receptor detects a change, it sends a message to the brain, which can then target the appropriate body systems and organs. These systems then respond to compensate for that change. The process of stopping and reversing a process in the body is usually called **negative feedback**, because the body is acting to negate or reduce the effect of the change.

Sweating when the external temperature increases is an example of negative feedback. The thermoreceptors in your skin detect that your body temperature is increasing, triggering the excretory system to release sweat. The sweat evaporates once it's released, which reduces the skin's temperature.

What are some of the different types of receptors in the body?

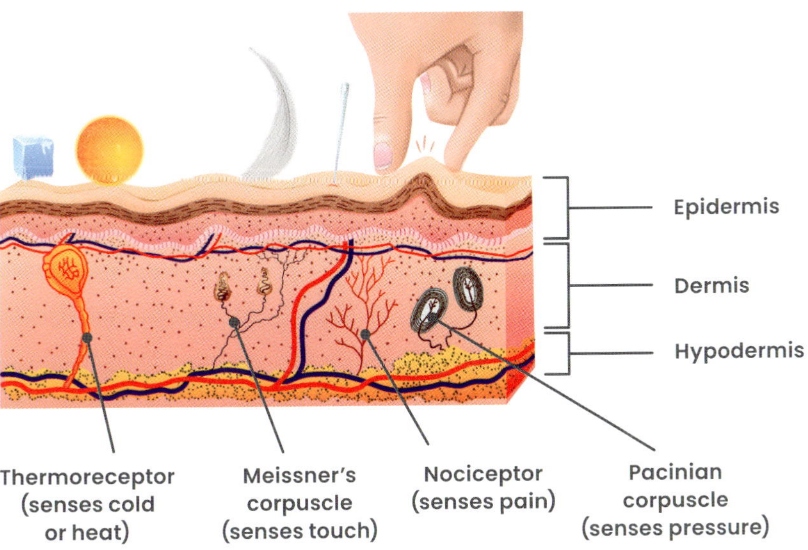

Figure 8.1 There are many different sensory receptors in the skin.

3 Homeostasis is the baseline state of the body

Whenever the body of a multicellular organism reacts to a change or stimulus with negative feedback, it is working to return to **homeostasis**. This is the 'normal' or baseline state of the body – the point at which every organ and system is working as it should.

When the human body is at homeostasis, vital body functions are maintained at the level that sustains life and good health. These functions include body temperature, blood pressure, glucose and oxygen levels in the blood, brain activity and many more. If an organism's body did not work to achieve homeostasis at all times, these functions would quickly cease.

Every organism has its own point of homeostasis, and every individual has a slightly different point of homeostasis than other organisms of the same type. Some people 'run hot', and have a slightly higher body temperature than others; some are more sensitive to changes in blood glucose. Regardless of your personal point of homeostasis, your body will always work to return to that point.

What is homeostasis?

INVESTIGATION 8.1A
Sensory receptors
INVESTIGATION 8.1B
Reaction time

CHECKPOINT 8.1

1 Explain what homeostasis is, and outline why it is important to survival.
2 Identify a chemical and physical change that an organism may react to.
3 Identify which receptors would be responsible in the following situations.
   a Your hand jerks away after you touch a hot plate.
   b You feel pain when you step on a sharp stone in bare feet.
   c You taste some sour milk.
   d A flower turns its head toward the Sun during the day.
4 Explain what a negative feedback response is, using an example that is not in the text.
5 Sweating helps to lower body temperature. Propose how shivering helps to increase it.

CHALLENGE

6 Undertake research to find out how the human body responds to changes in body temperature. Present your findings as a diagram or a flow chart.

SKILLS CHECK

- I can describe how physical and chemical changes can affect the body.
- I can identify the four different types of sensory receptors.
- I can describe negative feedback and homeostasis.

8.2 The coordination systems

At the end of this lesson I will be able to:

- **describe** the role of, and interaction between, the coordination systems in maintaining humans as functioning organisms.

KEY TERMS 

effector
a muscle, gland or organ that responds to a message sent by the nervous or endocrine system

gland
tissue that secretes hormones

hormone
a chemical secreted by a gland that triggers a response in certain cells

neuron
a specialised cell that makes up the nervous system

synapse
the gap between the axon and dendrite of two neighbouring neurons

LITERACY LINK 

Construct a table that compares and contrasts the structures and functions of the nervous and endocrine systems.

NUMERACY LINK 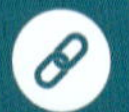

The motor neuron has an axon that is 1 m long, and a cell body that is 0.1 mm wide. How much bigger is the length of the neuron compared to its width?

When receptors respond to stimuli, they send messages to different body systems to trigger changes. These 'messages' are complex chemical and electrical signals, and the different systems must interact in multiple ways to respond effectively. This complex interaction is governed by two vital body systems, which we refer to as the coordination systems: the nervous system and the endocrine system.

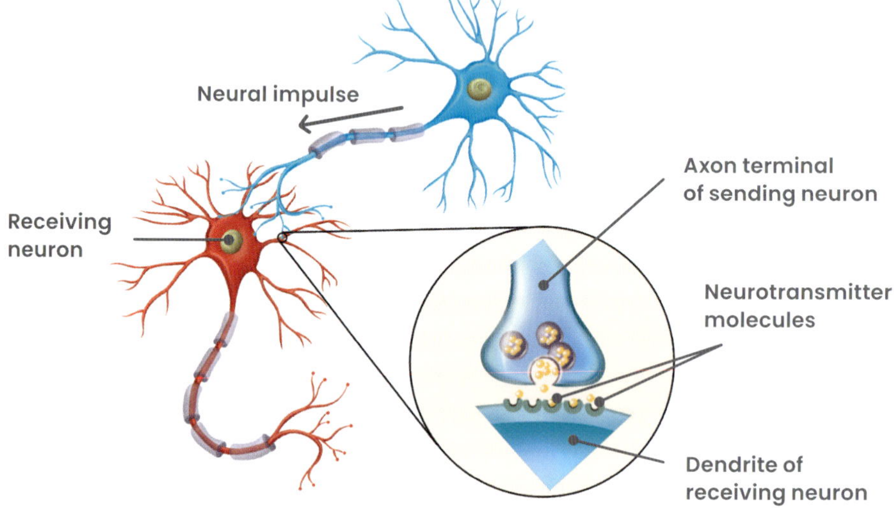

Figure 8.2 The axons and dendrites of neurons are separated by a gap called a synapse. The electrical impulse is translated into chemical neurotransmitters that cross the synapse and trigger the impulse to continue in the receiving neuron.

1 The nervous system transmits signals around the body

The nervous system is a network of cells and fibres that transmits fast messages between parts of the body. It consists of billions of cells called **neurons**, which form long nerve fibres. Those nerves connect the brain to muscles and organs by sending messages in the form of tiny electrical impulses.

The nervous system of vertebrates, such as human beings, has two parts. The *central nervous system* consists of the brain and the spinal cord. The *peripheral nervous system* is the network of nerves that runs through the rest of the body, connecting it to the central nervous system.

Neurons contain filaments called dendrites, which receive impulses, and long fibres called axons, which carry those impulses away from their cell body. Sensory neurons lead away from receptors, while motor neurons lead towards muscles, glands and other **effectors** that respond to signals.

Neurons do not actually connect to each other; they have tiny gaps between them, called **synapses**. When an electrical impulse reaches a synapse, it triggers the neuron to release chemical neurotransmitters. These cross the synapse and stimulate the next neuron, continuing the message.

How do signals pass along nerve cells?

2 The endocrine system releases hormones

The endocrine system consists of multiple **glands** – groups of cells that produce complex molecules called **hormones**. Once the gland is triggered, it secretes hormones into the bloodstream that travel to target cells and trigger a response. Endocrine responses are much slower than nervous system responses.

The pituitary gland is sometimes called the 'master gland' of the endocrine system. It sits at the base of the brain, just behind the bridge of the nose. While it's only the size of a pea, it secretes hormones that control many other endocrine glands. It also secretes hormones that regulate growth and reduce feelings of pain.

Hormones are often produced in pairs, one having the opposite effect to the other. For example, when blood sugar levels are too high, the pancreas produces insulin, which tells the liver to remove the excess glucose from the blood and store it. If blood sugar levels are too low, the pancreas releases glucagon, which tells the liver to return glucose into the blood.

How do hormones travel through the body?

3 The nervous and endocrine systems work together

The nervous and endocrine systems coordinate the body's responses to changes to its internal and external environment. They do this thanks to the brain, which plays a key role in both systems.

The brain consists of three main sections. The *cerebrum* is the largest part, consisting of two wrinkled hemispheres; this is where conscious thought takes place. The *cerebellum*, at the back of the brain, controls movement and balance. Finally, the *brain stem* connects the brain to the spinal cord, as well as controlling the automatic actions of the body.

The *hypothalamus* is a region in the centre of the brain that is responsible for interpreting signals from all over the body to ensure homeostasis. It does this through sending chemical signals to the pituitary gland, telling the gland to release hormones that will bring the body back into balance.

Which part of the brain coordinates the endocrine system?

Figure 8.3 A side view of the human brain

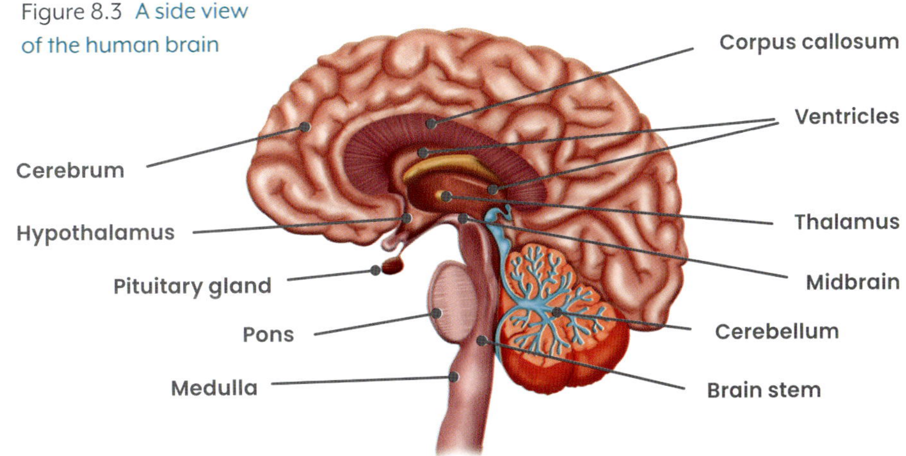

INVESTIGATION 8.2
Sheep brain dissection

CHECKPOINT 8.2

1 Identify the main types of neuron.
2 Distinguish between the central nervous system and the peripheral nervous system.
3 Describe how an impulse travels from one neuron to another, and identify the structures it passes through.
4 Suggest the benefit of hormones being produced in opposing pairs.
5 Identify the part of the brain that controls the endocrine system.
6 Suggest why it is important for the nervous and endocrine systems to work together.

CHALLENGE

7 Create a labelled model of neurons that shows how a signal is passed from one to another.

SKILLS CHECK

- I can describe the role of the nervous system.
- I can describe the role of the endocrine system.
- I can describe how the nervous and endocrine systems work together.

8.3 The cardio-respiratory system

At the end of this lesson I will be able to:

- **describe** how the coordinated function of internal systems in multicellular organisms provides cells with the requirements for life, including gases.

KEY TERMS 

cellular respiration
the process that all living things use to produce cellular energy from glucose and oxygen

diffuse
to move from an area of high concentration to an area of low concentration

gas exchange
the exchange of oxygen and carbon dioxide between an organism and the environment

LITERACY LINK 

Create a mnemonic to help you remember the structures of the respiratory and circulatory systems.

NUMERACY LINK 

A person has a resting heart-rate of 70 beats per minute. Each time their heart beats, it moves about 80 mL of blood through the heart. Calculate how much blood (in litres) is pumped through their body in 24 hours.

Every living thing needs energy to live, and that energy comes from **cellular respiration**: turning oxygen and glucose into energy for cells to use. In mammals, two body systems coordinate to support cellular respiration: the respiratory system, which exchanges oxygen and carbon dioxide, and the circulatory system, which transports these gases, along with nutrients and waste, around the body. These two systems are often considered together as the cardio-respiratory system.

1 The respiratory system is needed for gas exchange

The role of the mammalian respiratory system is to swap carbon dioxide from the blood with oxygen from the air. Its structure allows this **gas exchange** to happen very efficiently.

When you inhale, your diaphragm (a band of muscles under the lungs) moves down and your ribs move out, drawing air into the respiratory system through your nostrils or mouth. It goes through the trachea into two bronchi – one for each lung – before passing into smaller and smaller passages called bronchioles.

At the end of the bronchioles are tiny sacs called alveoli, which are surrounded by tiny capillaries. The oxygen in the air **diffuses** through the thin cell walls of the alveoli into the capillaries, where it bonds to haemoglobin, a protein carried in red blood cells. At the same time, the carbon dioxide in the blood plasma diffuses back out into the air within the alveoli. When you breathe out, your diaphragm moves up and your ribs move in, causing the air in your lungs to move back out.

In which structure does gas exchange take place?

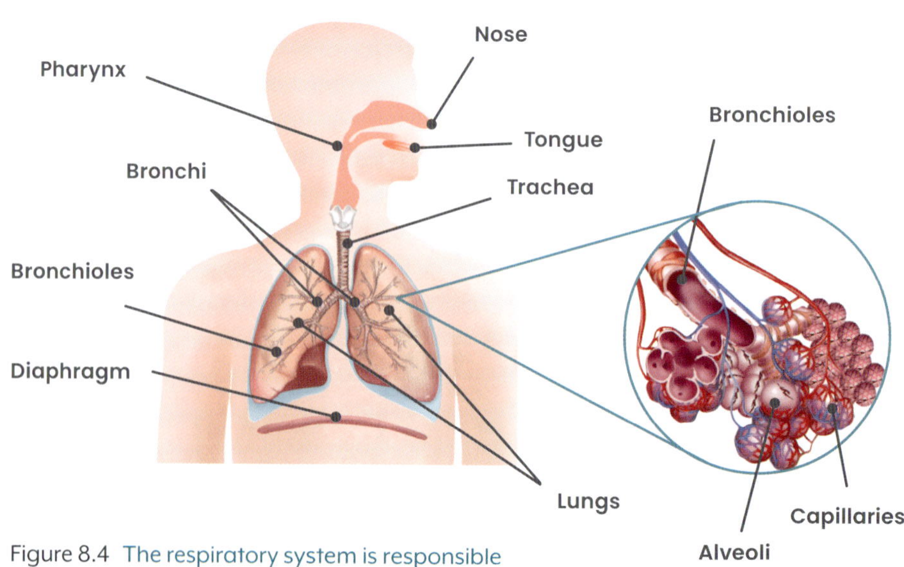

Figure 8.4 The respiratory system is responsible for the exchange of carbon dioxide and oxygen.

2 The circulatory system transports oxygen throughout the body

The mammalian circulatory system consists of the heart and blood vessels.

There are three main types of blood vessels. *Arteries* take blood from the heart to the body. They have thick muscular walls that expand and contract as the heart beats. *Veins* take blood from the body back to the heart. Their walls are not as thick as those of arteries, and they contain valves to keep the blood flowing in the right direction. *Capillaries* are the smallest blood vessels. They take blood right to the body's cells. They have walls that are just one cell thick – this allows nutrients and oxygen to easily move into the cells from the blood, and allows waste products to move out of the cells into the blood.

The heart is made up of four chambers; this structure allows oxygenated blood from the lungs to be pumped out to the body's cells, and deoxygenated blood returning from the body to be pumped to the lungs. Oxygenated blood comes from the pulmonary vein into the left atrium of the heart. It is then pumped into the left ventricle, before being pumped out of the heart through the aorta – the main artery that leads to the rest of the body. Deoxygenated blood from the body enters the right atrium of the heart from the vena cava – the main vein that leads from the body. It then moves into the right ventricle, before leaving the heart through the pulmonary artery to go to the lungs for gas exchange.

The circulatory system is not only important for transporting oxygen and carbon dioxide around the body, but for transporting nutrients from the digestive system (such as glucose) to the cells. It also moves waste products from the cells to the organs that process and remove them.

What are the three main types of blood vessel?

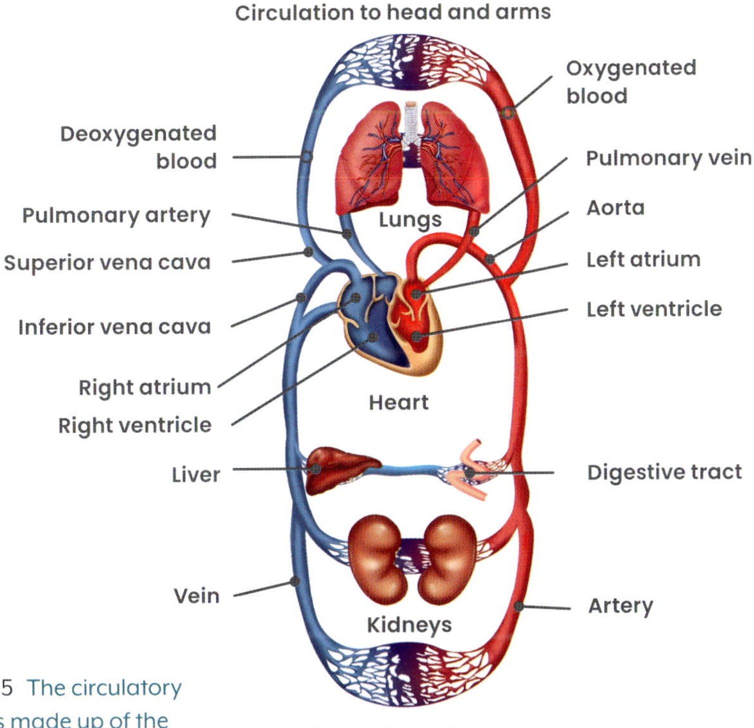

Figure 8.5 The circulatory system is made up of the heart and blood vessels.

INVESTIGATION 8.3A
Heart dissection

INVESTIGATION 8.3B
Heart rate, breathing rate and exercise

CHECKPOINT 8.3

1 What is the advantage of having lungs made up of thousands of tiny alveoli?

2 Explain why carbon dioxide in the blood needs to be exchanged with oxygen from the air.

3 Identify the structures of the circulatory system that a red blood cell moves through after it leaves the capillaries in the lungs, until it returns to the same position.

4 Identify which side of the heart contains oxygenated blood, and which side of the heart contains deoxygenated blood. What is the advantage of keeping them separated?

CHALLENGE

5 Find out about the structures of the respiratory and circulatory systems in other animals such as fish and amphibians. Use a table to compare them with those of a mammal.

SKILLS CHECK

- I can describe the structure and role of the respiratory system.
- I can describe the structure and role of the circulatory system.
- I can describe how the respiratory system and the circulatory system work together to support cellular respiration.

8.4 The digestive and excretory systems

At the end of this lesson I will be able to:

- **describe** how the coordinated functions of internal systems in multicellular organisms provide cells with the requirements for life, including nutrients and water, and removes cell wastes.

KEY TERMS 

amino acid
a simple molecule that is the basic unit of a protein

digestive enzyme
a chemical that acts to break down large chemical structures in food into smaller forms

nitrogenous
containing nitrogen

LITERACY LINK 

Write a story about how a hamburger and milkshake travel through the human digestive system.

NUMERACY LINK 

The volume of urine passed by a person varies from 0.250 to 0.400 L. Express these volumes in millilitres and then microlitres.

Mammals obtain nutrients and energy from the food that they eat. The nutrients in food are often not in the form required by the body's cells, so the body must process them. The digestive system breaks food down both physically and chemically so that the nutrients in it can be absorbed into the body.

Cellular activity, such as cellular respiration, creates waste products, and these must be removed from the body. The process that removes these wastes is known as excretion, and it is undertaken by organs in the excretory system like the kidneys, liver and even the skin and lungs.

1 The digestive system breaks down food

All living organisms require nutrients and water to survive, grow and sustain cellular processes. These nutrients include vitamins, minerals, simple sugars, fatty acids and **amino acids** (organic compounds that combine to form proteins). The role of the digestive system is to break down food into small molecules that the body can absorb. This happens in two ways: mechanical digestion, where the food is broken down into smaller pieces (for example, chewing), and chemical digestion, where **digestive enzymes** break the molecules in the food down into smaller molecules that can be absorbed by the body's cells.

In the mouth, food is broken into small pieces by the teeth and mixed with saliva, which contains enzymes that break down carbohydrates. When you swallow, this mix of food and saliva moves down into the oesophagus and then to the stomach.

Within the stomach, the food is broken down further into a soupy mixture of very small pieces and it mixes with gastric juices that contain enzymes and hydrochloric acid. This mixture is slowly released into the top section of the small intestine. Here it mixes with pancreatic juices and chemicals that neutralise the stomach acid droplets.

As this mixture moves through the rest of the small intestine, the nutrients begin to be absorbed. The walls of the small intestine are lined with tiny projections called villi, which contain capillaries. In the same way that oxygen molecules pass through the lung's capillaries into the bloodstream, the

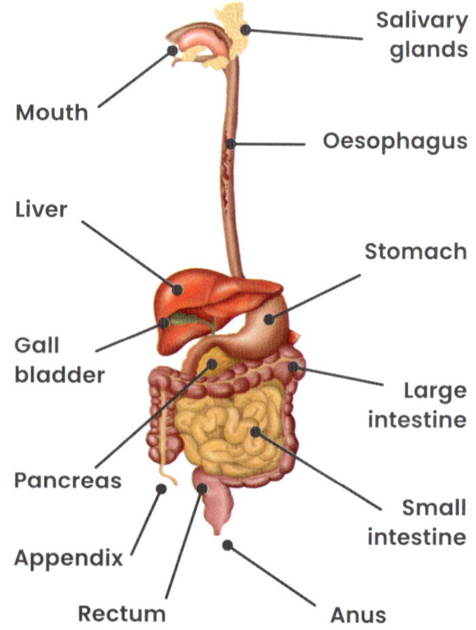

Figure 8.6 The role of the digestive system is to break down food into small molecules that the body can absorb.

nutrient molecules pass though the intestinal capillaries. The molecules enter the bloodstream, and the circulatory system moves them around the body to where they can be processed.

A lot of what we eat is left undigested as it passes through the small intestine; the nutrients are absorbed, but much material remains. This moves into the large intestine, also known as the bowel, where bacteria act on it, releasing some important vitamins and minerals that can be absorbed, producing gas as a by-product. Although water can be absorbed in the stomach and small intestine, the remaining absorption of water happens in the large intestine. The remaining solid waste collects in the rectum, and is finally expelled as faeces.

What is the role of the digestive system?

2 The excretory system removes waste

The chemical reactions in the body produce a variety of waste products that, if not removed, will build up in the body and cause harm. Excess carbon dioxide is removed by the lungs, and some excess salts and water are removed by the skin through sweat, but most water and other toxins are removed from the blood by the kidneys.

One major waste product is urea, a type of **nitrogenous** waste created by the liver. Urea is a by-product of reactions that break down proteins, amino acids and other molecules that contain nitrogen.

The kidneys are organs that remove waste products such as urea from the blood and ensure that the body retains the right amount of water. As blood moves into the kidneys through the renal arteries, microscopic structures called nephrons filter the blood, removing any waste products and excess water. The nephrons also make sure that substances the body needs, such as glucose, remain in the blood. The substances that have been removed from the blood are what we call urine, which leaves the kidneys through the ureters and is stored in the bladder until expelled through the urethra. The blood that has been filtered then returns to the circulatory system through the renal vein.

What is the structure in the kidney that filters blood?

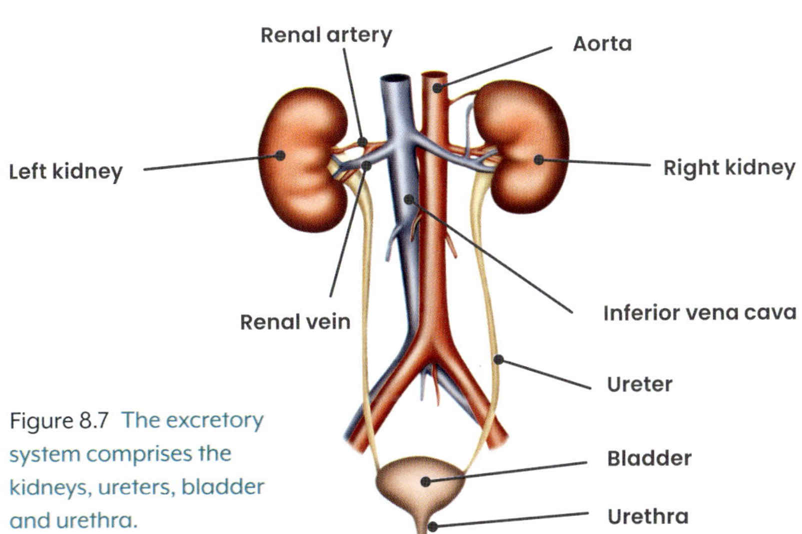

Figure 8.7 The excretory system comprises the kidneys, ureters, bladder and urethra.

CHECKPOINT 8.4 

1 What is the difference between mechanical and chemical digestion?
2 State the function of a digestive enzyme.
3 Identify the location where most absorption of nutrients occurs.
4 Describe the role of the kidneys, ureters, bladder and urethra in removing waste from the body.
5 Describe the role of the circulatory system in removing waste from the body.
6 Explain why it is important that the circulatory system coordinates with the digestive system.
7 The villi increase the surface area of the small intestine. Explain why this is important.

CHALLENGE

8 Carry out research to compare the digestive systems of a carnivore (such as a dog), a herbivore (such as a cow) and a human. How does diet relate to the different structures within the digestive systems of these mammals?

SKILLS CHECK

- I can describe how the digestive system provides the body with nutrients.
- I can describe how the excretory system removes wastes from the body.
- I can describe the role the circulatory system plays in transporting waste and nutrients around the body.

8.5 Responses to disease

At the end of this lesson I will be able to:

- **outline** some responses of the human body to infectious and non-infectious diseases.

KEY TERMS 

antibody
a protein that responds to a specific antigen

antigen
a substance that triggers the production of antibodies

lymphocyte
a type of white blood cell that produces antibodies in response to a pathogen

pathogen
an agent that causes disease

phagocyte
a type of cell capable of engulfing and destroying bacteria

LITERACY LINK 

Think of a metaphor for how the adaptive immune system works.

NUMERACY LINK 

The average red blood cell is 7.5 micrometres in diameter. A white blood cell is about 15 micrometres. Bacteria can vary from 0.5 to 2.0 micrometres. Draw a scale diagram of both types of blood cell plus a bacterial cell.

The systems of the body interact and coordinate to keep us active and healthy. Diseases and infections interfere with the normal functioning of the body. Fortunately, the immune system protects us from disease. Unfortunately, it doesn't always work instantly, which is why it takes time to recover when we do get sick.

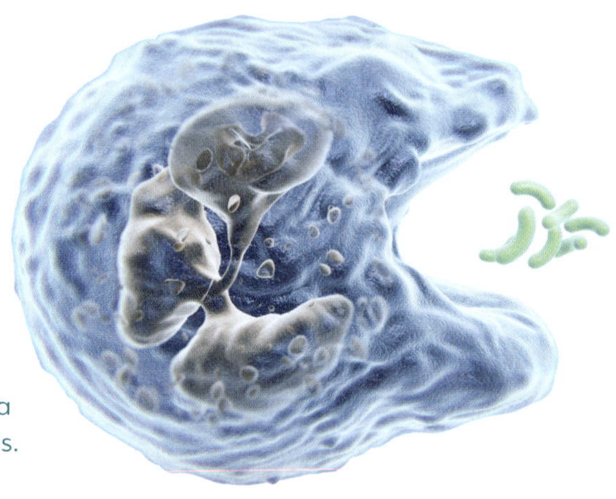

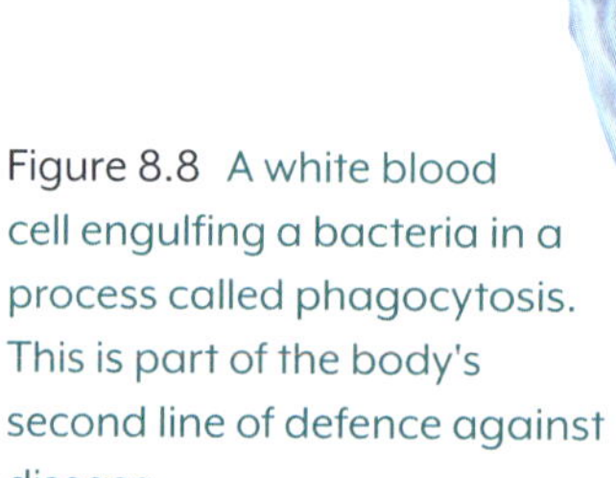

Figure 8.8 A white blood cell engulfing a bacteria in a process called phagocytosis. This is part of the body's second line of defence against disease.

1 Diseases may be infectious or non-infectious

Diseases cause damage to the body by preventing it from working as it should. Diseases can be classified as either infectious or non-infectious.

Infectious diseases can be transmitted (passed) from one person to another and are caused by agents called **pathogens**. Examples of pathogens are bacteria, viruses and fungi. Different pathogens can be transmitted in different ways, including skin-to-skin contact, breathing in droplets exhaled by an infected person, or the exchange of bodily fluids such as blood or saliva.

Non-infectious diseases aren't caused by pathogens and cannot be transmitted from one person to another. There are five main types of non-infectious diseases.

- Genetic diseases: conditions inherited from your parents, such as haemophilia (a disorder where blood doesn't clot)
- Lifestyle diseases: conditions caused by the way you live, such as dietary habits or smoking
- Parasitic diseases: conditions caused by parasitic organisms living on or inside your body. While these organisms do not spread an infectious disease, they may release toxins and their presence can still make you sick.
- Incorrect body function: conditions caused by organs or body systems not working the way that they should, such as diabetes or cancer
- Immunologic diseases: conditions caused by the immune system malfunctioning in some way

What is the difference between infectious and non-infectious diseases?

❷ The first defence against pathogens is keeping them out

The first 'line of defence' that the body can use against pathogens is preventing them from entering the body in the first place. Some examples of the first line of defence include the following.

- Your skin forms a barrier that keeps out pathogens. The circulatory system works quickly to seal any cuts in skin by forming clots and then a scab.
- Mucous-coated membranes line the nasal passage and the airways leading to the lungs. This mucous traps dust and any pathogens. Microscopic hairs called cilia move any foreign debris away from the lungs, so that it is either coughed or sneezed out, or swallowed.
- Tears and saliva contain antimicrobial substances that can kill pathogens.
- Acidic gastric juices can kill pathogens that are on or in food we eat.

What is the first line of defence against pathogens?

❸ The immune system fights pathogens and diseases

When pathogens make their way into the body, the immune system works to destroy them using the second and third lines of defence. These systems consist of a small number of organs and tissues, such as the thymus, lymph nodes and bone marrow, as well as the white cells in the blood.

The second line of defence involves cells called **phagocytes** that recognise foreign substances, engulf them and destroy them with the help of other cells. Fever and inflammation may occur when the second line of defence is active.

The third line of defence involves a specific, adaptive response of the immune system. In other words, it responds in a very specific way to different types of pathogens. A pathogen that enters the body has chemicals 'flags' on its outside, called **antigens**. When antigens are detected, specialised white blood cells called **lymphocytes** produce **antibodies** as a response. These are proteins that encourage cells of the immune system to kill off the pathogens, fighting off the infection.

After the pathogen has been destroyed, special memory cells remain, ready to make antibodies the next time the body is infected by the same pathogen. This way, the cells destroy the pathogen before it makes you sick, making you immune to it. Vaccines take advantage of this adaptive immune system.

What is a lymphocyte?

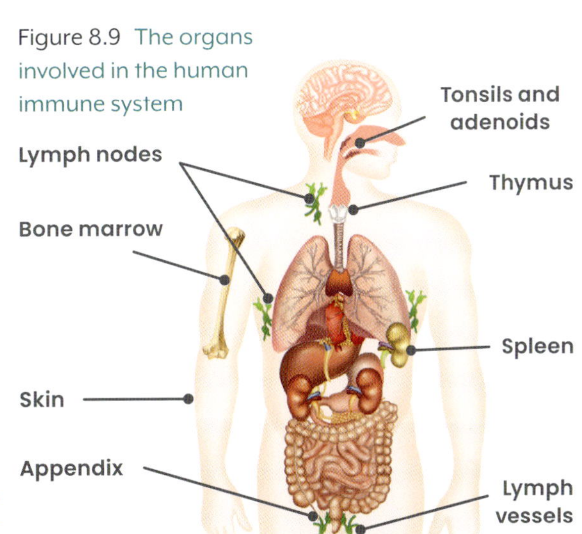

Figure 8.9 The organs involved in the human immune system

CHECKPOINT 8.5 

1 What is a pathogen?
2 Identify two ways that pathogens can be transmitted from person to person.
3 Using evidence from the text, explain why it is important to wash your hands regularly if you have a cold.
4 Describe one way that pathogens can be prevented from entering the body in the first place.
5 Describe the second line of defence against pathogens.
6 Explain how the immune system adapts to fight specific pathogens.
7 Edward Jenner is credited with creating the first vaccine. Carry out research to find out about his contributions to medical science.

CHALLENGE

8 Classify the following as infectious or non-infectious diseases.
   a chicken pox
   b haemophilia
   c tapeworm
   d influenza
   e measles
   f HIV
   g diabetes

SKILLS CHECK

- I can describe the difference between infectious and non-infectious disease.
- I can describe ways that the body prevents infection by pathogens.

8.6 Animal and plant diseases

At the end of this lesson I will be able to:

- **discuss**, using examples, how the values and needs of contemporary society can influence the focus of scientific research; e.g. the occurrence of diseases affecting animals and plants.

KEY TERMS 

contagious
able to spread from one organism to another

pustule
a small blister

quarantine
the isolation of people, plants, animals or objects that may have been exposed to biosecurity threats

LITERACY LINK 

Create a leaflet that can be used to inform the general public about the risks of FMD or guava rust.

NUMERACY LINK 

A farmer discovers 23 of his sheep have symptoms of FMD. In total, he has 446 sheep on his farm. What percentage of his sheep have the disease?

Every organism has a different immune system, and is vulnerable to different pathogens. Many diseases that affect animals can also affect other animals and humans. The diseases that affect plants usually don't affect humans or animals, but there are a few exceptions. Scientists develop treatments to keep animals and plants healthy, and to protect the global economy and ecosystems.

1 Foot-and-mouth disease affects animals with cloven hoofs

Foot-and-mouth disease (FMD) is a highly **contagious** disease that affects animals with cloven (divided) hoofs, such as cattle, sheep, goats and pigs. The disease is caused by a virus that can be transmitted through droplets in the air, saliva, faeces and milk. As the name suggests, the disease causes blisters and ulcers on the feet and mouth of affected animals, leaving them unable to walk or eat properly. It does not affect humans.

FMD is found in many parts of the world, particularly in Asia, Africa, South America and the Middle East. There have also been outbreaks of the disease in other parts of the world, such as in the United Kingdom in 2001 and 2007. There is currently no cure for FMD, although vaccines can reduce the chance that animals become infected.

The greatest threat of FMD is economic rather than medical. Because it cannot be cured, and because it's so contagious, infected animals have to be destroyed and their carcasses burned. This has an enormous impact on farms and businesses that sell or export animals or their meat or milk. The 2001 outbreak in the United Kingdom caused a loss to the British economy of almost 20 billion dollars.

The CSIRO is undertaking research to help prevent the spread of FMD in our neighbouring countries. This includes developing better techniques to quickly diagnose the disease, as well as vaccines that are effective against new strains of the virus.

What are the symptoms of FMD?

Figure 8.10 Foot-and-mouth disease is a highly contagious disease that is a significant threat to the agricultural industry.

❷ Guava rust affects many species of myrtle plants

Figure 8.11 Guava rust

Guava rust is a plant disease caused by a fungus that only infects plants belonging to the *Myrtaceae* family. This is a large family of more than 100 different types of plants, including eucalyptus trees, which is why the disease is also known as eucalyptus rust.

Guava rust is found throughout North America, South America, Asia and South Africa. It is primarily spread by spores of the fungus, which can be transported in the air or on the clothes and shoes of people working on trees. Spores can also survive on the surface of cut timber, wood packaging or infected plant material.

The disease attacks the leaves, stems and shoots of affected trees, and can also affect fruits and flowers. It causes tiny **pustules** to form, which soon erupt into yellow spores. As the infection spreads, it causes leaves to become deformed and stunts the plant's growth; it can even kill plants entirely. It is a significant threat to ecosystems, as well as to timber industries.

Scientists are undertaking research into what makes some plants resistant to guava rust and other similar diseases, in order to better understand the impact the fungus could have if released into the Australian ecosystem.

What kind of plants are affected by guava rust?

❸ Biosecurity laws help control animal and plant diseases

Neither FMD nor guava rust are found in Australia, which is a very good thing! An outbreak of FMD would have a massive and terrible impact on Australia's sheep and cattle industry. Similarly, if guava rust infected Australia's eucalyptus forests, it could ruin our logging industry – not to mention the damage it could cause to the ecosystems that involve those forests, for example to the koalas that eat eucalyptus leaves.

One vital protection we have against these diseases, and other dangerous plant and animal diseases, is that Australia is an island continent. Many diseases are airborne, travelling over national borders, but because Australia is surrounded by water, those diseases cannot arrive that way.

We're also protected by Australia's biosecurity laws. These laws set strict guidelines on the types of animal and plant materials (including meat and wood as well as live organisms) that can be brought into the country. Any suspect materials are placed in **quarantine** when they reach Australia's ports. Quarantined materials are tested and, if they carry a risk, they may be destroyed or sent back to their point of origin. These laws also apply within Australia – that's why you may not be allowed to bring fruit with you when you are travelling interstate.

Why does Australia have such strict biosecurity laws?

CHECKPOINT 8.6 

1 Identify the pathogen that causes FMD and ways it can be transmitted.

2 Suggest why vaccines are useful for preventing FMD.

3 Explain how the CSIRO's research into FMD is of benefit to the agricultural industry in Australia and neighbouring countries.

4 Identify the pathogen that causes guava rust and ways it can be transmitted.

5 Explain why it is important to better understand the fungus that causes guava rust.

6 What is meant by biosecurity?

7 Use evidence from the text to explain Australia's advantageous position with regards to controlling the spread of disease that can affect animals and plants.

CHALLENGE

8 The CSIRO plays an important role in undertaking research to help tackle national and international health and biosecurity challenges. Conduct your own research to find some examples of the work the CSIRO is undertaking. Select one project and present a summary of it to your class.

SKILLS CHECK

- I can describe how foot-and-mouth disease can impact the agricultural industry.
- I can describe how guava rust can impact native ecosystems and the forestry industry.
- I can describe the importance of biosecurity in Australia.

8.7 Human diseases and illnesses

At the end of this lesson I will be able to:

- **discuss**, using examples, how the values and needs of contemporary society can influence the focus of scientific research; e.g. an epidemic or pandemic disease in humans, or lifestyle related non-infectious diseases in humans.

KEY TERMS 

dengue fever
a mosquito-borne viral disease occurring in tropical and subtropical areas

epidemic
widespread occurrence of an infectious disease in a community

graft
a piece of living tissue that is transplanted surgically

pandemic
disease occurring within a whole country or around the world

NUMERACY LINK 

Too much cholesterol in the bloodstream increases the risk of heart disease or stroke. In 2017–18, 3.1% of the Australian population between 35–44 years had high cholesterol. This value increased to 6.8% for 45–54 years, 14.1% for 55–64 years and 21.2% for people over 65. Tabulate then graph this data.

As well as protecting plants and animals, Australia's geography and our biosecurity laws also protect us from diseases that affect humans. There are many dangerous diseases, such as rabies and *Ebola*, that are not found in Australia because of these protections. However, many other diseases make their way here – and when there is a global pandemic, it's impossible for us to remain untouched.

1 Diseases can spread in epidemics and pandemics

An **epidemic** is a major outbreak of a disease within one or more communities. Diseases spread very quickly during an epidemic, making it very difficult for doctors and medical officials to treat all sufferers. Australia's health system is effective in preventing epidemics. There were none recorded within Australia in the 20th century, and only one in the 21st century: a 2009 outbreak of **dengue fever** in Queensland.

Australia doesn't have the same level of protection against **pandemics** – disease outbreaks that affect global populations. The worst recorded pandemic in history was the Black Death, which killed millions of people in Europe during the 14th century. There have also been terrible pandemics of measles, smallpox, tuberculosis and other deadly diseases throughout history.

There are many pandemics currently affecting the world, and they're very hard to control when so many people travel internationally. Diseases such as influenza, measles and HIV are all global issues. Fortunately, many of these diseases can be prevented and treated relatively easily with vaccinations and modern medicines.

What is the difference between an epidemic and a pandemic?

Figure 8.12 The 1918 H1N1 influenza virus was the deadliest pandemic of the 20th century. It was known as the 'Spanish Flu' because it became more widely known after the disease spread from France to Spain.

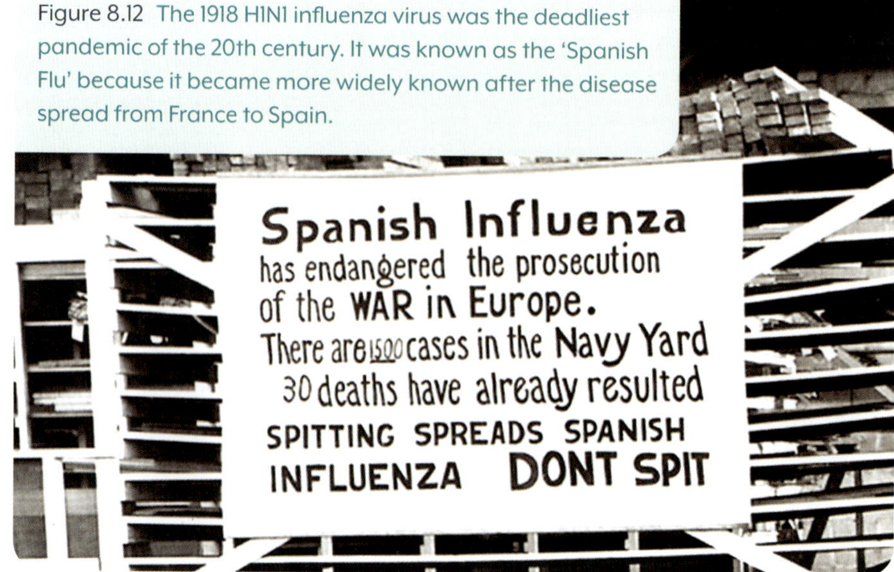

❷ Lifestyle diseases affect large numbers of Australians

While Australia has some protection against infectious diseases, Australians are just as vulnerable as people from other countries to non-infectious lifestyle diseases. A 2017 survey of records from Australian doctors and hospitals indicated that more than 12 per cent of Australians had hypertension, or high blood pressure, and more than 8 per cent had hyperlipidaemia, or high cholesterol. These are both diseases that can be caused or made worse by lifestyle factors, such as alcohol consumption and poor diet.

Diabetes is another non-infectious disease that is very prevalent in Australia – it's estimated that more than 1.7 million Australians suffer from some form of it. Diabetes is a disorder that affects the metabolism and is marked by high blood sugar levels. It can lead to cardiovascular disease, stroke, kidney disease, damage to the eyes and many other conditions.

Type 1 diabetes is a condition in which the pancreas no longer secretes insulin; it can be treated with regular insulin injections. Type 2 diabetes is a lifestyle disease, where insulin is still produced, but the target cells do not react to it. It is linked to obesity, high intake of sugar in the diet and lack of exercise.

What is a lifestyle disease?

❸ Medical scientists research cures and treatments

Health and medical research is one of the largest and most active areas of scientific research in the world today. This research is conducted by scientists such as biologists, doctors and biotechnologists.

Early medical researchers provided us with cures and treatments that are still in use today. Vaccines were invented at the end of the 18th century. They work by introducing a small amount of disease antigens to a patient, prompting their immune systems to develop antibodies and making them immune to that disease in the future. Antibiotics kill bacteria that cause diseases. Penicillin, the first antibiotic medicine, was invented in 1928.

As diabetes is a major issue in Australia, it's also a major focus of medical research. The Australian Foundation for Diabetes Research is currently developing a form of insulin **graft** – an artificial device that creates insulin in the body. The graft is a network of microscopic bubbles, created from seaweed, that are filled with insulin. Devices like this could revolutionise the treatment of Type 1 diabetes. Type 2 diabetes is better treated with lifestyle changes; the CSIRO has developed diet and exercise plans to help sufferers change their lifestyle and improve their own health.

How could insulin grafts change how Type 1 diabetes is treated?

Figure 8.13 Most people with diabetes must regularly check their blood sugar levels using a blood glucose meter.

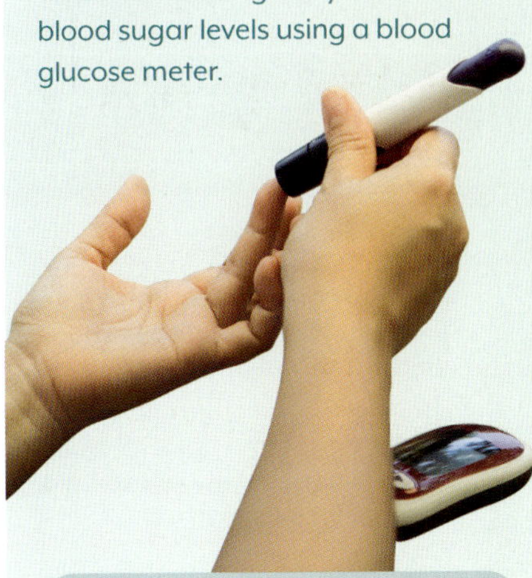

CHECKPOINT 8.7 

1 Suggest why international travel could cause pandemics.
2 Identify three lifestyle diseases that are common in Australia.
3 Explain the difference between Type 1 and Type 2 diabetes.
4 Suggest why an artificial device that creates insulin would benefit someone with Type 1 diabetes and not someone with Type 2 diabetes.

CHALLENGE

5 Funding for medical research is limited in Australia. Find out more about how the prevalence of disease influences where research money is allocated. Use this to determine some topics that can be debated by your class.

SKILLS CHECK

- I can describe the difference between an epidemic and a pandemic.
- I can describe some non-infectious diseases that cause problems in Australia.
- I can describe why diabetes is a major focus of medical research in Australia.

CHAPTER SUMMARY

Body systems coordinate to sustain life.

Nervous and **endocrine systems** coordinate the body's responses to changes in the internal and external environment.

Hormones are chemicals produced by glands of the endocrine system; they trigger an effect in target cells to help maintain homeostasis.

The **excretory system** is responsible for the removal of the waste from cellular processes.

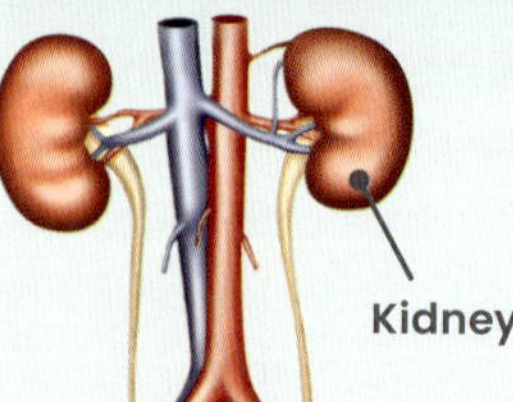

Homeostasis maintains bodily functions at levels that sustain life and good health.

Negative feedback is a response that compensates for changes in the environment.

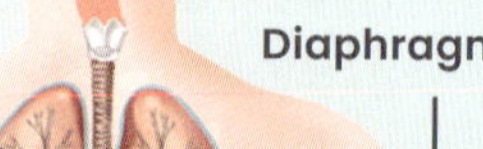

The **respiratory system** takes in oxygen and removes carbon dioxide.

The **circulatory system** is responsible for the transport of nutrients and wastes around the body.

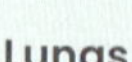

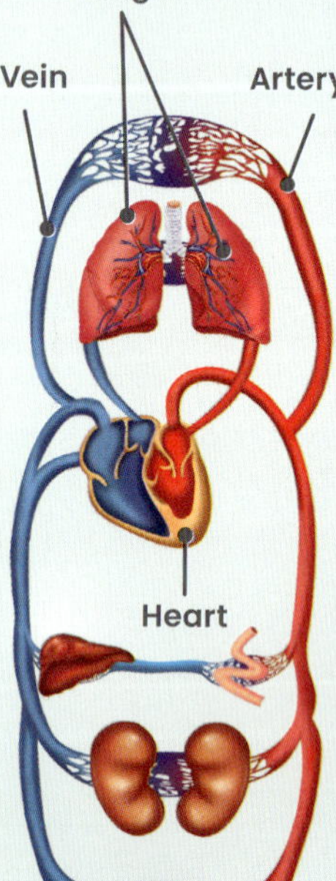

The **digestive system** is responsible for processing food into nutrients.

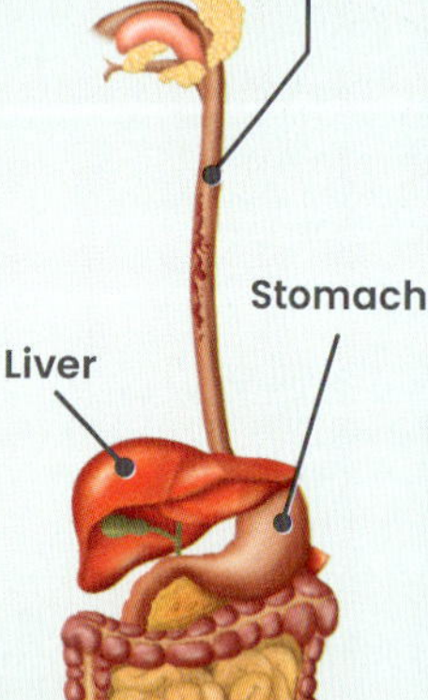

The **immune system** defends the body against pathogens.

Infectious diseases are caused by pathogens.

Non-infectious diseases are diseases not caused by pathogens.

An **epidemic** is a major outbreak of a disease within one or more communities.

A **pandemic** is a major outbreak of a disease that affects global populations.

◀ **Guava rust** and **foot-and-mouth disease** ▶

Organisations such as the CSIRO undertake research to help prevent the spread of infectious diseases.

FINAL CHALLENGE

1 Match each receptor to its correct stimulus.

Photoreceptor	Chemicals
Chemoreceptor	Light
Thermoreceptor	Changes in pressure or position
Mechanoreceptor	Temperature changes

LEVEL 1

50XP

LEVEL UP!

2 Identify which body system the following structures belong to.

a aorta
b bronchiole
c cerebrum
d kidney
e pituitary gland
f small intestine

3 Summarise the first, second and third lines of defence in the immune response to a bacterial infection.

4 Describe, using examples, the importance of biosecurity measures in Australia.

5 Explain what homeostasis is and give some examples of situations that require negative feedback in order for the body to remain in balance.

6 Compare the endocrine and nervous system with regard to:

a type of signal (chemical or electrical impulse)
b how the signal travels
c speed of the signal.

7 Explain the advantage of the respiratory system and circulatory systems working in coordination together.

8 Explain why your heart rate and respiration rate increase during exercise.

9 Identify what the lungs, skin and kidneys have in common.

10 Explain why diabetes is a growing disease in Australia and justify why the disease should or should not be a focus for scientific research.

9 Energy and matter in ecosystems

All living things and their environment are interconnected. If one thing changes so will everything else. The removal of wolves from Yellowstone National Park in the USA resulted in changes to the food web that affected how the rivers flowed – but this wasn't known until the wolves were returned.

By studying interactions between organisms and their environment as well as how matter and energy move through ecosystems, ecologists can help us learn how to manage and conserve ecosystems.

1 LEARNING LINKS

Brainstorm as much as you can remember about these topics.

Food chains and food webs include interactions between producers, consumers and decomposers.

Structural features and other adaptations of living things help them to survive in their environment.

The physical conditions of the environment affect the growth and survival of living things.

2 SEE-KNOW-WONDER

List three things you can **see**, three things you **know** and three things you **wonder** about this image.

3 CRITICAL + CREATIVE THINKING

In common: List five things that a coral reef has in common with a rainforest.

Five facts: List five facts, thoughts or opinions about the use of zoos to preserve endangered species.

What if ... a community of hippopotamuses were introduced to the Murray River. How would this affect the ecosystem?

4 THE MOST BIODIVERSE!

Biodiversity is the diversity of life in one particular place. The most biodiverse ecosystems are those close to the equator where there is high rainfall and it is warm all year round. On land, this means tropical rainforests; in the oceans, it means coral reefs. One in 10 species lives in the Amazon rainforest. It is home to more than 40 000 different plant species, more than 2000 species of birds, reptiles and mammals, and more than 2.5 million species of insects.

9.1 Communities in ecosystems

At the end of this lesson I will be able to:

- **recall** that ecosystems include communities of interdependent organisms.

KEY TERMS 

abiotic
non-living

biotic
living

ecosystem
a community of living things and their environment

community
a naturally occurring group of animals, plants and other organisms

consumer
an organism that gains energy by consuming other living organisms

producer
an organism that makes its own food using energy from the Sun

LITERACY LINK 

List as many verbs as you can that describe interactions between different organisms in an ecosystem.

NUMERACY LINK 

The population of a certain breed of frog in a small patch of rainforest was counted each day for a period of seven days. The numbers counted were as follows: 36, 47, 32, 29, 41, 43, 32.

Calculate the mean, median and mode for this data set.

An ecosystem is an area that contains living things and their environment. Ecosystems can be very small (for example, a pond) or very large (for example, an island). A community refers to all of the living things in an ecosystem, which rely on each other to survive and interact in different ways.

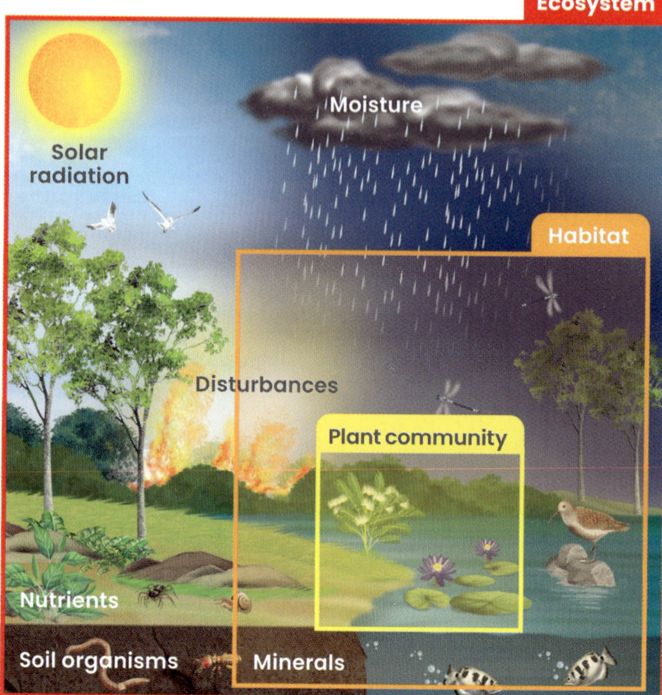

Figure 9.1
An ecosystem is an area that contains living things and their environment.

1 Communities are made up of different species

A **community** is made up of populations of all the different species that live in the ecosystem. All organisms within a community depend on each other for energy, nutrients and survival. Through the process of photosynthesis, **producers** convert the energy from the Sun into chemical energy that can then be used by other organisms. **Consumers** obtain their energy by eating other organisms.

Organisms such as bacteria and fungi are decomposers, recycling the nutrients from dead organisms and making these nutrients available to other species in the ecosystem.

The living things in an ecosystem are referred to as **biotic** factors. The non-living things, such as soil and water, are known as **abiotic** factors.

What is the difference between a consumer and a producer?

2 Ecological relationships maintain balance in communities

Organisms in an ecosystem can have more complex relationships than just predator–prey. These relationships maintain the balance within the community. A change in the population of one species will impact on the population of other species.

Table 9.1 Ecological relationships

Interaction	Effects on population	Example
Mutualism	The interaction is beneficial to both species.	A cassowary eating the fruit of plants, and distributing the seeds in its dung
Commensalism	One species benefits but the other is unaffected.	A staghorn fern growing on the branch of a tree
Competition	Both species are negatively affected because they compete for the same resource.	Sugar gliders competing with superb parrots for nesting hollows
Predator–prey	The predator benefits and the prey is harmed/killed.	A bronze whaler shark eating sardines
Parasitism	The parasite benefits and the host is harmed.	Mistletoe taking nutrients from a host tree

What is an example of an ecological relationship?

3 Each organism has an ecological niche

Each organism has a particular job within the ecosystem – their ecological niche. An organism's niche includes the interactions it has with other organisms, the type of food it eats, where it lives and how it reproduces. Only one organism can fill a specific niche in an ecosystem. Many different species can co-exist within an ecosystem because no two species have exactly the same niche. Introduced species cause problems when they take over a niche – often they will out-compete the native animal, leading to localised extinction.

If organisms that fill a particular niche disappear, then there will be flow-on effects throughout the entire community. If a pollinator disappears, then plants cannot reproduce, which limits the amount of food available to the herbivores, which in turn limits the amount of food available for carnivores.

What is an ecological niche?

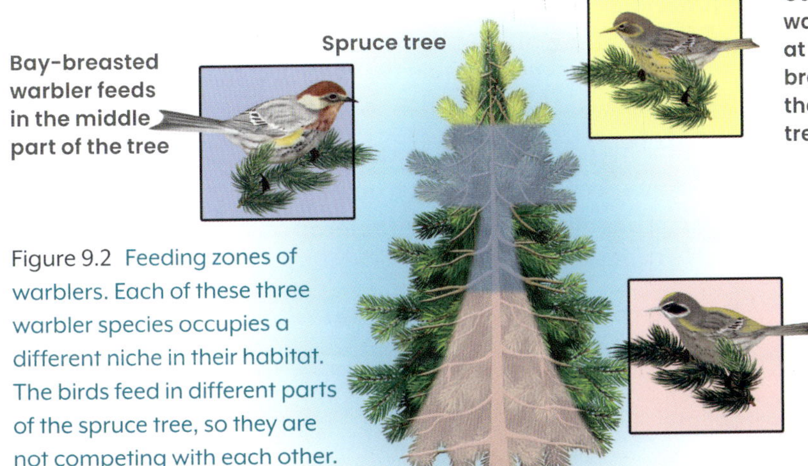

Figure 9.2 Feeding zones of warblers. Each of these three warbler species occupies a different niche in their habitat. The birds feed in different parts of the spruce tree, so they are not competing with each other.

INVESTIGATION 9.1
Identifying community members in a terrestrial ecosystem

CHECKPOINT 9.1

1 Copy and complete these sentences.

A community is made up of ________ of all the different ________ that live in the ecosystem. All organisms within a community ________ on each other for ________, ________ and ________.

2 Classify the following interactions.

a Anemones attach to the claws of boxer crabs. The anemones provide the crab with protection in return for food.

b Bandicoots feed on insects in your backyard.

c Dung beetles feed on kangaroo dung. The kangaroos are unaffected.

d A tick sucks blood from a wallaby.

e A cleaner shrimp removes parasites from a sea turtle.

3 Explain, using an example, what an ecological niche is.

CHALLENGE

4 Competition can be between different species or members of the same species. Using a local ecosystem, find examples of both types of competition.

SKILLS CHECK

- I can define what an ecological community is.
- I can identify at least three ways that organisms in a community can interact.

9.2 Abiotic factors in ecosystems

At the end of this lesson I will be able to:

- **recall** that ecosystems include abiotic components of the environment.

KEY TERMS 

adaptation
a feature that enables an organism to survive in its environment

zone of tolerance
the range of an abiotic factor that an organism can survive in

LITERACY LINK 

An adjective is a word that describes or clarifies a noun. Think of five adjectives that could be used to describe the abiotic factors of an ecosystem.

NUMERACY LINK 

A 5 g sample of soil was analysed in an area close to the beach. It was found to contain 230 mg of salt. Calculate the percentage of salt in the soil.

Abiotic factors are just as important as the biotic or living factors. Factors such as water availability, soil type, temperature and availability of nutrients (for example, nitrogen) affect what living things will be in the ecosystem. If the abiotic factors change outside of their normal level, there can be major impacts on the rest of the ecosystem.

1 Abiotic factors affect what can live in an ecosystem

Abiotic factors are chemical and physical factors within an ecosystem. Abiotic factors include:

- rainfall and water availability
- soil type
- temperature
- pH of soil or water
- sunlight
- nutrient availability in soil or water
- gas (oxygen and carbon dioxide) availability
- salinity of soil or water.

The abiotic factors within an ecosystem will affect what can live there. Plants require a certain amount of sunlight, nutrients and water to grow, so these factors determine the types of plants that live in certain areas of an ecosystem. This, in turn, affects what consumers can live there.

What is an abiotic factor?

2 Organisms have adaptations to abiotic factors

Organisms have evolved **adaptations** to help them survive in their environment. The adaptations can be:

- structural – a feature of an organism's body that helps it survive (for example, the sharp claws of a koala helps it climb trees)
- behavioural – the way an organism responds to its environment to help it survive (for example, the huddling of emperor penguins to stay warm)
- physiological – a process inside an organism's body that helps it survive (for, example, the ability of a camel to retain water by producing very little urine).

Even though organisms have evolved to adapt to their ecosystems, they can still only tolerate specific levels of each abiotic factor. If the level of an abiotic factor moves out of this **zone of tolerance**, the organism will not survive.

What are the three types of adaptation?

3 Coral bleaching happens when coral is stressed

Coral bleaching is an example of what happens when abiotic factors move out of the zone of tolerance.

Coral live in a mutualistic relationship with an alga called zooxanthellae. The algae live inside the structure of the coral. The coral provides the algae with a home and the algae photosynthesise and provide the coral with 90% of its energy requirements. The algae are also responsible for the coral's colour.

If the water temperature or pollution levels increase to outside the coral's zone of tolerance, the coral expels the algae and becomes bleached. If temperatures and pollution levels return to normal, the coral allows the algae to return and so it will recover. If the levels remain too high for a long time, then the coral will die because it is not receiving the energy it needs.

The optimum water temperature range for most reef-building corals is 20–32°C, although different species have their own zones of tolerance. Increasing temperature levels in Australian waters have led to coral bleaching all along the Great Barrier Reef.

What is coral bleaching?

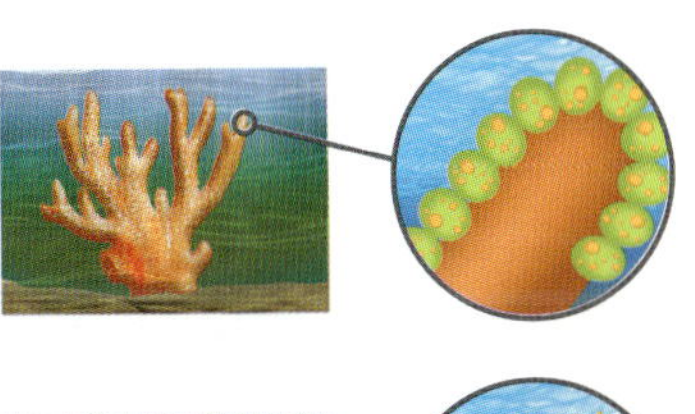

Healthy
Algae live inside the coral.
They depend on each other to survive.
Algae give the coral its colour.

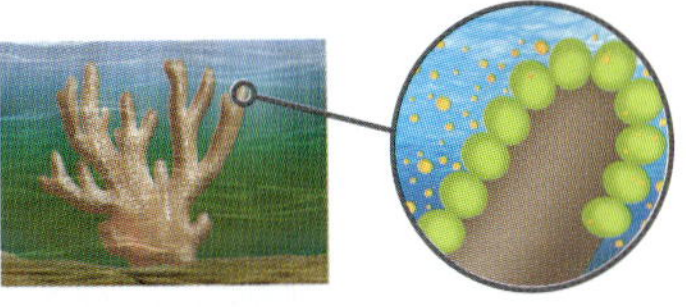

Stressed
If the coral gets stressed by rising temperatures or pollution levels, the algae leave the coral.

Bleached
Without the algae, the coral does not receive enough food. It turns white and is more likely to get diseases and die.

Figure 9.3 The stages of coral bleaching

Figure 9.4 Coral before, during and after a bleaching event. The dead coral becomes covered with algae.

INVESTIGATION 9.2
Measuring abiotic factors

CHECKPOINT 9.2

1 Identify some abiotic factors that are important for a:
   a plant
   b saltwater fish.

2 Define *adaptation*.

3 Explain, using a specific example, what is meant by an organism's tolerance level to an abiotic factor.

4 Predict what would happen if pollution caused the salinity level of a freshwater pond to increase.

CHALLENGE

5 Research to compare the abiotic factors in the Daintree Rainforest (Queensland) with those of the Simpson Desert (central Australia). Find out about rainfall (water availability), temperature range and nutrient availability in the soil. How do these factors influence what organisms live in these ecosystems?

SKILLS CHECK

- I can explain what an abiotic factor is.
- I can identify at least three examples of abiotic factors.
- I can explain how abiotic factors can determine what organisms live in an ecosystem.

9.3 Matter cycles in ecosystems

At the end of this lesson I will be able to:

- **outline** using examples how matter, such as nitrogen, cycles through ecosystems.

KEY TERMS 

atmospheric nitrogen
nitrogen that is found in the atmosphere

mutualism
a relationship in which both organisms benefit

LITERACY LINK 

Write a story describing what happens to an atom of nitrogen as it cycles through the atmosphere, soil and living things.

NUMERACY LINK 

The nitrogen content of wheat is 1.68%; for barley, it's 1.62%; and for broad beans, it's 4.0%.

Plot a column graph of this data.

Matter cannot be created or destroyed. On Earth, matter cycles through different parts of ecosystems, including the atmosphere, soil and different organisms. The atoms that make up your body were once part of the food that you consumed, and the air that you breathed.

If you eat a steak, the atoms in that steak have come from the plants that the cow has eaten, and before that from the carbon dioxide, water and nutrients that were used by the plants.

1 Nitrogen is an important element for life

Nitrogen (N) is an important element that cycles through ecosystems. Nitrogen is important because it is found in amino acids, the building blocks of proteins, which are essential for all life. Nitrogen is the most abundant element in Earth's atmosphere, making up approximately 78% of air. In the atmosphere, nitrogen is mostly found as molecular nitrogen (N_2).

Plants are a very important part of the nitrogen cycle because they can take up inorganic nitrogen (nitrogen from non-living sources) through their roots from the soil. Hence nitrogen enters the biosphere and cycles through living things. Bacteria are also vital to the nitrogen cycle and different species play different roles – converting nitrogen from one form into another.

Why is nitrogen important for living things?

2 Nitrogen cycles from the atmosphere

Although there is a lot of nitrogen in the air, plants cannot use this **atmospheric nitrogen**. Plants need to obtain nitrogen from the soil (through their roots) in the form of nitrate (NO_3^-). They use this nitrogen to make amino acids.

Most nitrogen is transformed by different species of bacteria into forms that plants can use. This usually happens in several steps, each involving a different species of bacteria. Some of these bacteria live in the roots of legumes (e.g. peas and beans). This relationship is an example of **mutualism** because both the bacteria and the plant benefit. Nitrogen-fixing bacteria convert nitrogen from the air in the soil into a form that the plant can use. There are also denitrifying bacteria, which release nitrogen back into the atmosphere.

When a herbivore eats a plant, the animal produces proteins from the amino acids that the plant has made. When a herbivore is consumed by a higher-order predator, the amino acids are then taken in and used by the predator.

Figure 9.5 Soy bean roots contain nodules of bacteria that take nitrogen from the air in the soil and convert it into a form that the plant can use.

Decomposers such as bacteria and fungi play a key role in the nitrogen cycle. They break down the amino acids in dead matter, releasing nitrogen back into the soil so that it can be reused by plants.

How is nitrogen used by plants?

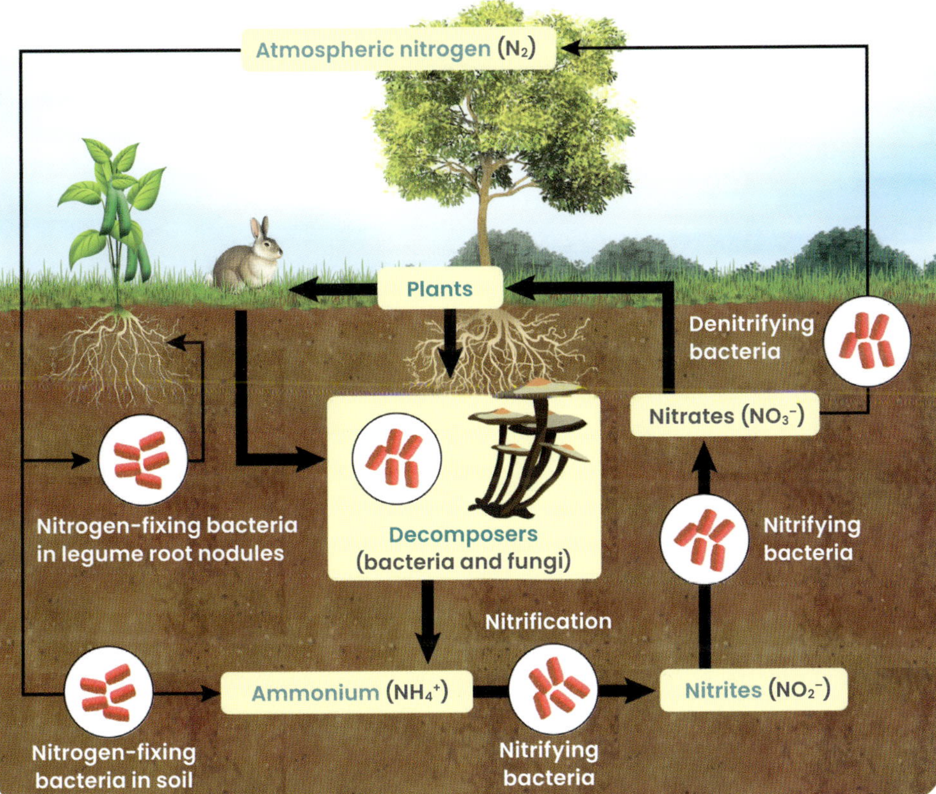

Figure 9.6 The nitrogen cycle

3 Other elements also cycle through ecosystems

Carbon, oxygen, hydrogen and phosphorus are other important elements that cycle through ecosystems. All of these elements are important for the formation of molecules that are vital for life.

Carbon, oxygen and hydrogen cycle through ecosystems through the processes of photosynthesis and cellular respiration. They move from inorganic molecules, such as carbon dioxide, molecular oxygen and water, into organic molecules, such as glucose and carbohydrates, which are then passed up through the food chain.

Phosphorus plays an important role in the formation of DNA. Plants take up phosphorus from soils and convert it into molecules that are then passed up through the food chain, returning to the inorganic forms through excrement and decomposition.

What other elements cycle through ecosystems?

CHECKPOINT 9.3 

1 Explain what is meant by a 'matter cycle'.

2 In what form is nitrogen found in the atmosphere?

3 Identify the two types of organisms that are important for the nitrogen cycle. What role do they play?

4 Predict what would happen to the nitrogen cycle in an ecosystem:

a where there were no decomposers

b that became too cold for the nitrogen-fixing and nitrifying bacteria to survive.

5 Carnivorous plants catch insects not to gain energy, but to extract nutrients such as nitrogen. Suggest what this means about the soils that they live in.

6 Identify two processes that allow carbon, oxygen and hydrogen to cycle through ecosystems.

CHALLENGE

7 Phosphorus is another element that cycles through ecosystems. Use the internet to find out more about the phosphorus cycle.

SKILLS CHECK

- I can outline the steps involved in the nitrogen cycle.
- I can explain the importance of the nitrogen cycle for living things.

9.4 Energy flows through ecosystems

At the end of this lesson I will be able to:

- **use** food webs to describe how energy flows through ecosystems.

KEY TERMS 

photosynthesis
the chemical reaction between carbon dioxide and water in the presence of sunlight and chlorophyll that produces glucose and oxygen

trophic level
the position of an organism in a food chain

LITERACY LINK 

Photosynthesis and chemosynthesis are processes used by organisms to convert energy from one form to another. Suggest what these words mean, then do some research to find out.

NUMERACY LINK 

If the energy content of the first trophic level is 140 000 joules and only 10% of energy is passed on to the next trophic level in a food chain, how much energy will an organism in the third trophic level consume? How much energy will have been lost?

Energy enters most ecosystems as solar energy from the Sun. Producers use this energy to make glucose. Glucose is then used as an energy source by all of the other living things in the ecosystem, being passed on as one organism consumes another. All organisms lose energy to the environment as heat, in excrement and in reproduction. This means that higher-order consumers need to consume a larger number of lower-order consumers to meet their energy requirements.

❶ Plants bring energy into ecosystems

Energy enters most ecosystems as light energy from the Sun. Producers convert light energy into chemical energy during **photosynthesis**. In photosynthesis, plants use light energy to produce glucose ($C_6H_{12}O_6$) from carbon dioxide (CO_2) and water (H_2O).

In plants, photosynthesis happens in chloroplasts. These cell organelles are packed with chlorophyll, a green pigment that, when sunlight hits it, starts the reactions of photosynthesis.

Where does the energy in ecosystems originate?

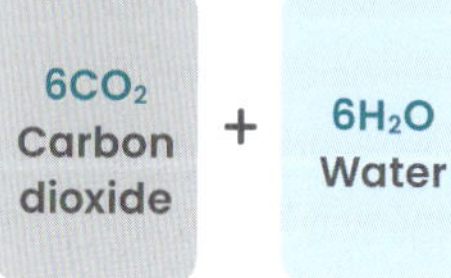

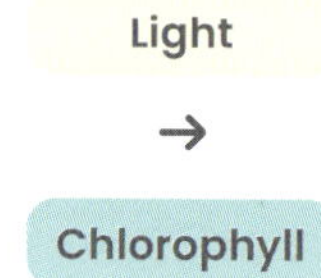

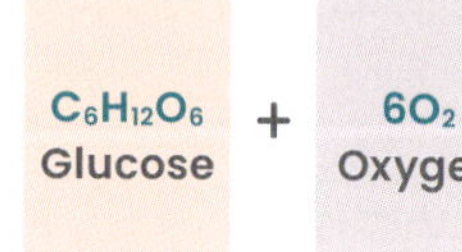

❷ Energy flows up the food web

A food web shows feeding relationships within an ecosystem and how energy flows through the ecosystem. The arrows in a food web indicate the direction that energy is flowing; from the producer, to the primary and secondary consumers, and lastly to the tertiary consumers.

Food webs organise the organisms into **trophic levels**. The trophic level of an organism refers to its position within a food chain. Producers are at the bottom and the highest-order consumer is at the top. Energy passes to the higher trophic levels through consumption.

What do the arrows in a food web indicate?

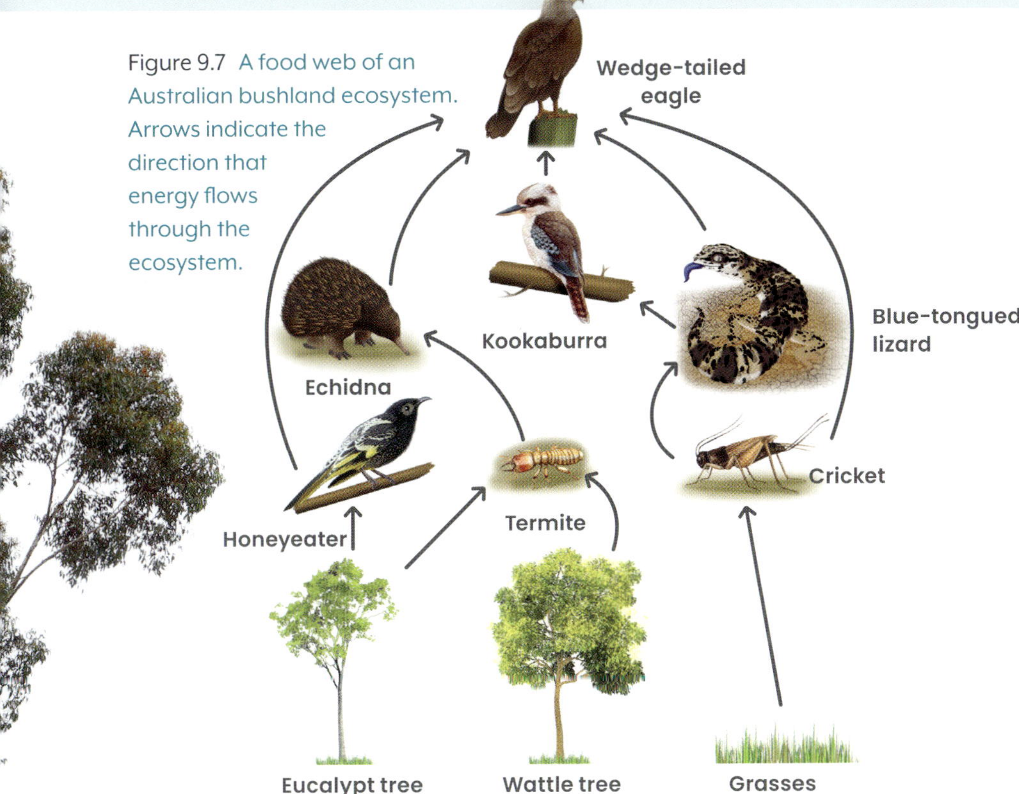

Figure 9.7 A food web of an Australian bushland ecosystem. Arrows indicate the direction that energy flows through the ecosystem.

INVESTIGATION 9.4
Algal balls

CHECKPOINT 9.4 

1 Why are plants known as producers?
2 Identify where photosynthesis takes place in a plant.
3 How is energy passed between organisms?
4 What can a food web be used to show?
5 Identify where most of the energy goes that an organism consumes.
6 Predict what would happen if a disease wiped out the major producer species in an ecosystem.
7 Predict what would happen if a large population of fourth-order consumers was released into an ecosystem.

CHALLENGE

8 In some unusual ecosystems, bacteria use a process called chemosynthesis to convert chemical energy into a resource that other organisms can use. These communities have evolved around structures called 'black smokers'. Use the internet to find out about the food webs in black smoker communities.

❸ Energy is lost at each trophic level

Energy is lost at each trophic level. Every organism loses energy to the environment through metabolic heat loss. Energy is also used up through reproduction and in the production of waste such as faeces. Therefore, this energy is not available when that organism is consumed.

Only about 10% of the energy in a trophic level is passed onto the level above. This is why a high number of producers and lower-level consumers are required to support a small number of higher-order consumers. This relationship can be represented in an energy pyramid.

What percentage of energy is available to the next trophic level?

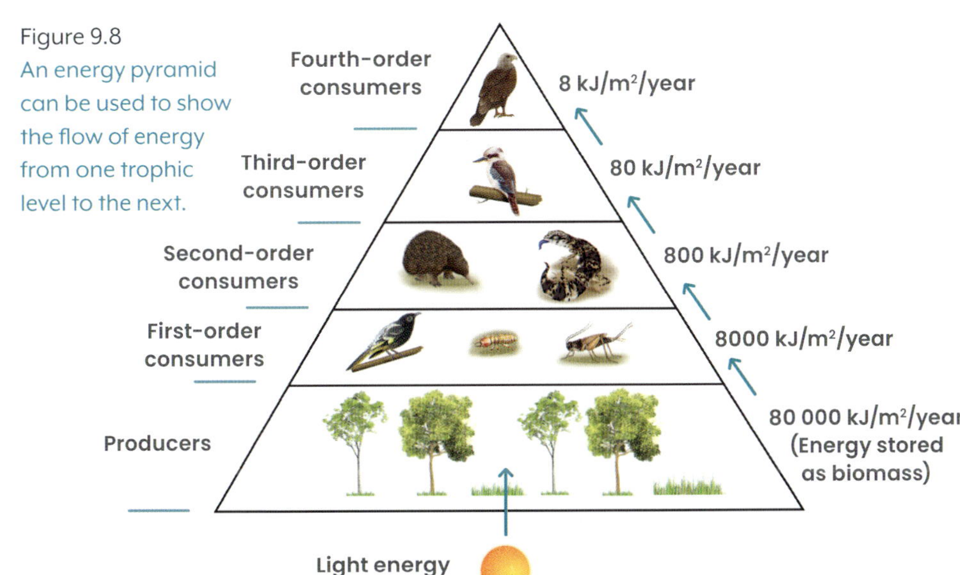

Figure 9.8 An energy pyramid can be used to show the flow of energy from one trophic level to the next.

SKILLS CHECK

- I can identify where energy in a food web originates.
- I can identify how energy is passed from one organism to another in a food web.
- I can use a food web to describe how energy flows through an ecosystem.

9.5 Changes to ecosystems

At the end of this lesson I will be able to:

- **analyse** how changes in some biotic and abiotic components of an ecosystem affect populations and/or communities.

KEY TERMS 

biodiversity
the variety of species in an ecosystem

cellular respiration
the chemical reaction between glucose and oxygen that produces energy, and carbon dioxide as a waste product

eutrophication
the process in which nutrient levels increase in a waterway, resulting in increased algal growth and decreased dissolved oxygen levels

keystone species
a species that plays a crucial role in its ecosystem

NUMERACY LINK 

The population of foxes in a particular ecosystem suddenly increases. Sketch a graph predicting what would happen to the population of rabbits and foxes in the area over time. (Use a different colour for each). Extend the graph to show what would happen to both populations over a number of seasons.

Within an ecosystem, everything is linked. Therefore, if one thing changes, so will others. Changes can happen to the biotic factors; for example, a species may be introduced or removed. Changes can also happen to the abiotic factors, such as a change in the nutrient availability or temperature. These changes can affect other abiotic factors as well as causing changes to the ecosystem's community if species cannot adapt to the changes.

1 Increased nutrient levels in waterways cause algal blooms

It is important that levels of nutrients, such as nitrogen, remain balanced in ecosystems. Producer organisms require these nutrients to grow. If there is too little of a nutrient, then the producers will be unable to grow. If too many nutrients flow into waterways from pollution or stormwater run-off, then it can cause other problems.

Eutrophication occurs when nutrient levels in a waterway increase so much that algae reproduce rapidly and algal blooms grow on the surface. This decreases the oxygen levels in the water. Different aquatic species require different oxygen levels. As the oxygen levels start to drop, those organisms that require the most oxygen die first.

Oxygen levels in the water decrease when:

- more algae use the oxygen dissolved in the water for **cellular respiration** – especially overnight when there is no photosynthesis to replace it
- algae block sunlight from reaching other producers deeper in the water and stop the producers from photosynthesising
- the increased numbers of algae die and decompose, so that more decomposers grow and increase their cellular respiration.

What is eutrophication?

Figure 9.9 This green water in the Condamine River in Queensland shows there is excessive algal growth and eutrophication.

2 Keystone species are essential to ecosystems

Keystone species are organisms that play essential roles in their ecosystems. No other species in the ecosystem can fill its niche. If a keystone species is removed, the whole ecosystem can collapse. The numbers of species falls, causing a loss of **biodiversity** as more than just the keystone species becomes extinct.

Southern cassowaries (*Casuarius casuarius johnsonii*) are a keystone species in the north Queensland tropical rainforest ecosystem. They play a key role in the dispersal of the seeds of up to 150 plant species. Cassowaries eat the fruit and then deposit the seeds in their dung throughout their territory.

The southern cassowaries are endangered because of habitat destruction and other human impacts. Without the cassowaries, some rainforest plants could also become extinct. If these plants were to disappear, the rainforest would, too, because more extinctions would follow at the different trophic levels.

What is a keystone species?

3 Introduced species compete with native species

The introduction of species into an ecosystem can have wide-ranging impacts on the community. Introduced species can compete with native species for the same resources, remove species through predation or toxins, and even damage soils and plant life through trampling, burrowing and digging.

In Australia, most ecosystems have been affected by the introduction of species. Foxes and cats are effective predators and have hunted many species to extinction. Rabbits out-compete small marsupials for food and other resources because they reproduce at a much faster rate. Brumbies (feral horses) and camels trample fragile soils and plants that have not evolved to cope with hard-hooved animals.

Why do rabbits out-compete native marsupials?

Figure 9.10 The southern cassowary is a keystone species in Queensland's tropical rainforests.

CHECKPOINT 9.5 

1 Identify one way an ecosystem would be affected by a change to:
  a an abiotic factor
  b a biotic factor.

2 Explain why increasing levels of nutrients such as nitrogen in lakes and rivers can be detrimental for the organisms living there.

3 Explain why the loss of one species can lead to the loss of others.

4 Create a Daintree Rainforest food web that contains a cassowary, rainforest plants, amethystine python, estuarine crocodile, white-tailed rat, musky rat kangaroo and insects.

5 Identify three ways that an introduced species can impact on an ecosystem.

CHALLENGE

6 Climate change is having a significant impact on Australian ecosystems because it is changing abiotic factors such as temperature and water availability. Research and select one case study of an Australian ecosystem being affected by climate change and present your findings to your class.

SKILLS CHECK

- I can identify some examples of changes that can occur to abiotic and biotic components of an ecosystem.
- I can give an example of how a change to an abiotic factor could affect a population or community.
- I can give an example of how a change to a biotic factor could affect a population.

9.6 Indigenous management of ecosystems

At the end of this lesson I will be able to:

- **assess** ways that Aboriginal and Torres Strait Islander peoples' cultural practices and knowledge of the environment contribute to the conservation and management of sustainable ecosystems.

KEY TERMS 

euryhaline
the ability to survive in water with various salinity levels, from fresh to very salty

evolve
change over many generations to adapt to the environment

sustainably
to use a resource in a way that avoids depletion and maintains balance

LITERACY LINK 

The Dreaming stories of Aboriginal and Torres Strait Islander peoples were one way of passing knowledge between generations. Find one story from your local area and summarise the message that it contains.

NUMERACY LINK 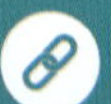

The small sawfish usually has 23–34 teeth on each side of the 'saw'. A fisherman catches five sawfish in his net. Calculate the approximate number of teeth in total.

Aboriginal and Torres Strait Islander peoples have been managing the land **sustainably** for tens of thousands of years. Their ecological knowledge has been passed from one generation to the next.

Today, scientists work closely with Indigenous land managers to learn more about the conservation of ecosystems. The National Environmental Science Program is one such initiative in which Indigenous Australians partner with scientists and governments to undertake environmental and climate research.

1 Ecosystems can be managed with fire

By lighting relatively low-intensity, slow-burning and controlled fires at specific times of the year, Indigenous Australians managed fuel levels and encouraged the growth of key plant species. Because of this practice, some Australian plant species **evolved** to require fire as part of their life cycle. These plants need fire to open seed pods or to clear away competition, and fire also adds nutrients back into the soils. Smoke also helps some seeds to germinate.

Indigenous Australians also used fire to flush out animals so they could be easily hunted.

Since the European colonistion of Australia, the use of fire to manage ecosystems has declined. This has caused a loss of biodiversity in some ecosystems. However, burning practices are being used again in many regions, which has improved the biodiversity and led to a return of some species.

Why do some Australian plants need fire?

Figure 9.11 Indigenous fire management practices are now being reintroduced to some Australian ecosystems to restore biodiversity.

Figure 9.12 Banksia seed pods will only open after being burnt by a relatively cool fire. The seeds can then germinate in the nutrient-rich soil, free from competition.

❷ Indigenous knowledge is helping to manage sawfish populations

Sawfish are some of the world's most endangered marine fish species. Four of the world's five species are found in the estuaries of northern Australia. An estuary is an area where a freshwater river meets the ocean resulting in brackish, or slightly salty, water. Sawfish are a **euryhaline** species, which means that they can live in water that varies in salinity from fresh to very salty. Very few sharks and rays can do this.

The sawfish populations in Australia's northern estuaries and rivers have declined over the last 50 years, and the species are now protected. Threats to the population include fishing, and barriers across rivers, such as road crossings and dams, that limit the migration of the sawfish up and down the river.

Many Indigenous groups have strong cultural connections to sawfish. They know where and when to find the different species and have noted their decline. Scientists can collaborate with these groups to determine ways of protecting threatened sawfish populations. Indigenous knowledge of past sawfish distribution combined with recent population studies helps to determine strategies that will protect these species from extinction.

What is one threat to sawfish populations?

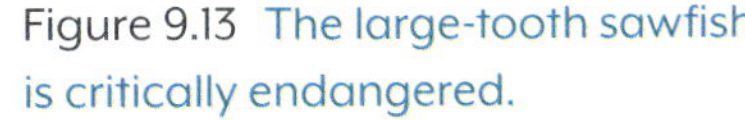

Figure 9.13 The large-tooth sawfish is critically endangered.

❸ Indigenous knowledge is being used to manage ecosystems

There have been many successful strategies involving Indigenous peoples' ecological knowledge to manage ecosystems.

- The Great Barrier Marine Park Authority works closely with the traditional owners of the land to manage and protect the Great Barrier Reef.
- Indigenous communities are working with scientists to study the effect feral cats have on ecosystems and to determine eradication methods.
- Indigenous researchers are working to conserve biodiversity in urban areas.

What is one way that Indigenous ecological knowledge is used to manage ecosystems?

CHECKPOINT 9.6

1 Is fire a biotic or an abiotic factor?

2 List four benefits to an ecosystem of controlled fires.

3 Identify two impacts on ecosystems that do not experience regular slow-burning fires.

4 Explain how returning to a regular slow burning fire regime benefits Australian ecosystems.

5 How is a sawfish different from other sharks and rays?

6 Predict the impact on an ecosystem if sawfish were removed.

7 Explain why Indigenous knowledge is useful to scientists who are studying current populations of sawfish.

CHALLENGE

8 Find out more about another collaboration between one group of Aboriginal and Torres Strait Islander people and scientists to help sustainably manage ecosystems. Include an overview of the research and evaluate how Indigenous peoples' knowledge is contributing to it.

SKILLS CHECK

- I can identify two ways that Indigenous Australians' knowledge and/or practices can be used to help sustainably manage ecosystems.

9.7 Conservation strategies

At the end of this lesson I will be able to:

- **evaluate** some examples of strategies used to balance conserving, protecting and maintaining the quality and sustainability of the environment with human activities and needs.

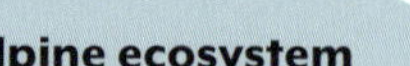

KEY TERMS

alpine ecosystem
an ecosystem existing at high altitudes

conservation
protection and maintenance

omnivore
an organism that eats both plants and animals

sustainable
able to be maintained long term

LITERACY LINK

Create a slogan that could be used by a national park to educate visitors about reducing their impact on the park ecosystem.

NUMERACY LINK

The entire area of NSW is 809 444 km^2. If the area of land devoted to national parks is around 70 000 km^2, calculate the percentage of the state that is devoted to national parks.

Human activities can significantly impact on ecosystems. **Conservation** measures aim to minimise the effect of human activities and protect ecosystems. Australia has thousands of national parks and reserves where a large variety of native landscapes and ecosystems are managed and protected. More than 7 million hectares or more than 9% of New South Wales is protected and managed by the NSW National Parks and Wildlife Service (NPWS). NPWS manages **sustainable** tourism in these areas by considering the conservation of plants and animals and protecting sites of cultural and heritage significance.

1 We need to balance the needs of visitors with protecting national parks

Kosciuszko National Park was established in 1967 and more than 1.5 million people visit every year. It covers an area of 6900 km^2 and includes Australia's highest peak (Mt Kosciuszko), snow fields, unique **alpine ecosystems** and several endangered species.

One of the endangered species is the mountain pygmy possum (*Burramys parvus*). The known habitat of this tiny **omnivore** totals only 8 km^2, because it lives in rocky areas and boulder fields at elevations above 1300 m. Habitat removal and climate change are the major threats to the mountain pygmy possum's survival.

What is Kosciuszko National Park home to?

Figure 9.14 The boulder field habitat of the mountain pygmy possum. Artificial boulder fields and corridors across ski slopes have been constructed to help preserve the species.

Figure 9.15 The mountain pygmy possum is an endangered species.

❷ Building ski resorts involves removing habitat

Even though they currently take up less than 1% of the area of Kosciuszko National Park, ski resorts have caused significant damage through pollution from rubbish, emissions and sewage and by removing important habitat. When the ski runs were constructed, many of the boulder fields that were important mountain pygmy possum habitat were removed, as they were dangerous to skiers.

The NPWS works to repair the damage to the ecosystem, including possum habitats, through careful management of the park. Visitors are encouraged to stay within the resorts. Artificial boulder fields have been built away from the ski runs, and small corridors have been built across the ski runs so that the possums can move from one area to another. This is very important during their mating season.

What area of the Kosciuszko National Park do the ski resorts take up?

❸ Formed walkways help limit damage to plant life

NPWS has built formed walking tracks along the most popular routes in Kosciuszko National Park. For example, the track from Thredbo to the top of Mt Kosciuszko has been constructed as a paved or raised metal walkway in order to protect the surrounding vegetation. If visitors only use the formed walkways, then damage to fragile alpine plant life is limited to a very small area.

One ecosystem the walkways protect is the alpine sphagnum bogs. Sphagnum moss is an important species because it helps to retain water when the snow melts, slowly releasing it into waterways that eventually flow into the Snowy River. Trampling causes significant damage to the sphagnum bogs, and formed walkways can reduce the risk of tourists trampling through the bogs.

What does a formed walkway aim to prevent?

Figure 9.16 Hiking on formed tracks protects fragile alpine plant life, such as sphagnum moss.

CHECKPOINT 9.7 

1. What is the purpose of a national park?
2. What type of unique ecosystems are present in Kosciuszko National Park?
3. How can ski resorts damage natural ecosystems?
4. a What impact do ski resorts have on the habitat of the mountain pygmy possum?
   b How is this being addressed?
5. What is the benefit to the ecosystem of constructing walkways in popular areas?
6. Ski resorts take up less than 1% of the area of Kosciuszko National Park. Is it worth making efforts to reduce the damage to such a small fraction of the park? Justify your response.

CHALLENGE

7. The Snowy Hydro-Electric Scheme was constructed to produce hydroelectricity and provide irrigators on the western side of the Great Dividing Range with water. Carry out research to make a list of the pros and cons of the scheme.

SKILLS CHECK

- I can identify two examples of where national parks conserve the environment but still allow human activities.
- I can explain the costs and benefits of their strategies to the ecosystem.

CHAPTER SUMMARY

An **ecosystem** is an area that contains living things and their environment.

Ecosystem
Moisture
Solar radiation
Habitat
Disturbances
Plant community
Nutrients
Soil organisms
Minerals

Biotic components of an ecosystem include:
- animals
- plants
- fungi
- bacteria.

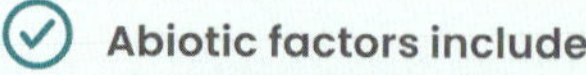

Abiotic factors include:
- water availability
- temperature
- soil.

A **community** is made up of populations of all the different species that live in an ecosystem.

All **organisms** within a community depend on each other.

Sustainable tourism strategies help protect wildlife and habitats.

Keystone species play essential roles in their ecosystems. Extinction of keystone species will result in the extinction of other species.

Knowledge of the past distribution of endangered species helps to determine strategies to protect them from extinction.

Energy flows through ecosystems. Food webs can be used to describe how energy flows through ecosystems.

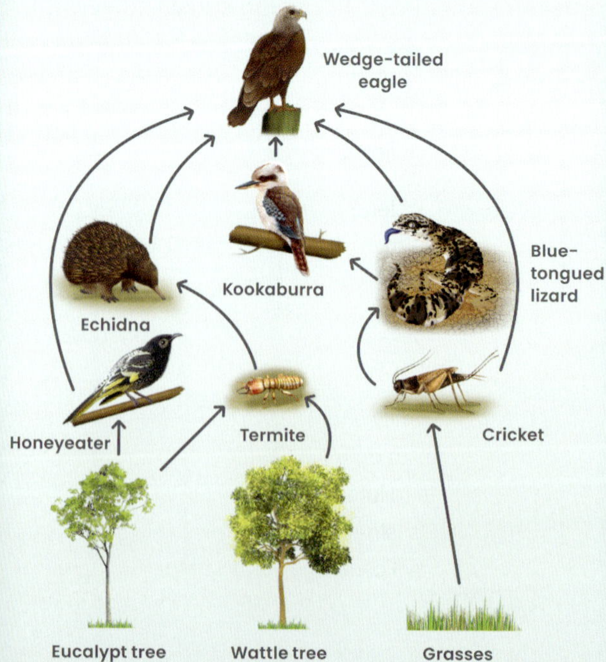

FINAL CHALLENGE

1. Match these terms with their definitions.

abiotic	a naturally occurring group of animals, plants and other organisms
biotic	the number of a species in a particular area
community	non-living
ecological relationship	living
ecosystem	an interaction between two organisms
population	a community of living things and non-living things

LEVEL 1

50XP

LEVEL UP!

2. Identify five different abiotic factors.
3. Identify three ways that organisms in a community can depend on one another.
4. Use a flow chart to show how nitrogen cycles through an ecosystem.
5. Use a flow chart to show how energy flows through an ecosystem.

6. Describe an example where a change to an abiotic factor in an ecosystem affected others.
7. Describe an example where traditional Indigenous conservation techniques have contributed to the conservation of an ecosystem.

8. Explain the significance of the directions of the arrows in a food web.
9. Explain why bacteria and other microorganisms are a vital part of any ecosystem.

10. As a class, identify a local example of a conservation strategy that balances human activities and needs. Create a report that evaluates the success or failure of this strategy. You could consider habit loss due to infrastructure, species loss, human requirements for recreational and commercial spaces and access to essential human services.

10 Genetics

The greatest leaps forward in science might be found in the smallest of places. In nearly every cell of your body, there is a tiny molecular blueprint called DNA. Scientific advances mean that DNA mapping that once took 15 years and cost $5 billion can now be done rapidly for less than $100. We can find out what diseases we are most at risk of, what medicines work best for us individually and what cultural heritage we may have but otherwise would never have known about. These advancements in biotechnology bring with them the potential for both miracles such as cell regeneration and terrors such as biological warfare.

1 LEARNING LINKS

Brainstorm as much as you can remember about these topics.

The reproductive system has an important role in humans.

Living things are made of cells.

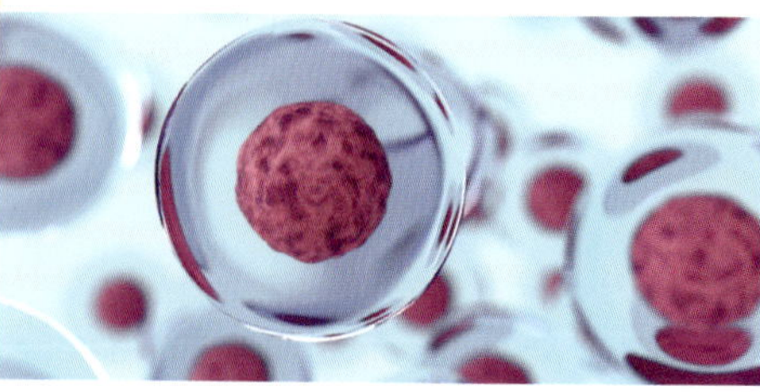

Developments in technology have helped us find solutions to current health issues.

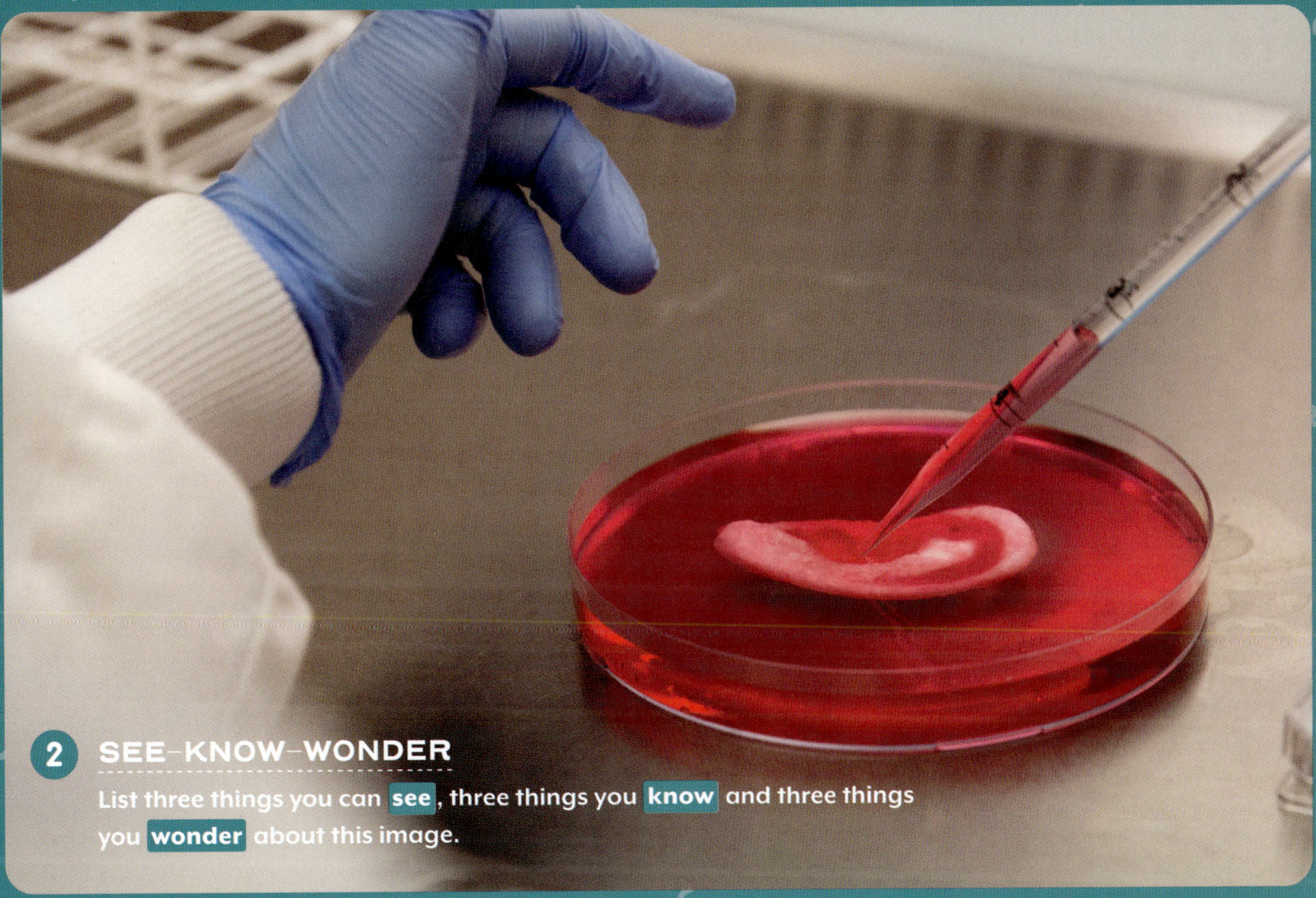

2 SEE-KNOW-WONDER

List three things you can **see**, three things you **know** and three things you **wonder** about this image.

3 CRITICAL + CREATIVE THINKING

Five questions: Write five questions that have the answer 'DNA'.

What if ... parents could choose the characteristics of their baby before it was born?

The BAR: Suggest one thing you could make **B**igger, **A**dd and **R**eplace on a human being.

4 THE BIGGEST PROJECT

In 1990, the Human Genome Project began. This was an international research project with the goal of mapping the entire human genome (all of human DNA), of more than 3 billion base codes. Scientists from all over the world worked for 15 years on research estimated to cost more than $5 billion. Because of the wonderful work on the Human Genome Project, scientists were able to identify and locate all the genes in human DNA.

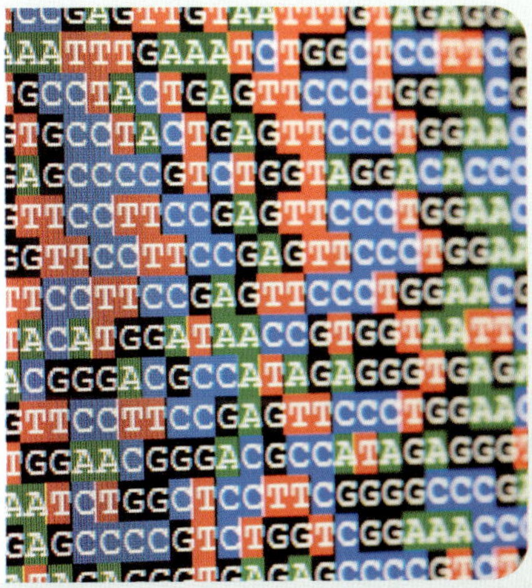

10.1 The human reproductive system

At the end of this lesson I will be able to:

- **relate** the organs involved in human reproductive systems to their function.

KEY TERMS 

gamete
a sex cell – an ova or sperm

menstruation
monthly discharge of blood and tissue from the lining of the uterus

ovum
a female sex cell, also known as an egg

semen
a protective fluid that contains the sperm

sperm
a male sex cell

LITERACY LINK 

Identify three terms from this section that you are unfamiliar with. Write a definition for each term.

NUMERACY LINK 

The human egg is about 0.1 mm in diameter and a sperm cell is about 0.05 mm in length. How many times bigger is the egg cell than a sperm cell?

The male and female human reproductive systems work hard nearly all of the time. Men produce about 1500 sperm every second, while women release an egg every month. To make a baby, these sperm must survive and traverse the female reproductive system. Only one of the sperm will fertilise an egg to begin the process of creating new life.

1 The female reproductive organs

The female reproductive organs are the vagina, uterus, fallopian tubes and ovaries. There are two ovaries, one on either side of the uterus. Ovaries are oval-shaped glands that produce the reproductive hormones oestrogen and progesterone. Oestrogen is one of the hormones involved in the development of the secondary sex characteristics at puberty – the widening of the hips, growth of pubic hair and breast development.

Ovaries also contain ova (eggs). These are the **gametes** (sex cells) of women and are also known as oocytes. Girls are born with all the eggs they will ever have, about 2 million.

When a girl starts menstruating, an egg (or **ovum**) is released every month down her fallopian tubes (the left and right alternatively). The fallopian tubes act as transport tunnels for the eggs to travel from the ovaries to the uterus. If an ovum is fertilised by a sperm, it develops and embeds in the wall of the uterus, which acts as a 'nest' for the developing embryo. The uterus wall is lined with tissue, blood and blood clots. If no fertilised egg is present, this unused material is shed out of the uterus, through the cervix and the vagina, and is observed as **menstruation**.

What are the organs of the female reproductive system?

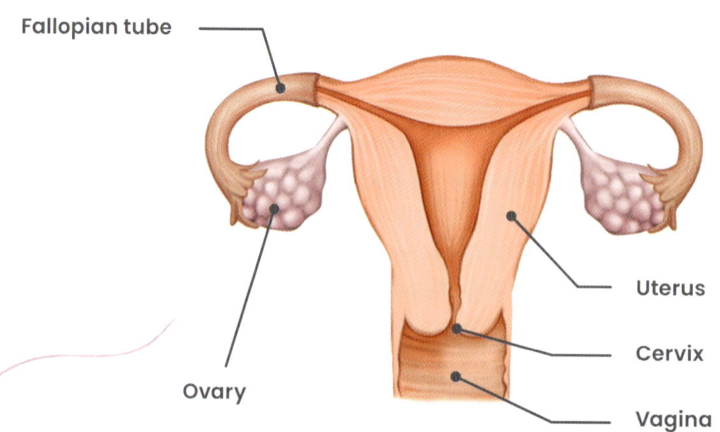

Figure 10.1 The female reproductive system

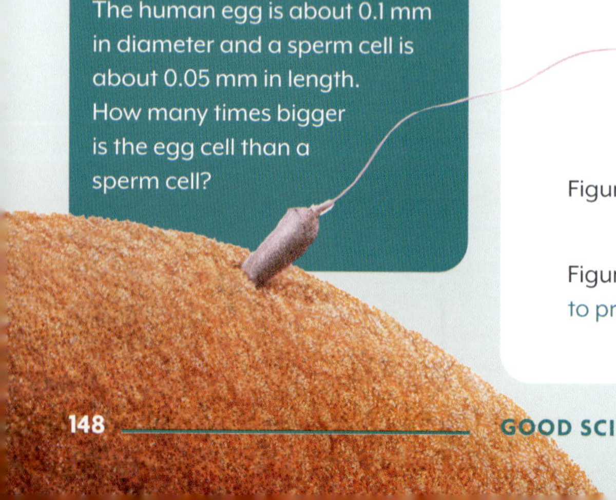

Figure 10.2 The sperm penetrates an ovum (egg) to produce a new cell.

2 The male reproductive organs

The male reproductive organs are the penis, testicles, epididymis, vas deferens and prostate gland. The testicles are outside the body in sacs called the scrotum. The testicles are glands that produce sperm and the male sex hormone testosterone.

The gametes of men are called **sperm**. The sperm travel from the testis to the vas deferens through a tube called the epididymis. The vas deferens is a duct that then transports the sperm to the ejaculatory duct. The penis has a single opening called the urethra, and both sperm and urine can travel down this passage.

When aroused, the penis floods with blood, stretching and becoming erect. This blocks off access to urine so that only semen can travel down during ejaculation. The prostate gland's function is to secrete the fluids that are expelled with sperm during ejaculation. These fluids and sperm together are referred to as **semen**. The semen provides the sperm with a medium that they can swim through and lubricates the urethra.

What are the organs of the male reproductive system?

Figure 10.3 The male reproductive system

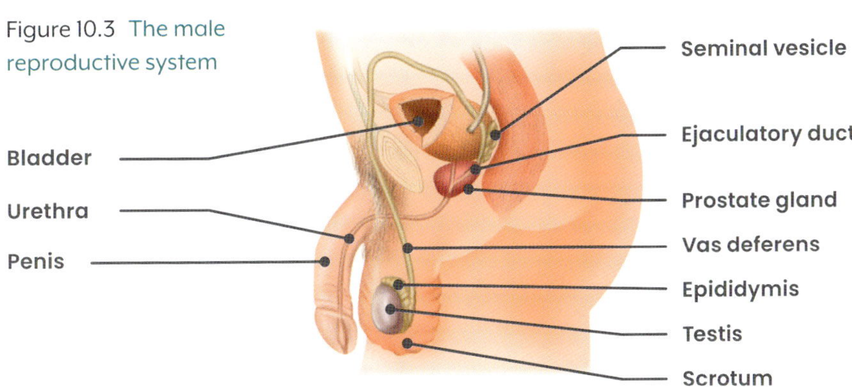

3 Ova are much bigger than sperm

The ova are the biggest cell types in the human body (about 0.1 mm in diameter), visible to the naked eye and dwarfing the tiny sperm cell (0.05 mm). The ovum has a nucleus that contains the genetic material. Around the nucleus is a cell plasma (yolk). In the cell plasma are all the substances required by the ova to develop if it is fertilised by a sperm cell.

Sperm cells have a head that contains a nucleus with the genetic material inside. Sperm also have a long tail (or flagellum), which propels the sperm forward using a wave-like motion and is used to dig into and penetrate the female ovum.

What are some differences between sperm and ova?

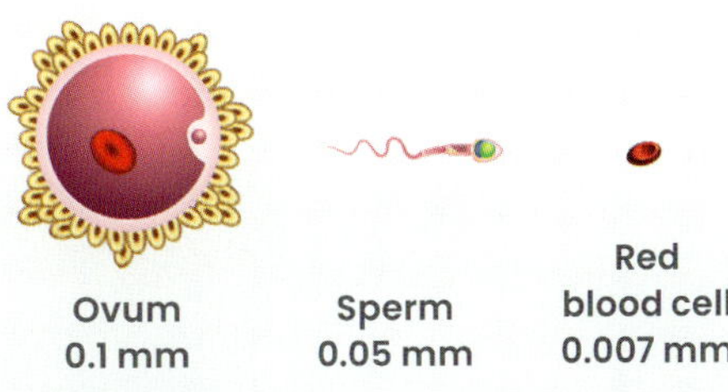

Figure 10.4 The female sex cell, the ovum, is one of the largest cells in the body.

CHECKPOINT 10.1 

1 Identify two organs of the female reproduction system and describe their function.
2 Identify two organs of the male reproductive system and describe their function.
3 Explain what menstruation is.
4 Are semen and sperm the same thing? Explain your answer.
5 What are the male and female sex cells called?
6 What is the difference between the vagina and the uterus?
7 The female ovum is a very large cell. Suggest why it is so big.
8 What is the male version of the ovaries called?
9 Identify the functions of these reproductive organs.
   a ovary
   b testicles
   c prostate gland
   d fallopian tubes

CHALLENGE

10 Carry out research and then compare human and plant reproductive systems using a Venn diagram.

SKILLS CHECK

- I can name the organs of the female and male reproductive systems.
- I can describe the function of the key organs in the male and female reproductive systems.

10.2 Heredity

At the end of this lesson I will be able to:

- **identify** that during reproduction the transmission of heritable characteristics from one generation to the next involves DNA and genes.

Your DNA is a blueprint completely unique to you. It contains all the instructions for your genetic traits, passed down from your mother and father. These characteristics are passed down on genes – sections of DNA. Some genes are common (such as those for brown eyes) and some are rarer (such as for red hair). Genes are passed down on chromosomes; humans typically have 46 chromosomes.

1 DNA is made up of nucleotides

Deoxyribonucleic acid (DNA) is the blueprint that contains all our biological instructions. Found in the nucleus of nearly every cell, DNA is in the shape of a twisted ladder called a **double helix**. The sides of the 'ladder' are the backbone of DNA and are made up of units called **nucleotides**.

Each nucleotide is made up of a phosphate group, a five-carbon sugar and a nitrogenous base. There are four different nitrogenous bases – adenine, thymine, guanine and cytosine – which pair together like the rungs of a ladder. The bases only partner in specific pairs – adenine pairs with thymine and guanine pairs with cytosine. (See more about DNA in Unit 10.4 on page 154.)

Where is DNA found?

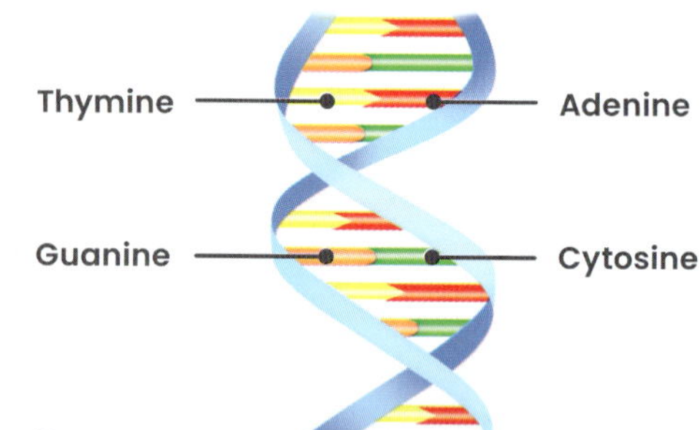

Figure 10.5 DNA has a double helix shape with a sugar–phosphate backbone and nitrogenous base pairs. Adenine pairs with thymine and guanine pairs with cytosine.

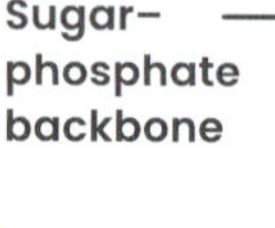

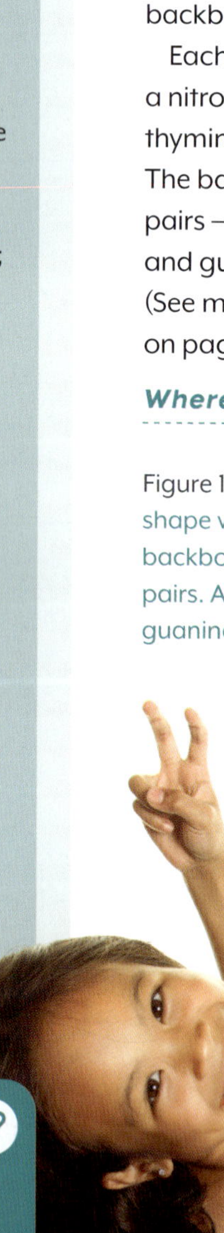

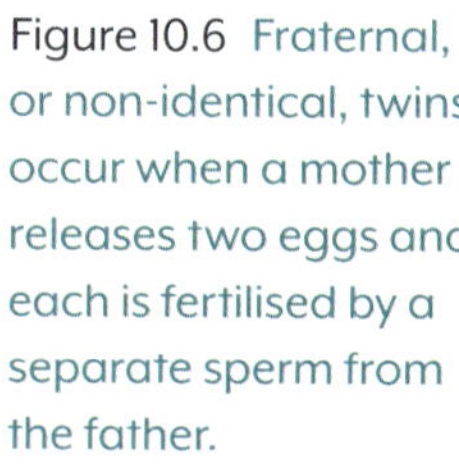

Figure 10.6 Fraternal, or non-identical, twins occur when a mother releases two eggs and each is fertilised by a separate sperm from the father.

KEY TERMS

chromosome
a tightly coiled strand of DNA

deoxyribonucleic acid (DNA)
found in the nucleus of a cell; the carrier of genetic information

double helix
the structure of a DNA molecule; a double-stranded spiral

gene
a segment of DNA, the basic functional unit of heredity

genome
an organism's entire set of DNA

genotype
the genetic code for a gene or an organism's entire genome

heredity
the passing on of traits from parents to their offspring

nucleotide
building block of DNA (adenine, guanine, thymine, or cytosine)

phenotype
how the genotype is physically expressed

NUMERACY LINK

If you stretched the DNA from one human cell all the way out, it would reach approximately 1.8 m long. Convert this value into mm.

❷ Genes are segments of DNA

One section of DNA is called a **gene**. A gene is the basic functional unit of **heredity**. Genes are how traits are passed down from mother and father to their offspring. Every human has two copies of each gene, one from their mother and one from their father.

Some genes code for proteins. This is very important because proteins:

- are used to repair and make new tissue
- make up much of our skin, hair, muscle and bone
- make up the chemical messenger system – hormones, blood and antibodies.

Other genes code for a variety of molecules with different functions. Genes can vary in size – they can be a few hundred bases long or millions of bases long. Your genes make up your **genotype** – the genetic code for a specific gene or genes – and determine your unique genetic identity (your **genome**). Your genotype codes for your **phenotype** – how those genes will be expressed. This can be through visible traits, such as hair or eye colour, or characteristics that aren't visible, such as blood type.

Gene

Figure 10.7 A gene is a segment of DNA.

What is a gene?

INVESTIGATION 10.2
Extracting DNA from strawberries

❸ Heritable characteristics are passed down on genes

Genes are passed down from parents to offspring in organised little packages called **chromosomes**. Human cells typically have 46 chromosomes; 23 are inherited from the mother and 23 from the father. During fertilisation, when the 23 chromosomes from the ovum fuse with the 23 from the sperm, the zygote (first cell) has a complete set of 46 chromosomes. These chromosomes contain all the genes the individual will ever have, a biological instructional manual for how to make a unique person.

This combination of genes from your mother and father determines your characteristics. Some traits, such as height, involve many genes.

Figure 10.8 The sperm and ovum contain 23 chromosomes each, which make a full set of 46 chromosomes in the fertilised egg.

Mother
Ovum (egg) 23 single chromosomes

Father
Sperm 23 single chromosomes

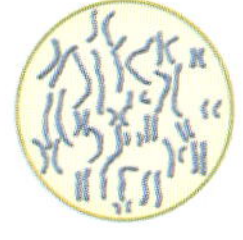

Fertilised egg 46 paired chromosomes

How are heritable characteristics passed down from parents to offspring?

CHECKPOINT 10.2

1 What does DNA stand for?
2 Explain the differences between DNA and genes.
3 Which DNA base pairs with adenine?
4 How many chromosomes are typically found in:
   a a sperm?
   b an ovum?
   c a zygote?
5 What is the backbone of DNA made up of?
6 What is the difference between phenotype and genotype?
7 Copy and complete these sentences.
   a A ___________ is a section of DNA.
   b DNA is shaped like _________
   c Genes are passed down from parents to offspring by _________

CHALLENGE

8 Research how DNA can be used to solve crimes.

SKILLS CHECK

- I can explain the difference between DNA and genes.
- I can describe how hereditary characteristics are passed from parents to offspring.

10.3 Chromosomes

At the end of this lesson I will be able to:

- **identify** that genetic information is transferred as genes in the DNA of chromosomes.

KEY TERMS 

allosome
a sex chromosome (1 pair in humans)

autosome
a non-sex chromosome (22 pairs in humans)

diploid
the number of chromosomes found in most cells

haploid
the number of chromosomes found in the sex cells (sperm and ova)

karyotype
a picture of an individual's full set of chromosomes

LITERACY LINK 

Explain what the term *aneuploidy* means. Give an example and use it in a sentence.

NUMERACY LINK 

A house-fly has a haploid number of 6. What is its diploid number?

A leopard frog has 26 chromosomes in its body cells. How many chromosomes are there in its sex cells?

Dogs have a diploid number of 78. What is their haploid number?

Chromosomes store DNA in neat packages during important stages of a cell's life cycle. The number of chromosomes differs across species – humans have 46 chromosomes, whereas chickens have 78 and peas have just 14. Problems with chromosomes can occur, and some people have extra chromosomes or are missing chromosomes. For example, Down syndrome is a genetic disorder in which people have an additional chromosome.

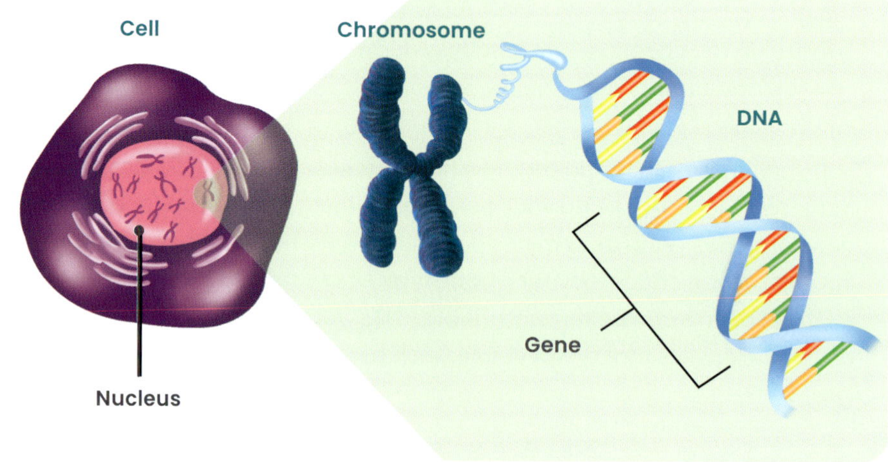

Figure 10.9 Chromosomes contain DNA, which in turn is made up of genes.

❶ Condensed chromosomes contain tightly coiled DNA

DNA often exists in cells as disorganised long pieces. Under a microscope, DNA can look like a bunch of squiggles. At various stages (especially during cell division) the DNA in chromosomes becomes tightly coiled and packed.

Each chromosome is made up of two identical chromatids. The two chromatids are joined so that they have short and long arms. Each chromatid contains one long DNA molecule that has been tightly coiled and packed into the small structure. In order to ensure the DNA is effectively coiled and packed, it is wrapped around little protein balls called histones.

What is a chromosome made of?

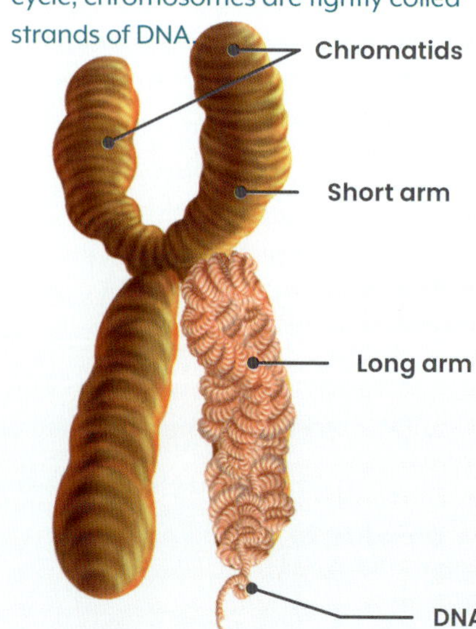

Figure 10.10 At some stages of the cell cycle, chromosomes are tightly coiled strands of DNA.

INVESTIGATION 10.3
Genetic trait survey

❷ Sex chromosomes are different from non-sex chromosomes

Typically, humans have 46 chromosomes in each cell. These are often described as 23 pairs because they can pair up according to the location of genes, size and shape. A full set of chromosomes (in humans 46) is known as the **diploid** number. Sperm and ova only have half this number (23), which is called the **haploid** number. Of the 23 pairs of chromosomes in humans, 22 look very similar to each other – like little X's. These are the 22 non-sex chromosomes, also known as **autosomes**.

The 23rd pair of chromosomes is the sex chromosomes, also known as **allosomes**. These determine the sex of the individual – a female has two X's, whereas a male has an X and a Y chromosome. Males produce sperm that have either an X or a Y chromosome, so they have a 50:50 chance of passing on each chromosome. Females will always pass on an X.

What are autosomes?

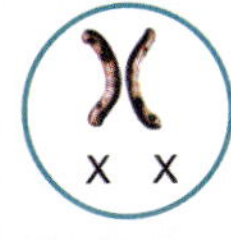

Figure 10.11 A female has two X chromosomes whereas a male has an X and a Y chromosome.

❸ Karyotypes are an individual's full set of chromosomes

Karyotypes are pictures of an individual's chromosomes. In order to get this picture, cells containing condensed chromosomes are stained and examined under a microscope. An image of the chromosomes is taken and each chromosome is cut out, placed next to its pair and numbered. The largest chromosome is chromosome 1. By observing the appearance and number of chromosomes, scientists can check for chromosomal abnormalities.

In some cases, the number of chromosomes is different from the usual number. This is called *aneuploidy*. If a person has an additional chromosome number 21, then they have a genetic disorder called Down syndrome. Other differences that can occur are people with more than two sex chromosomes, only one sex chromosome or additional chromosomes 13 or 18.

What is a karyotype?

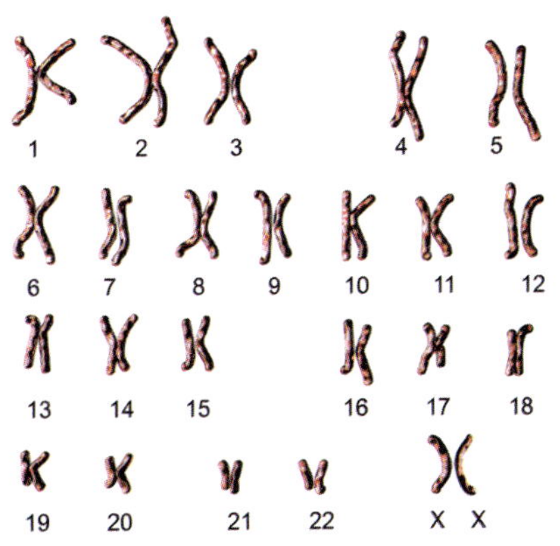

Figure 10.12 A female human karyotype. The final XX chromosome pair shows that the person is female. Males have an XY chromosome pair instead, but the Y chromosome is missing here.

CHECKPOINT 10.3 

1 Describe how genetic information is transferred from one generation to the next.
2 Do all species have the same number of chromosomes? Explain your answer.
3 Which is bigger, a strand of DNA or a gene? Justify your answer.
4 Draw and label a diagram of a chromosome.
5 Describe the difference between an allosome and an autosome.
6 What is the difference between a diploid and haploid number of chromosomes?
7 Do X and Y chromosomes actually look like X's and Y's? Explain your answer.
8 What are karyotypes used for?
9 What are the male and female sex chromosomes?
10 Redraw the karyotype shown in Figure 10.12 for a human male.

CHALLENGE

11 Research a genetic disorder and create a short PowerPoint presentation on it.

SKILLS CHECK

- I can explain what chromosomes are and describe their structure.
- I can explain how chromosomes transfer genes from parents to offspring.

10.4 DNA replication

At the end of this lesson I will be able to:

- **outline** how the Watson–Crick model of DNA explains the exact replication of DNA and changes in genes (mutations).

KEY TERMS 

mutation
an error in DNA caused by a substitution, an insertion or a deletion of a base

nitrogenous
containing nitrogen

LITERACY LINK 

Write a numbered step-by-step process outlining how DNA is replicated.

NUMERACY LINK 

Three bases (a base triplet) on a strand of DNA codes for an amino acid. How many different ways can you arrange a base triplet using the 4 bases?

American biologist James Watson and British physicist Francis Crick presented the famous Watson–Crick model of DNA in the early 1950s. Although DNA had been discovered decades earlier, Watson and Crick were the first to present a model that described the components of DNA and how they were arranged.

Figure 10.13 James Watson and Francis Crick with their famous model of DNA

1 The Watson–Crick model shows DNA as a double helix

The Watson–Crick model of DNA is the model currently accepted and used throughout the world. The model is a double-stranded twisted helix. The two strands of DNA are antiparallel (parallel, but in the opposite direction) to each other and have a sugar–phosphate backbone. The four **nitrogenous** bases (adenine, thymine, cytosine and guanine) connect the two backbones and exist in pairs held together by hydrogen bonds. DNA sequences are written out using letters to represent the bases – A, T, C and G.

At the time, many scientists around the world were competing to present a model of DNA. However, Watson and Crick had access to the work of a young chemist called Rosalind Franklin. Franklin worked with powerful X-rays and was able to obtain photographic evidence that helped Watson and Crick discover the structure of DNA. Her contribution to the Watson–Crick model is often overlooked.

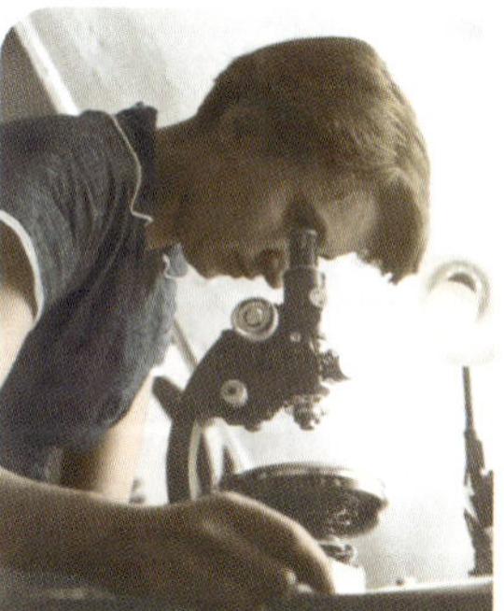

Figure 10.14 Rosalind Franklin

What is the Watson–Crick model of DNA?

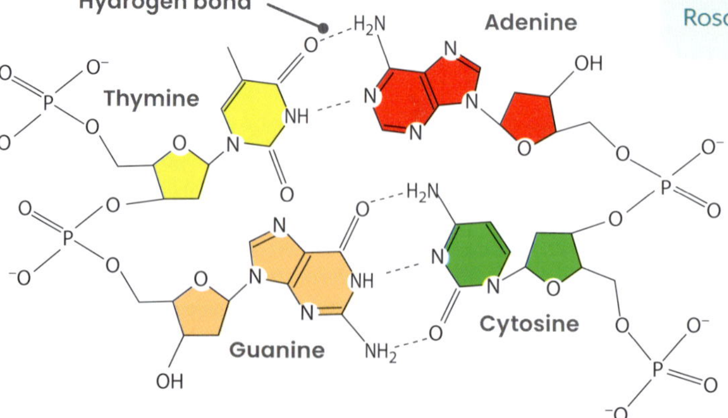

Figure 10.15 The structure of the nucleotides, adenine (A), cytosine (C), guanine (G) and thymine (T). They are made up of a sugar, a phosphate group and a nitrogenous base.

❷ DNA replication produces a complementary strand of DNA

DNA is replicated whenever a new cell is formed. In this process, one strand of old DNA acts as a template from which a new strand is produced. The new DNA molecule is made of one old strand and one new strand.

The old double-stranded DNA molecule is 'unzipped' by an enzyme called DNA helicase and prepared for replication by a molecule called RNA. Once prepared by RNA, the nitrogenous bases are ready to act as a template for a strand of complementary new DNA. The new DNA strand matches up along the old strand, so thymine is only paired with adenine and guanine is only paired with cytosine.

The new DNA grows one base at a time in a process called elongation. More enzymes then 'proofread' the new strand to find any mistakes, and then again to finish and complete the process. The new DNA molecule is referred to as a 'daughter molecule'.

How does DNA replicate?

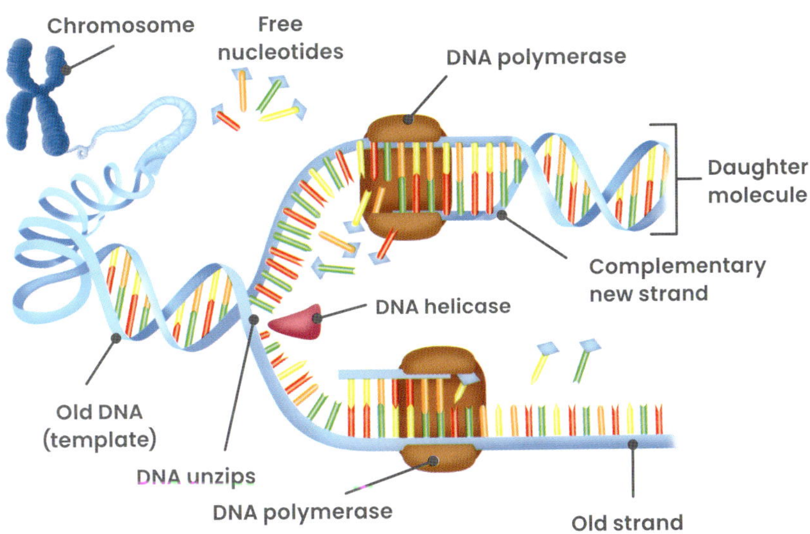

Figure 10.16 In DNA replication, an old strand of DNA acts as a template to make a new complementary one.

❸ DNA mutations are errors that happen during replication

Even though the DNA replication process is very tightly controlled, mistakes can happen when the bases are being read or synthesised. These mistakes result in changes in the DNA being produced and are known as **mutations**.

There are three main types of mutation – substitution, insertion and deletion. In substitution, a different base is incorrectly inserted; for example, thymine is placed where cytosine should go. In insertion, one or more extra bases is inserted somewhere by mistake. In deletion, one or more bases is removed.

Genetic mutations can cause cell death or disease to organisms, but many are not harmful. In some cases, mutations can even be beneficial.

What are mutations?

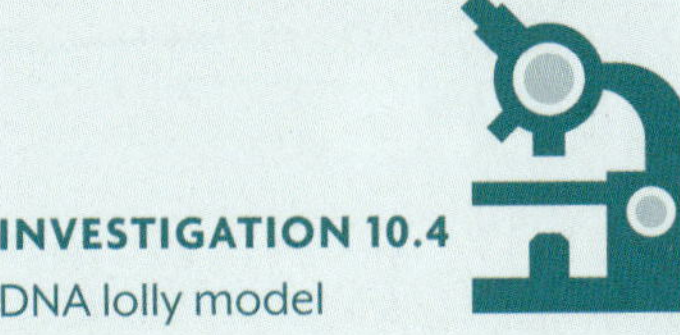

INVESTIGATION 10.4
DNA lolly model

CHECKPOINT 10.4

1 Copy and complete these sentences.
DNA is replicated whenever a new ________ is formed. To replicate, one strand of ________ acts a template from which a new ________ strand is produced. This results in one ________ strand of DNA and ________ strand of DNA and one ________ strand.

2 Give the complementary DNA bases for the following DNA sequence: GCCAGCTTACG.

3 Describe the Watson–Crick model of DNA.

4 Explain why DNA needs to be replicated.

5 What are the three types of mutation?

6 What role does RNA play in DNA replication?

7 Support or reject this statement, justifying your choice. 'Enzymes are not important in DNA replication.'

CHALLENGE

8 Research the chemist Rosalind Franklin, whose work helped Watson and Crick with their model of DNA. Prepare a short essay about her life.

SKILLS CHECK

- I can describe the Watson–Crick model of DNA.
- I can describe the process of DNA replication.
- I can explain how mutations occur.

10.5 Developments in technology

At the end of this lesson I will be able to:

- **describe**, using examples, how developments in technology have advanced biological understanding; for example, vaccines, biotechnology, stem cell research and in vitro fertilisation.

KEY TERMS 

biotechnology
the use of living organisms to develop products

IVF (in vitro fertilisation)
fertilisation of the sperm and egg that occurs outside the human body

stem cell
a cell type that is unspecialised and can develop into other types of cell

surrogacy
an arrangement in which a woman becomes pregnant and gives the baby to another person

LITERACY LINK 

Write an informative fact sheet for couples considering IVF. Include what to expect as well as any advantages or disadvantages.

NUMERACY LINK 

Testicles are able to make about 1500 sperm cells per second. Calculate how many sperm cells are produced per minute.

Recent developments in science and medicine have led to some incredible new technologies. However, many of these developments have been accompanied by controversy.

Figure 10.17 Dolly the sheep was the first cloned mammal and had three 'mothers'.

1 In vitro fertilisation takes place outside the body

IVF (in vitro fertilisation) is a type of assisted reproductive technology in which a female ovum is fertilised by a sperm outside the human body. The ova are harvested from the woman when she ovulates. Then, in a laboratory, the ova are immersed in a liquid with the sperm for fertilisation. After fertilisation, the mixture is left for a few days to allow the embryo to develop, and it is then implanted into the wall of the uterus.

IVF was first successfully completed in 1978. The process is expensive and invasive, so it is often a last resort for couples who cannot conceive naturally. IVF can also be used for **surrogacy** – in which a woman becomes pregnant and gives the baby to another person, sometimes in exchange for financial reward. The embryo is implanted into the woman (who is not the biological mother) and she carries the baby to term.

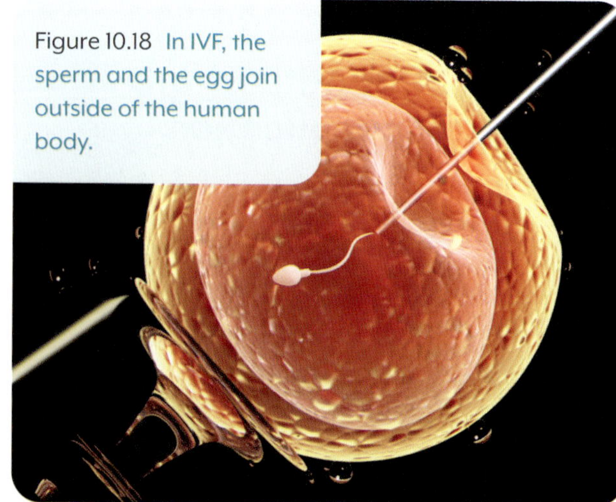

Figure 10.18 In IVF, the sperm and the egg join outside of the human body.

How does IVF work?

❷ Stem cells can develop into different types of cells

Figure 10.19 Stem cells have the potential to become many different types of cells.

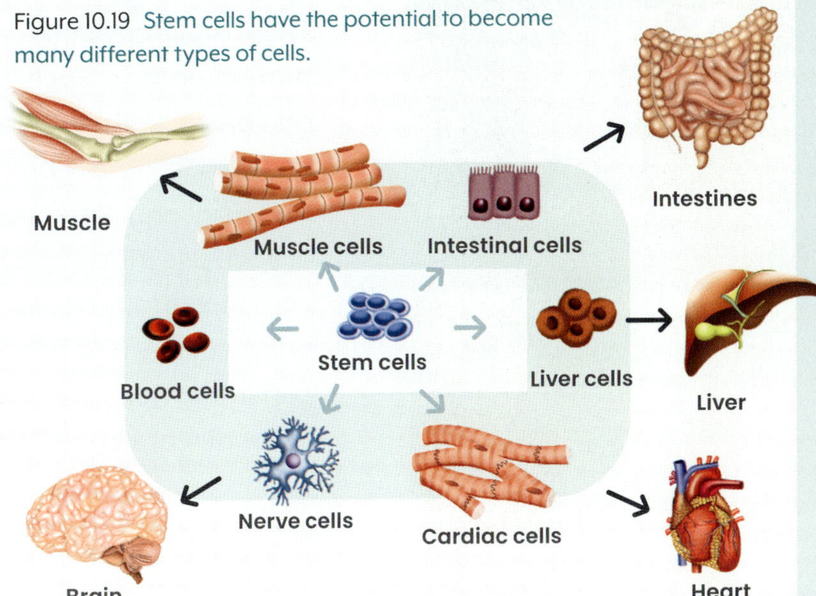

Stem cells are cells that have the potential to become many different types of specialised cells and can keep dividing to produce more cells. There are two main types of stem cells – adult stem cells (from sources such as bone marrow and intestines) and embryonic stem cells (from embryos). Adult stem cells can only develop into limited types of cells and cannot keep dividing indefinitely.

Embryonic stem cells can develop into nearly any type of cell, which provides almost limitless possibilities for new treatments in medicine. Imagine being able to create new nerve cells for someone with a spinal cord injury, or growing a new heart for someone who is waiting for a transplant, or repairing tissue in the eye so someone without vision can see again. Stem cell science is still unfolding. Currently, stem cells are only used in bone marrow transplants but their potential is exciting.

What are stem cells?

❸ Biotechnology produces products from living things

Biotechnology is any use of living things to develop products or change existing ones. Biotechnology has been happening for hundreds (if not thousands) of years in industries such as farming and medicine. Farmers have cultivated plant species by selecting traits over time. For example, carrots used to be purple, but after many years of careful cultivation, now most carrots are orange.

Newer examples of biotechnology include genetic engineering (alterations to DNA) and cloning. Cloning is the process of producing genetically identical organisms. Vaccines are a type of biotechnology that use a dead or weak variety of microorganism to make a person immune to them. Biotechnology is used to produce antibiotics and genetically modify food, to increase crop production and develop crops that are resistant to pests. Biological warfare is a negative development that has resulted from developments in biotechnology.

Biotechnology has also led to the development of genetic testing and screening, where individuals can have their entire genome mapped. Genome mapping often tells the person of their cultural heritage and risks of certain health problems. It can also inform doctors about what medicines would work best for that person.

What are some examples of biotechnology?

CHECKPOINT 10.5 

1 Give three examples of how developments in technology have advanced biological understanding.

2 Compare IVF to 'natural' pregnancy. What are the similarities and differences?

3 What is the difference between adult and embryonic stem cells?

4 The use of genetically modified organisms is controversial. Suggest why this might be.

5 Explain how biotechnology can be viewed as both a new and very old science.

CHALLENGE

6 Research another example of biotechnology and create a pamphlet that could be used to advertise it.

SKILLS CHECK

- I can explain what biotechnology is.
- I can give at least three examples of biotechnology.

10.6 Social and ethical considerations of biotechnology

At the end of this lesson I will be able to:

- **discuss** some advantages and disadvantages of the use and applications of biotechnology, including social and ethical considerations.

KEY TERMS

biotechnology
the use of living organisms to develop products

ethical consideration
a factor that takes into account what is right and what is wrong

GM
genetically modified; an organism that has had alterations to its DNA

social consideration
a factor that affects society and the people within society

LITERACY LINK

Create an informative poster about one type of biotechnology.

There are many applications of **biotechnology**. Each has advantages and disadvantages. Biotechnology also has a huge impact on society and so the social and ethical considerations of each use must be discussed. **Social considerations** are factors that affect society and people and **ethical considerations** are factors that take into account what is right and what is wrong.

Figure 10.20 Anthrax infection is caused by a bacteria. It causes a range of symptoms, including blisters and ulcers, nausea and fever. Anthrax has been used as a biological weapon.

1 In vitro fertilisation (IVF)

Advantages

Some people are unable to conceive a child naturally because of their age, health or other factors. IVF allows them to have children.

Disadvantages

IVF is extremely expensive. Each round of treatment costs thousands of dollars. It is also invasive; the couple must undergo a number of procedures and the woman requires regular injections of hormones.

Social considerations

Not everyone can afford IVF, so it is not an equitable treatment. Should some people miss out on having a baby just because they can't afford IVF?

Ethical considerations

It has been reported that some IVF doctors have been pressured to achieve high success rates and implant several fertilised eggs into a woman, instead of just one. This has resulted in multiple births, which is less safe for the mother and the babies.

What are some advantages of IVF?

Figure 10.21 When multiple embryos are implanted, IVF has a much higher chance of multiple births.

2 Stem cells

Advantages

Stem cells may be able to cure many different medical diseases and disorders, from diabetes to Parkinson's disease.

Social considerations

Stem cell research is a new science, so the long-term impacts for people and society are unknown.

Disadvantages

Stem cell science is still new, so currently stem cells are only used fully and completely in bone marrow transplants.

What are some disadvantages of using stem cells?

Ethical considerations

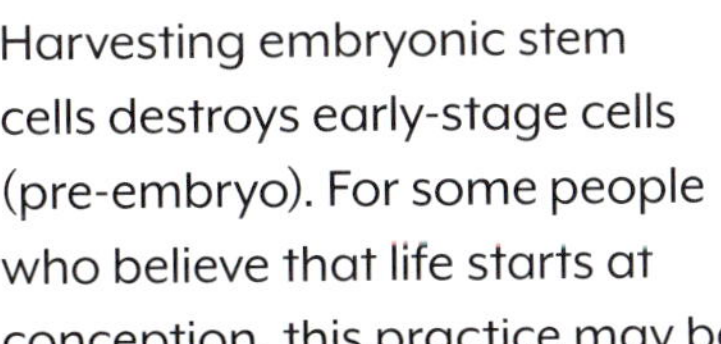

Harvesting embryonic stem cells destroys early-stage cells (pre-embryo). For some people who believe that life starts at conception, this practice may be considered unethical.

Figure 10.22 Stem cell therapies are a potential cure for diseases such as Parkinson's disease.

CHECKPOINT 10.6

1 Explain the difference between social and ethical considerations.
2 Describe some of the advantages of stem cell science.
3 Outline some of the ethical concerns about the use of embryonic stem cells and then present your own opinion.
4 A social consideration of IVF is equity. Explain why.
5 Consider biological weapons (such as anthrax). Outline some social and ethical considerations about their use.

CHALLENGE

6 Research a type of biotechnology that is not mentioned in this chapter. Provide a summary of the biotechnology and include a discussion of any social and ethical considerations of its use.

3 Genetically modified crops

Advantages

Crops can be genetically modified (**GM**) to contain more nutrients, take up less farming land and use less water, pesticides, fertilisers and soil. GM crops also tolerate herbicides, so farmers can spray and remove weeds without damaging the crop.

Social considerations

GM crops can contaminate non-GM or organic crops on neighbouring farms. This could prohibit a neighbouring farmer from labelling their products as 'certified organic' and limit their ability to earn an income.

Disadvantages

GM crops are pollinated just like any other crop. Some farmers are concerned that pollen from GM crops will be carried over into their non-GM crop and cross-pollinate the crop.

Ethical considerations

Most core foods are now genetically modified in some way but are not labelled as GM. This means that consumers cannot make an informed choice about buying and consuming them.

What are some social and ethical considerations of genetically modified crops?

SKILLS CHECK

- I can give examples of the advantages and disadvantages of biotechnology.
- I can describe some social and ethical considerations of biotechnology.

CHAPTER SUMMARY

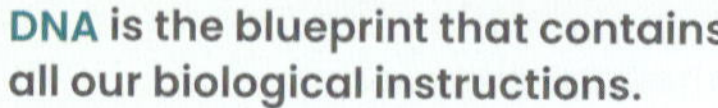

DNA is the blueprint that contains all our biological instructions.

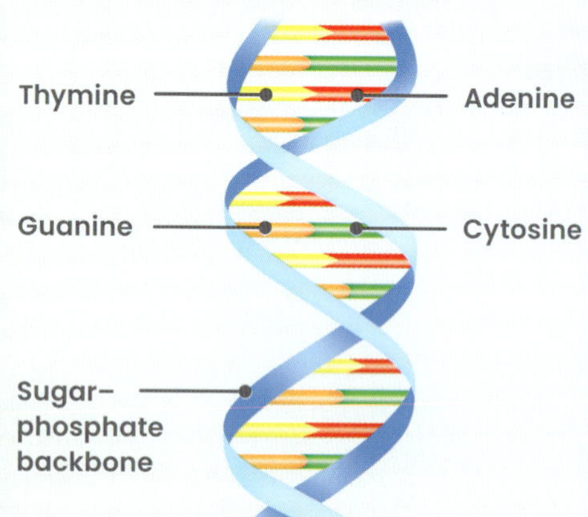

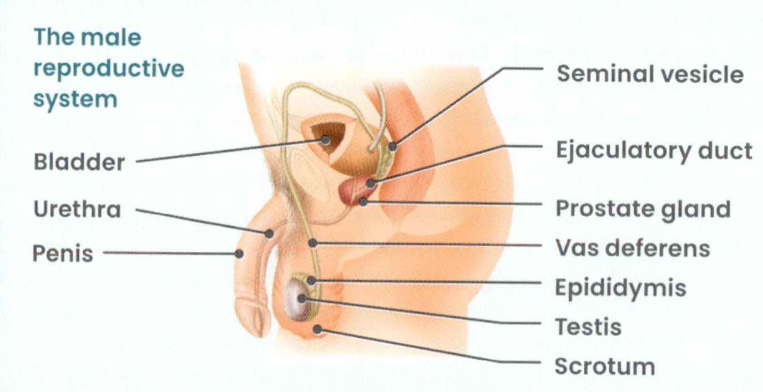

A gene is a section of DNA.

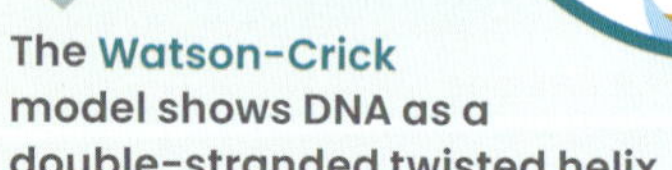

The Watson-Crick model shows DNA as a double-stranded twisted helix.

Genes are passed down in chromosomes.

A fertilised human egg has 46 paired chromosomes.

There are many social and ethical considerations of biotechnology.

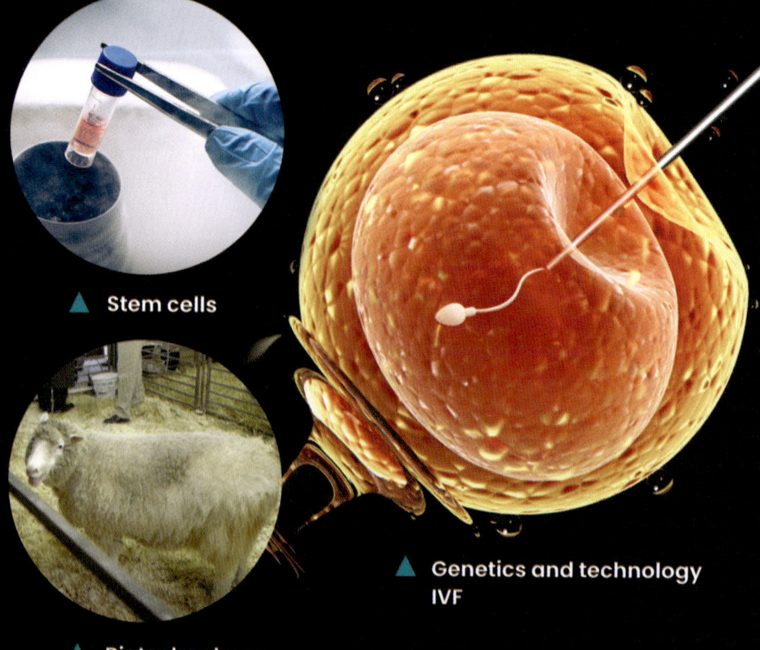

▲ Stem cells

▲ Biotechnology

▲ Genetics and technology IVF

FINAL CHALLENGE

1. How many chromosomes do humans typically have?
2. Which DNA base pairs with:
   a adenine?
   b cytosine?
3. Explain what karyotypes are used for.

4. Explain what biotechnology is. Give three examples of biotechnology.
5. Describe three types of mutation that can happen to DNA.
6. Explain the difference between a gene and a genome.

7. Complete these sentences.
   a Allosomes are ______________.
   b Autosomes are ______________.
   c A karyotype is ______________.
8. Describe some advantages and disadvantages of using stem cells in science and medicine.
9. How old are the eggs of a 35-year-old woman? Explain your answer.

10. Identify the functions of these reproductive organs.
    a testicles
    b penis
    c fallopian tubes
    d ovaries
    e prostate gland
11. Outline some ethical considerations of in vitro fertilisation.

12. Describe how DNA is replicated.
13. Describe the Watson–Crick model of DNA and include a diagram.
14. Explain with the use of an annotated diagram how genes are passed from one generation to the next.

11 Evolution

Evidence suggests Earth is more than 4.5 billion years old. After 1 billion years, signs of life appeared on Earth. Tiny cells in the ocean gave rise to simple marine plant life, followed by land plants, and then small animals. Billions of years later, complex organisms live on Earth. These complex organisms can reproduce sexually to form genetically complex and intricate life, such as human beings. Evolution is the change in characteristics of living things over successive generations, often as a result of natural selection.

1 LEARNING LINKS

Brainstorm as much as you can remember about these topics.

Fossils require particular conditions to form.

Structural features and other adaptations of living things help them to survive in their environment.

The physical conditions of the environment affect the growth and survival of living things.

2 SEE–KNOW–WONDER

List three things you can **see**, three things you **know** and three things you **wonder** about this image.

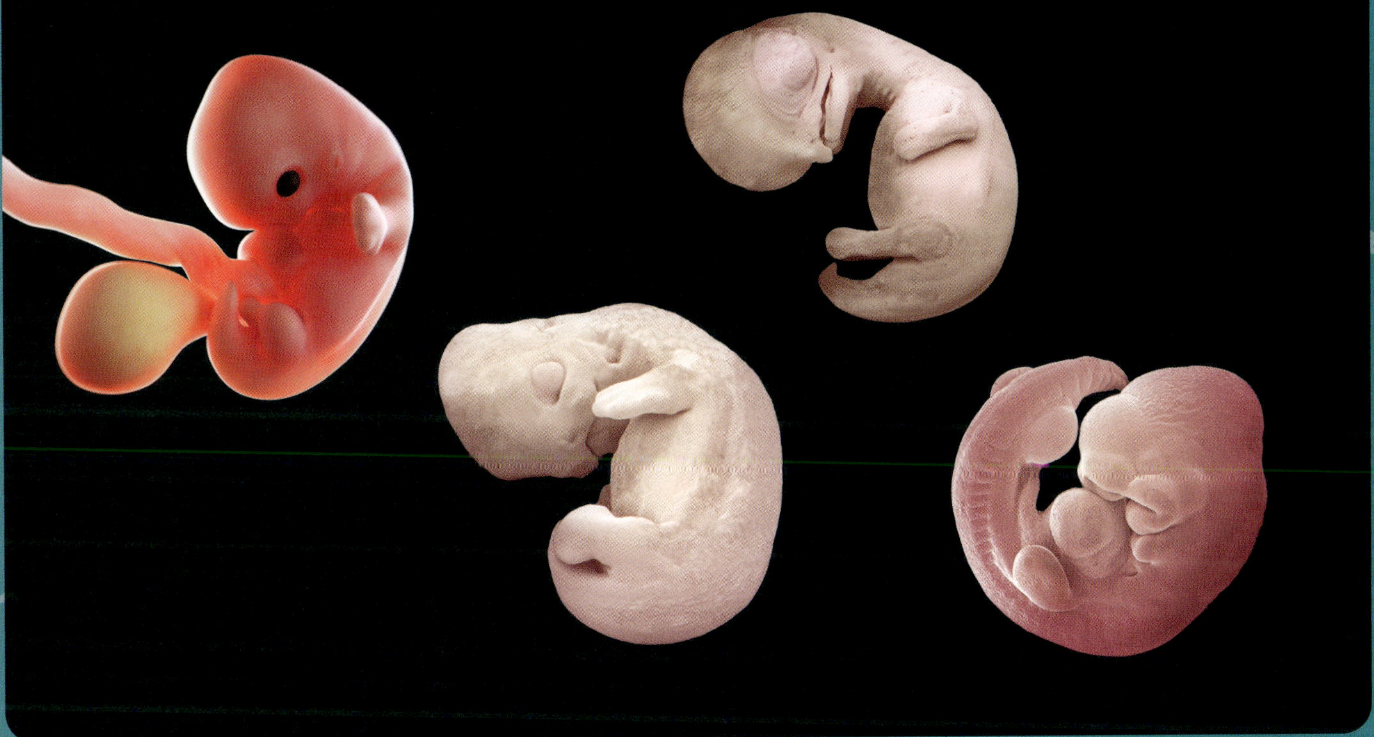

3 CRITICAL + CREATIVE THINKING

Prediction: What will a human look like in a million years?

Interpretations: A fossil is found of a dinosaur with a human baby on its back. Write some explanations for this.

Mismatches: How could you solve the problem of crop-eating insects that develop a resistance to pesticides by using a fishing net, a kettle and a box of matches?

4 THE FLUFFIEST MONSTERS!

Scientists studying ancient fossils have made the rather surprising discovery that many dinosaurs (including *Tyrannosaurus rex*) probably had feathers. *T. rex* didn't use the feathers to fly; instead, the feathers probably helped them keep warm, socialise or even attract mates. Some dinosaurs' feathers are thought to have been all the colours of the rainbow.

11.1 Evidence of evolution

At the end of this lesson I will be able to:

- **describe** scientific evidence for present-day organisms having evolved from organisms in the past.

KEY TERMS 

biogeography
the study of the past and present distribution of living organisms

comparative anatomy
the study of features of different species to look for evidence of a common ancestor

evolution
the change in characteristics of living things over time

homologous structures
parts of organisms that show evidence of a common ancestor

vestigial structure
a part of an organism that has lost some or all of its original function

LITERACY LINK 

Suggest how to break down the word *biogeography* in order to make meaning from the term.

NUMERACY LINK 

In adults, the femur (thigh bone) is said to be approximately ¼ of a person's height.

Measure the length of your femur and measure your height. Determine your femur-to-height ratio. Record class results and calculate the average.

One of the closest relatives to humans is thought to be the bonobo, a great ape found in regions of the Congo in Africa. Bonobos live to about 40 years of age and communicate vocally and through facial expressions.

How do we know that humans are related to the bonobo? There is much scientific evidence that humans and bonobos have evolved from a common ancestor. This evidence includes fossils, comparative anatomy, genetic evidence and biogeographical evidence.

1 Comparative anatomy looks at common features in different species

Evolution is a gradual change in the physical characteristics of organisms over many generations. The first definite clues of evolution came by comparing the anatomy (especially the bones) of species to find similarities and differences. Scientists using **comparative anatomy** have found evidence to show that different species have evolved from the same ancestor. Parts of the body in different species that show evidence of common features are called **homologous structures**.

Vestigial structures are another type of evidence. These are parts of organisms that have lost their usefulness as a species evolved. They remain in the organism, sometimes even causing problems. Vestigial structures in humans include wisdom teeth, the appendix, the tailbone and nipples on men.

How are homologous structures evidence of evolution?

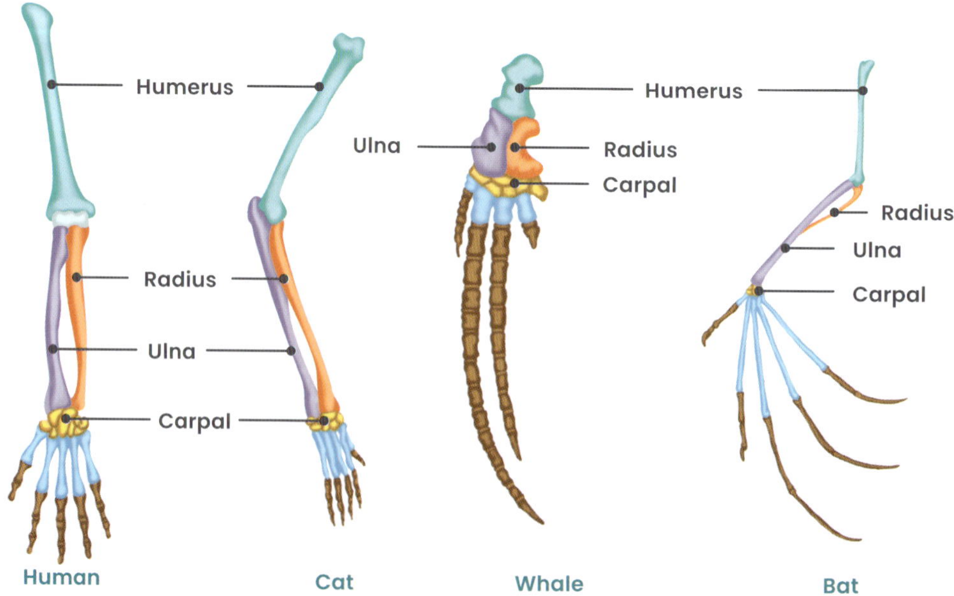

Figure 11.1 The human arm, cat leg, whale fin and bat wing are homologous structures that show evidence of a common ancestor.

❷ Genetic evidence shows how closely related different species are

DNA (deoxyribonucleic acid) is the molecule that codes for genetic instructions. DNA is a long molecule that contains all the biological information for humans and nearly every other living organism and is found in nearly every cell of your body. The molecule forms a double helix – a twisted ladder of genetic information.

From 1990 to 2003, teams of scientists from all over the world worked together to completely map the DNA of humans in a huge endeavour called the Human Genome Project. Now we can compare the DNA of different species and identify how similar they are. DNA mapping shows that humans and bonobo great apes share more than 98% of their genes, providing evidence that we had a common ancestor.

How can DNA be used as evidence of evolution?

❸ Biogeography is used as evidence for evolution

Biogeography is the study of the past and present distribution of living organisms. By studying where plants and animals are or were found on islands and continents, scientists can start to piece together a complex and ancient puzzle.

Much of the evidence for biogeography comes from fossils. If fossils of one species are found in both South America and Africa, then this is compelling evidence that the two continents were once connected.

Australia has been an isolated island for many millions of years, which is why Australian marsupials and mammals are unique. However, even after being separated from South America for millions of years, Australian marsupials still have some common features with those in South America. This means that they probably evolved from a common ancestor millions of years ago.

How can biogeography be used as evidence of evolution?

Figure 11.2 DNA is contained in a structure like a twisted ladder called the double helix.

INVESTIGATION 11.1
Comparative anatomy

CHECKPOINT 11.1

1 Describe evolution.
2 What is biogeography and what can it tell us about the evolution of species?
3 Explain how the Human Genome Project led to breakthroughs in using genetics to show evidence of evolution.
4 Outline the evidence that suggests that Australia has been an isolated island for many millions of years.
5 Describe the difference between homologous and vestigial structures.
6 Which of biogeography, genetics or comparative anatomy do you think gives the most scientific evidence of evolution? Justify your response.

CHALLENGE

7 Research some additional vestigial structures in humans. Explain what the function of each structure was and why it is not useful any more.
8 Embryology is the study of embryos. Embryos of many different kinds of animals look very similar. How might the study of embryology provide evidence of evolution?

SKILLS CHECK

- I can give specific examples of scientific evidence that supports evolution.
- I can describe what is meant by comparative anatomy, DNA and biogeography.

11.2 The fossil record

At the end of this lesson I will be able to:

- **relate** the fossil record to the age of Earth and the time over which life has been evolving.

KEY TERMS 

fossil
the geologically altered remains of an organism that was once alive

fossil record
a record of all fossils found on Earth

palaeontologist
a scientist who studies fossils

LITERACY LINK 

Write a job advertisement for a palaeontologist. Your advertisement should include a description of a typical day in the life of a palaeontologist.

NUMERACY LINK 

Use the chart in Figure 11.5 to construct a pie chart showing the relative durations of each of the different eras: Caenozoic, Mesozoic and Palaeozoic.

When you hold a fossil, you are holding a piece of ancient history, a remnant from millions of years ago when Earth was young and nearly unrecognisable.

Many things can be turned into a fossil, but the conditions must be perfect, which is why fossils are so rare. Fossils contain clues to the history of life on Earth and are important in the scientific study of comparative anatomy, including our own complex heritage.

Figure 11.3 Fossils form when layers of sediment cover an organism. After millions of years, the organism can become an impression in solid rock.

❶ Fossils form over millions of years under the right conditions

Fossils are the remains, impressions or traces of long dead organisms left in rock and kept safe by geological processes. Fossils can be preserved bones or other remnants, shells, footprints, nests and eggs. Fossils are rare because they only form under certain conditions – the temperature, amount of oxygen and even soil acidity must be just right.

First, the organism must be protected once it dies so that its bones (or seeds and bark) stay together and in a relatively good condition. This happens much more easily under water because the organism sinks to the bottom and is covered by mud or silt. Over many years, layers of sediment cover the organism.

Over millions of years, the sediments harden and become rock, and the organism is then preserved as a fossil.

Fossils can also form when the bones completely decay to leave a hollow mould. Other minerals then seep in and harden, making a three-dimensional replica of the original bones.

How do most fossils form?

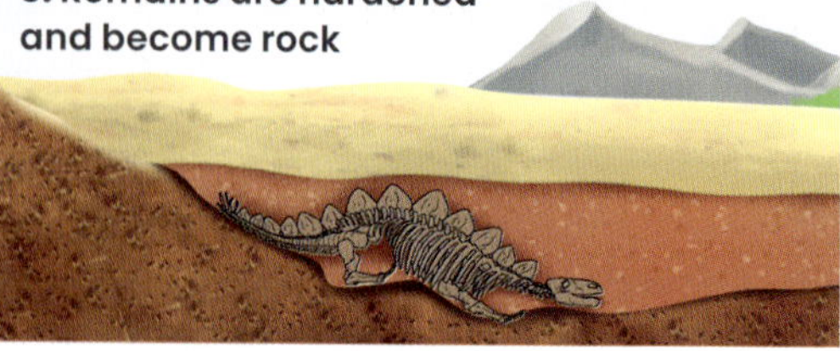

Figure 11.4 Fossils form over millions of years, usually in sedimentary rock.

❷ The fossil record provides evidence for evolution

Scientists who study fossils and the fossil record are called **palaeontologists**. The **fossil record** provides evidence for evolution because it shows how organisms have changed through geologic time from the simplest single-cell organisms to human beings.

Fossils have been found and dated from all around the world but it is thought many species are missing from the record, or have not yet been discovered. It is hard to get an exact picture of all life forms from the fossil record because of the gaps. Also, it is hard to know whether a particular fossil is a good representation of a species.

What is the fossil record?

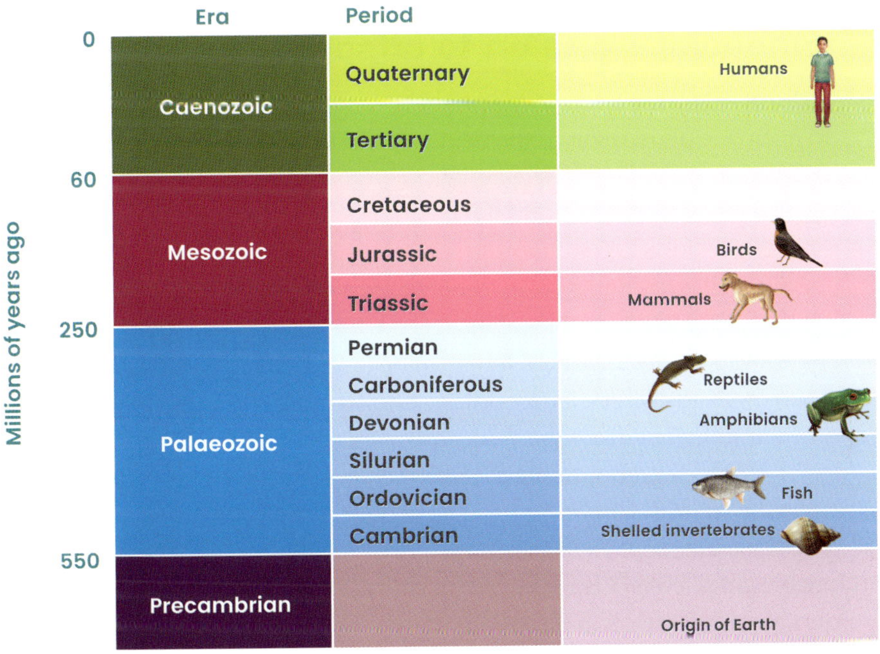

Figure 11.5 This sequence of animal fossils begins with 500-million-year-old fossils of simple invertebrates (animals without backbones) and ends with human beings.

❸ Fossils tell us about the age of Earth and evolution

Scientists date fossils in two main ways – relative dating and absolute dating.

Relative dating gives an approximate age of a fossil by comparing it to other fossils or by studying the layers and ages of the surrounding stone.

Absolute dating is commonly done by studying the radioactive minerals in the rock. Scientists can calculate the age of the rock by investigating how much the radioactive elements in them have decayed.

Older fossils are of simpler organisms than younger fossils. This evidence supports the theory of evolution because it suggests that much more complex organisms evolved from simpler ones.

How can fossils be used as evidence of evolution?

INVESTIGATION 11.2
Dating fossils

CHECKPOINT 11.2

1 Give two reasons why the fossil record is incomplete.
2 Explain the difference between relative and absolute dating.
3 The oldest fossils discovered are of the simplest organisms. Explain why this is the case.
4 What is the fossil record?
5 Create a flow chart showing the process of fossilisation.
6 Explain how fossils can be used as evidence for evolution.
7 Outline the two main ways for establishing the age of a fossil.
8 Fossils occur mostly in sedimentary rock. Suggest why this might be the case.
9 Give some reasons why the fossil record may be viewed as an 'incomplete puzzle'.

CHALLENGE

10 Research the biggest fossil discoveries made by scientists. Choose one that you think is the most significant scientifically and explain why.
11 Research the horse fossil record and explain how it provides evidence for evolution.

SKILLS CHECK

- I can describe how a fossil is formed.
- I can explain why the fossil record can be used as evidence of evolution.

11.3 Natural selection

At the end of this lesson I will be able to:

- **explain**, using examples, how natural selection relates to changes in a population; for example, when bacteria develop resistance to antibiotics and insects develop resistance to pesticides.

KEY TERMS 

antibiotic
a medicine that kills bacteria or stops them reproducing

hereditary
traits passed by parents to their offspring

natural selection
the process where organisms that are better suited to their environment tend to survive and reproduce, passing their traits on to further generations

pesticide
a substance that kills pests, such as insects or weeds

NUMERACY LINK 

A student measures the number of red bugs and green bugs on a bush as follows:

Time (days)	Red bugs	Green bugs
0	22	22
15	15	26
30	9	31
45	3	35

Graph this data and make a conclusion in terms of natural selection.

Have you ever wondered why giraffes have such wonderfully long necks? Their long necks let them eat the highest leaves that other animals can't reach. Most scientists now think that giraffes got their long necks through a process called natural selection. A giraffe with a longer neck than other giraffes was much more likely to find food and survive than other giraffes. So, giraffes with long necks would be more likely to reproduce and pass their longer necks on to the next generation.

1 Natural selection involves passing beneficial traits on to offspring

Natural selection occurs when beneficial traits are passed on from one generation to the next and is often referred to as 'survival of the fittest'. The beneficial traits are hereditary – they are passed on through the genes. However, factors other than 'fitness' affect natural selection – even luck.

Imagine the members of an insect species can be green or brown. Brown insects are better camouflaged on the bark of trees. Green insects are much more easily seen by predators, and so more green insects die. Over time, fewer green insects are left to reproduce, and eventually no green insects are left in the species – this is natural selection.

Natural selection was first proposed in the 1850s by celebrated scientist Charles Darwin, who put his ideas about natural selection and evolution into his book *On the Origin of Species.*

What is natural selection?

Figure 11.6 How many moths can you see here? The paler moth is camouflaged and so is much more likely than the black moth to survive and pass its genes on.

❷ Galapagos finches helped Darwin develop his theory of natural selection

In 1831, Charles Darwin set sail from England. On his journey, Darwin found enormous fossils and many species of animals that forced him to consider new theories for the origins of life on Earth.

During Darwin's 5-year expedition, he discovered a number of species of finches in the Galapagos Islands, off the coast of Ecuador. Although the finches looked very similar, their beaks were different lengths and shapes, which reflected the different food sources on each island. Darwin theorised that the finches had a common ancestor, but after they separated geographically and had different food sources, the finches evolved different beaks through natural selection.

What did Darwin learn about evolution from the finches in the Galapagos Islands?

Figure 11.7 The finches of the Galapagos Islands demonstrated the evolution of different beaks in response to the food sources available to them.

❸ Antibiotic resistance is a result of natural selection in bacteria

Natural selection can affect modern humans and society. For example, mosquitoes carry deadly diseases such as malaria. An effective way of reducing the incidence of malaria is to spray populated areas with **pesticides** (chemicals that control pests). However, in some areas mosquitoes are becoming resistant to pesticides. Natural selection means resistant mosquitoes are more likely to survive and pass on the resistance to their offspring. This results in a population resistant to pesticides that could also be carriers of malaria.

Another example of natural selection happening is the increasing resistance of bacteria to **antibiotics** as a result of the overuse of antibiotics. The resistant bacteria have more chance of surviving and reproducing and antibiotic resistance increases throughout bacterial populations. Scientists are in a constant battle to stay one step ahead of the bacteria, synthesising new antibiotics – even though these can become ineffective very quickly.

What are some problems caused by natural selection?

INVESTIGATION 11.3
Natural selection in practice

CHECKPOINT 11.3

1 Define *natural selection*.
2 Who was Charles Darwin and what did he contribute to modern science?
3 There are many important benefits to natural selection but there are also some modern disadvantages to humans; describe one.
4 Explain why natural selection is producing antibiotic-resistant bacteria.
5 How long do you think it would take for species to differentiate and evolve as Darwin's finches did? Justify your answer.
6 Natural selection is when beneficial traits are passed on to the next generation. Can you think of a situation where this wouldn't happen?

CHALLENGE

7 Create a timeline of Darwin's life, including major breakthroughs and expeditions.
8 Consider the diagrams of Darwin's finches in Figure 11.7. Create a summary of why the form of each beak suits the function (accessing each food source).

SKILLS CHECK

- I can define natural selection.
- I can give specific examples or case studies that show my understanding of natural selection.

11.4 Factors affecting survival

At the end of this lesson I will be able to:

- **outline** the roles of genes and environmental factors in the survival of organisms in a population.

KEY TERMS 

adaptation
a change in the structure or function of an organism that has evolved over generations that makes it better suited to its environment

gene
a unit of heredity passed from parents to offspring; made of DNA

genetic variation
the differences in genes within individuals of a population

LITERACY LINK 

Write a letter to the editor of a newspaper. In your letter, outline how habitat destruction affects the survival of plants and animals in Australia and suggest a way to combat habitat destruction.

NUMERACY LINK 

A scientist wants to measure the fish population in a 50 m^2 pond. He chooses five different areas of 1 m^2 each. His results are: 6, 4, 9, 3, 11.

Calculate the approximate number of fish in the entire pond.

Many factors affect the survival of organisms in a population, including an individual's genetics (and corresponding physical traits) and the environment it is in.

Figure 11.8 Angler fish have many genetic adaptations for survival. Many environmental factors affect them too.

The angler fish (Figure 11.8) has many physical traits that have evolved over time and that allow the fish to survive in some of the darkest environments on Earth. Food is scarce in these environments, so the fish lures its prey with a little light projected in front of its huge mouth. Female angler fish release powerful chemicals so that potential mates can find them in the pitch-black waters.

❶ Survival depends on genetic diversity in populations

DNA is the blueprint for your biology. DNA is found in the nucleus of most cells. Segments of DNA are called **genes** and can determine different physical characteristics. For example, your height and bone structure are partly determined by your genes, which were passed down to you from your mother and father.

Natural selection acts to produce a population that is well suited to an environment. Over time it can reduce genetic variation, with organisms becoming more genetically alike. **Genetic variation** describes the variation in DNA sequences that is present in individuals of the same species. Genetic variation is critical to the continuation of life on Earth.

A population is a group of the same species living in a specific area and interbreeding. When populations lack genetic diversity, they are vulnerable to changes in their environment. They may be unable to adapt to changes because they don't contain the necessary genetic options.

For example, the Tasmanian devil lacks genetic diversity because of its low numbers, isolation and generations of inbreeding. When a highly contagious facial tumour spread through the population in the mid-1990s, no individuals had genetic resistance to the condition. The disease now affects more than 80% of Tasmanian devils.

Figure 11.9 The Tasmanian devil has low genetic diversity and so is susceptible to contagious facial tumours.

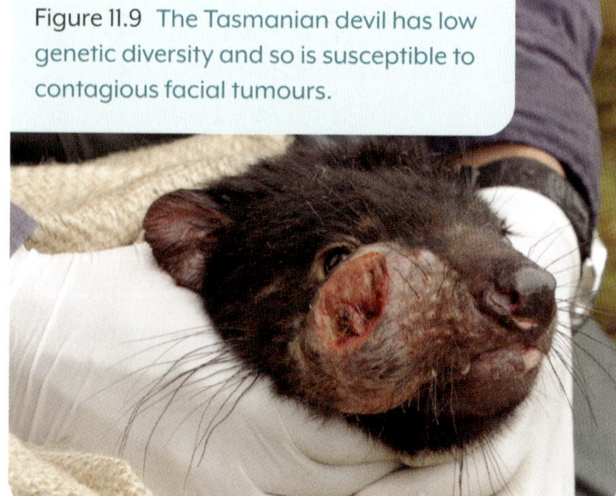

How can genetics influence survival?

2 Environmental factors affect survival

Many environmental factors influence survival; for example, the amount of sunlight, temperature, wind, soil quality, pH and water. Most organisms can only survive in a fairly narrow temperature range. For example, polar bears thrive in freezing temperatures but would perish in tropical climates. Polar bears have several **adaptations** to their cold environment, including a thick coat and a layer of fat, as well as white fur that helps them blend in with their icy environment.

Other factors that affect survival are the availability of resources and mates, as well as the presence of predators and population density. Population density is a measure of the number of individuals living in a particular area and is very important to the survival of organisms. If the population density is too high, then individuals have to compete for food, water and reproductive mates. However, if the population density is too low, the population can lack genetic variation and it can be hard for individuals to find a mate.

How can the environment influence the survival of organisms?

3 Human activity affects the survival of other species

Human civilisation has had a devastating effect on the natural environment and, thus, the survival of plants and animals. Habitat destruction has increased as humans have cleared land for farming, buildings and roads. Humans use fossil fuels, cut down trees for buildings or burning, build dams and change or destroy waterways.

Habitat destruction leaves plants and animals without the resources they need to live and reproduce. The World Wildlife Fund estimates that since 1970, humans have caused the death of approximately 60% of all mammal, bird, reptile and fish species.

Conservation management can do much to assist species to avoid extinction. National parks provide protected places for plants and animals. Reducing deforestation, using natural resources sustainably and protecting habitats will safeguard some of the most vulnerable species.

How do human impacts on the environment affect the survival of organisms?

Figure 11.10 Habitat destruction has a devastating effect on the survival of plant and animal populations.

INVESTIGATION 11.4
Environmental and genetic factors affecting survival

CHECKPOINT 11.4

1 What is genetic variation?
2 Explain what population density is and how it can impact the survival of organisms in a population.
3 What role do genes play in the survival of an organism in a population?
4 List some environmental factors that affect the survival of organisms.
5 Explain why a lack of genetic variation can be very detrimental to the survival of a species.
6 Describe habitat destruction and how it can affect survival.
7 Suggest some ways to control and reduce habitat destruction.
8 Choose an animal and describe the genetic adaptations (physical traits) that improve its chances of survival.

CHALLENGE

9 Select and carry out research on an endangered species. Write a short report summarising the threats to its survival.

SKILLS CHECK

- I can explain the role of genes in the survival of organisms.
- I can explain the role of environmental factors in the survival of organisms.

CHAPTER SUMMARY

Evidence for evolution

 Comparative anatomy

 Genetics

   Biogeography

▼ **Fossils** form over millions of years.

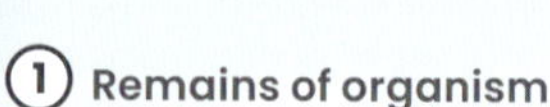
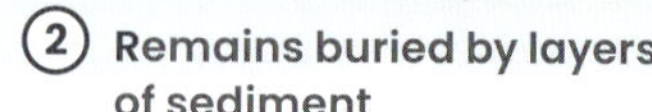

1. Remains of organism in water
2. Remains buried by layers of sediment
3. Remains are hardened and become rock
4. Remains become exposed on surface as fossil

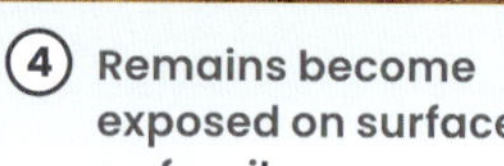

Survival relies on:

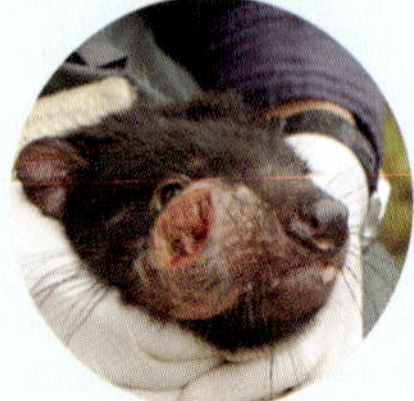

▲ genetic diversity

▲ human activity

▼ environment

▼ The **fossil record** provides evidence for evolution.

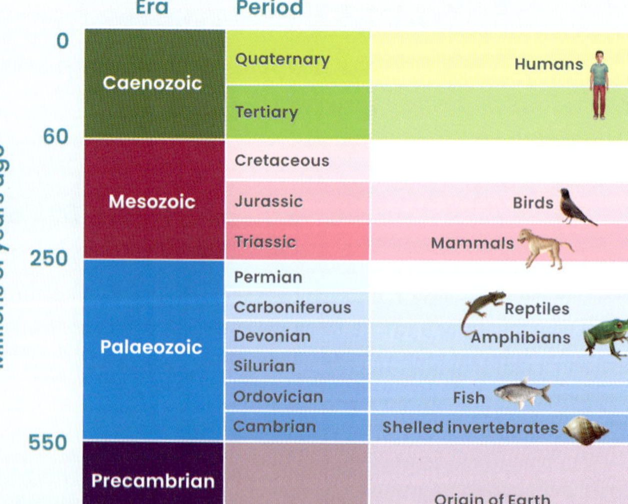

Millions of years ago	Era	Period	
0	Caenozoic	Quaternary	Humans
		Tertiary	
60	Mesozoic	Cretaceous	
		Jurassic	Birds
		Triassic	Mammals
250	Palaeozoic	Permian	
		Carboniferous	Reptiles
		Devonian	Amphibians
		Silurian	
		Ordovician	Fish
		Cambrian	Shelled invertebrates
550	Precambrian		Origin of Earth

Natural selection

'Survival of the fittest' → Beneficial traits passed down → Populations become better suited to their environment → 'Survival of the fittest'

The Galapagos finches demonstrate evolution by natural selection, developing different beaks according to the food sources available.

Large ground finch (seeds)

Cactus finch (cactus fruit and flowers)

Vegetarian finch (buds)

Woodpecker finch (insects)

★ FINAL CHALLENGE ★

1. Define *evolution*.
2. Describe how fossils are formed.
3. Explain how environmental factors affect the survival of individuals and populations.

LEVEL 1
50XP
LEVEL UP!

4. Explain how competition is an environmental factor that affects survival.
5. Describe natural selection and give an example.
6. Explain how Darwin's finches demonstrate principles of natural selection.

LEVEL 2
100XP
LEVEL UP!

7. Fossils are quite rare. Suggest why.
8. Explain how fossils and comparative anatomy are linked to our understanding of evolution.
9. Explain how natural selection has resulted in bacteria that are resistant to antibiotics.
10. Suggest ways to stop or slow habitat destruction.

11. Analyse the usefulness of the fossil record in providing evidence for evolution.
12. Which do you think is more important to the survival of organisms – genes or environmental factors? Give evidence to support your opinion.
13. Write three journal entries from Charles Darwin's perspective during his 5-year expedition.

14. Define the following terms and outline the evidence they provide for evolution.
    a. comparative anatomy
    b. genetics
    c. biogeography
    d. fossils

LEVEL 5
300XP

12 Inside atoms

What are things made of? What is the smallest part of matter? Why do some elements have unique properties? These are all questions that, over many centuries, and through scientific collaboration, led scientists to discover tiny particles that eventually became known as 'atoms'. Through the ages, as technology progressed, different scientists discovered different parts of the atom, such as electrons, protons and neutrons. Recently, using a specialised tool called a scanning tunnelling microscope, scientists have been able to take an image of an atom.

1 LEARNING LINKS

Brainstorm as much as you can remember about these topics.

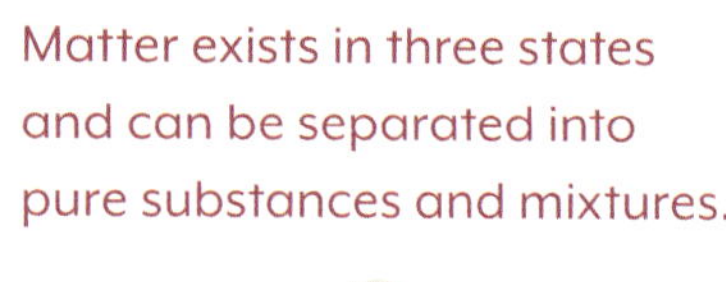

Matter exists in three states and can be separated into pure substances and mixtures.

Atoms are made up of protons, neutrons and electrons.

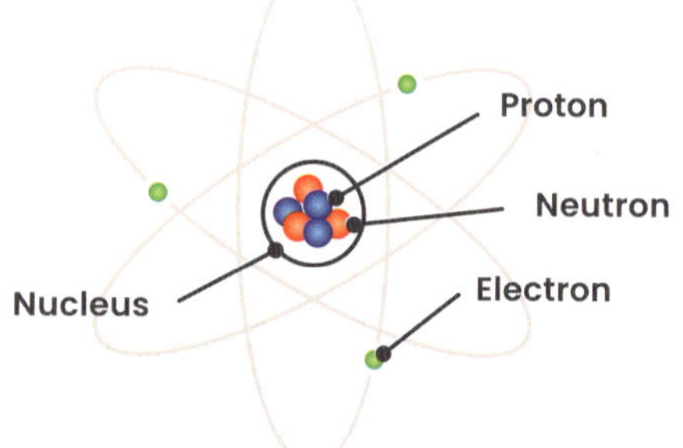

Fossil fuels are burnt for energy.

2 SEE–KNOW–WONDER

List three things you can **see**, three things you **know** and three things you **wonder** about this image.

3 CRITICAL + CREATIVE THINKING

What if ... 50% of all matter in the universe disappeared overnight?

The reverse: Think of five things that are not made of atoms.

The BAR: Think of one thing you would make **B**igger, one thing you would **A**dd and one thing you would **R**eplace in a nuclear power plant.

4 THE ENERGETIC DEBATE!

Scientists have been able to split the atom to release a large amount of nuclear energy. This energy can be used to generate electricity and for diagnosing and treating diseases. Although nuclear energy has many amazing benefits, some people are concerned about its safety because a nuclear accident could have catastrophic consequences. In 2011, the Fukushima nuclear reactor in Japan was damaged after an earthquake and tsunami. This led to dangerous radioactive material leaking into the environment.

12.1 Matter is made of atoms

At the end of this lesson I will be able to:

- **identify** that all matter is made of atoms, which are made up of protons, neutrons and electrons
- **describe** the structure of atoms in terms of the nucleus, protons, neutrons and electrons.

Everything that you see around you is made of matter. Imagine you are sitting on a beach observing your surroundings. You observe three states of matter – liquid water, gas as air and clouds, and solid sand.

You feel the sand grains and wonder how tiny each grain is. You wonder if there is anything smaller than this. The answer is yes. The atom is the basic unit of matter, made up of tiny subatomic particles called protons, neutrons and electrons.

1 Matter takes up space and has mass

Matter is anything that takes up space and has mass.

Matter can be divided into pure substances and impure substances.

Pure substances include elements (made up of one type of atom) and compounds. Compounds are made of two or more types of elements chemically joined together into particles that cannot be physically separated.

Impure substances include mixtures, which can be physically separated into their different components. Air, milk and salt water are all mixtures.

What is matter?

Matter

Pure substances

Impure substances (mixtures)
- Can be physically separated
- For example, air, milk, blood, salty water

Elements
- One type of atom
- Cannot be physically separated
- For example, gold (Au), nitrogen (N_2)

Compounds
- Two or more types of atoms
- Cannot be physically separated
- For example, salt (NaCl), carbon dioxide (CO_2)

KEY TERMS 

atom
the smallest particle of matter, made up of electrons orbiting a nucleus of protons and neutrons

electrically neutral
having an equal number of protons (positively charged) and electrons (negatively charged)

element
a pure substance made up of one type of atom

matter
particles that make up all physical substances

molecule
two or more atoms chemically bonded to each other

nucleus
the central part of an atom, containing protons and neutrons

LITERACY LINK 

Create a mind map using the key terms and at least two additional terms of your choice.

2 Matter is made of atoms

All matter is made of atoms. **Atoms** are the basic building blocks of matter. There are more than 90 different elements that exist naturally and some that have been synthesised by scientists in the laboratory.

Atoms are **electrically neutral** and are made up of three small subatomic particles.

- Protons have a positive charge.
- Electrons have a negative charge.
- Neutrons have no charge.

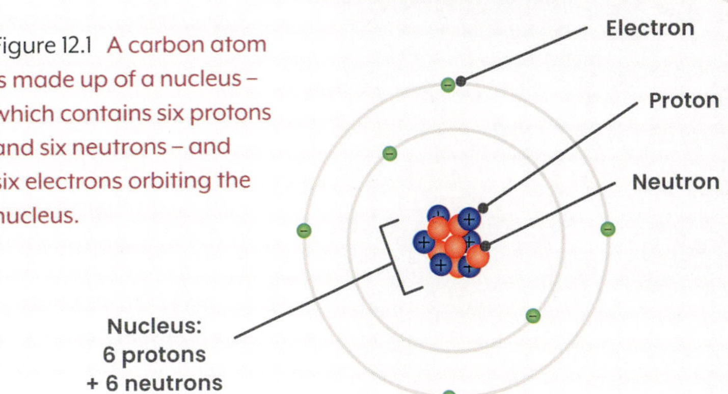

Figure 12.1 A carbon atom is made up of a nucleus – which contains six protons and six neutrons – and six electrons orbiting the nucleus.

Different atoms can bond to form **molecules**. For example, a glucose molecule is made up of six carbon atoms, 12 hydrogen atoms and six oxygen atoms, bonded together. Identical atoms can also bond to form molecules. For example, two oxygen atoms combine to form an oxygen (O_2) molecule.

What three subatomic particles is an atom made up of?

3 Atoms are made up of protons, neutrons and electrons

Scientific experiments in the 20th century proved the existence of atoms and their components – electrons and a nucleus, which contains protons and neutrons. The electrons are negatively charged and orbit the nucleus in clouds. Electrons are much smaller than protons and neutrons. An electron has only $\frac{1}{1840}$ the mass of a proton or neutron.

This means that almost all of the mass of the atom is in the nucleus, a tiny, dense core made up of positive protons and neutral neutrons. The protons and neutrons are held together by a strong nuclear force.

Atoms are held together by a strong attraction between the protons and electrons because they are oppositely charged, like opposite poles of a magnet. The electrical charge on a proton is exactly equal and opposite to the charge on an electron. An atom contains equal numbers of protons and electrons, so an atom has no overall charge.

What is the nucleus made up of?

	Electron	Proton	Neutron
Symbol:	e^-	p	n
Charge:	1^-	1^+	0
Relative mass:	$\frac{1}{1840}$	1	1
Location:	Electron cloud around the nucleus	Nucleus	Nucleus

Figure 12.2 Properties of the subatomic particles electrons, protons and neutrons

CHECKPOINT 12.1

1 Describe matter.
2 Draw a simple diagram of an atom and label the three main parts.
3 Explain why protons and electrons must have opposite charges.
4 Describe the difference between atoms and molecules.
5 Elements and compounds are different. Explain how, using evidence from the text.
6 Describe how the particles in an atom are held together.

CHALLENGE

7 Atoms are made up of protons, neutrons and electrons but protons and neutrons are made up of even smaller particles. Research to find out what these are.

SKILLS CHECK

- I can describe what matter is and what it is made of.
- I can describe with diagrams or in words the structure of atoms, including the subatomic particles.

12.2 Atomic theories

At the end of this lesson I will be able to:

- **outline** historical developments in atomic theory to demonstrate how models and theories have been contested and refined over time through a process of review by the scientific community.

KEY TERMS 

alpha particle
a positively charged particle that is four times larger than a proton

cathode ray
a stream of electrons observed in a high-vacuum tube

Dalton's atomic theory
the theory that states that all matter is made up of tiny particles

LITERACY LINK 

Write a creative story about a day in the life of one of the scientists mentioned in this section.

NUMERACY LINK 

Construct a timeline showing the development of atomic theory, including names of scientists and their contributions. Ensure the timeline is scaled correctly.

Early ideas developed by scientists described atoms as tiny spheres that could not be split into anything simpler. This idea was difficult to test because atoms were not visible. Over time, many scientists have collaborated and devised clever experiments to develop the atomic model that we use today.

1 Democritus proposed that everything is made of atoms

Democritus was an ancient Greek philosopher who lived around 400 BCE. His greatest contribution to science was to suggest that all substances are made of small particles that are indivisible (cannot be divided) and indestructible. He called these particles *atomus*, meaning 'cannot be cut or divided'.

What did Democritus propose?

2 Dalton proposed that compounds are made of two or more different types of atoms

In 1808, English chemist John Dalton proposed his **atomic theory**. He proposed that all matter is made up of atoms, and atoms of a given element are identical in size, mass and other properties. For example, every carbon atom is exactly the same as every other carbon atom, in every way. Dalton concluded that each element has unique atoms and that atoms of different elements combine with each other in simple ratios to form new substances, called compounds.

What did John Dalton's atomic theory state?

3 Thomson discovered electrons

At the beginning of the 20th century, English physicist J.J. Thomson studied the effect of passing cathode rays through gases. A **cathode ray** tube is a glass tube that fires a beam of negatively charged particles. By observing how the cathode ray beam behaved, Thomson identified the particles as electrons.

Thomson proposed that if every atom contains negatively charged particles, then there must also be positively charged material in every atom to balance the electrical charge. His model was described as the plum pudding model in which the atom consists of a positively charged sphere with electrons embedded in it (Figure 12.3).

How did Thomson discover electrons?

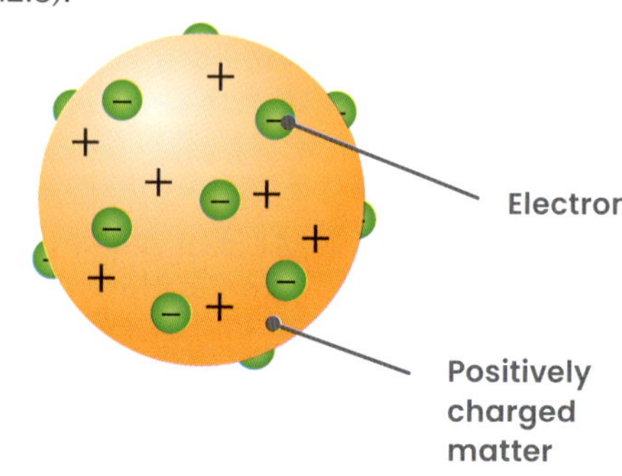

Figure 12.3 Thomson's plum pudding model describes the atom as a positively charged sphere (the 'pudding') with negatively charged electrons (the 'plums') embedded in it.

4 Rutherford proposed the nuclear model of the atom

INVESTIGATION 12.2A
Making an atom

INVESTIGATION 12.2B
Flame test

Ernest Rutherford was a New Zealand-born physicist. Working in England around 1910, he tested Thomson's plum pudding model by bombarding a piece of gold foil with **alpha** (α) **particles**, which are small, positively charged particles.

According to the plum pudding model, Rutherford would have expected all of the alpha particles to pass through the gold foil with little or no deflection. This is because the positive charge in the plum pudding model was assumed to be spread out throughout the entire volume of the atom, so the overall charge would be too weak to significantly affect the path of the relatively massive and fast-moving alpha particles.

However, although most of the alpha particles went straight through the atoms, a small percentage were deflected at large angles. This could be explained by the atom having a very small, dense, positively charged nucleus at its centre. This led Rutherford to conclude that the atom is mostly empty space with a tiny, dense, positively charged nucleus and electrons orbiting the nucleus. Rutherford later identified that the positive components of the nucleus were protons.

What did Rutherford discover?

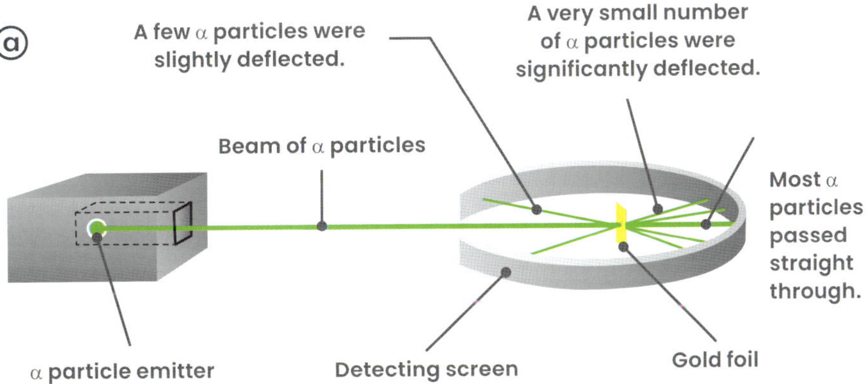

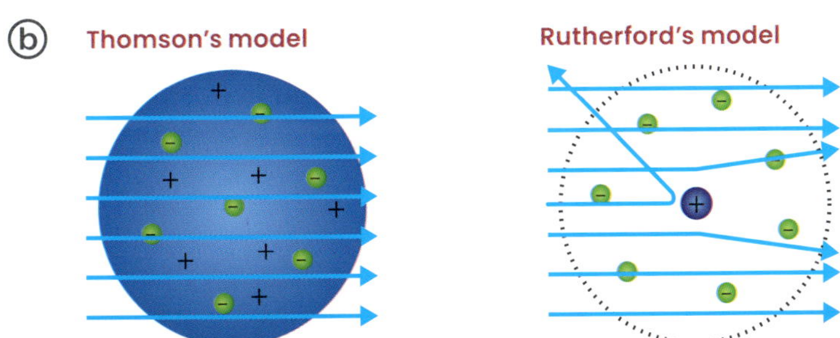

Figure 12.4 (a) Rutherford's famous experiment bombarded a sheet of gold foil with positively charged alpha particles. It showed that the atom is mostly empty space, with a small, dense, positively charged nucleus. (b) According to Thomson's plum pudding model, there was nothing dense or heavy enough in the atom to deflect the α particles. However, Rutherford's observations could only be explained by the presence of a dense, positively charged nucleus.

12.2 continued...

12.2 continued...
Atomic theories

KEY TERMS 

electron shell
space around the nucleus where electrons circulate

spectral line
dark or bright line in an otherwise uniform spectrum

5 Bohr proposed the concept of electron orbits

In 1913, Danish physicist Niels Bohr worked out mathematically that the electrons in Rutherford's model must move around the nucleus in orbits of fixed sizes and energy levels, like planets orbiting the Sun. This energy was explained by emission spectra – atoms absorb and emit a fixed amount of radiation when the electrons move between the orbits or **shells**, which are at specific distances from the nucleus. This produces a series of regular **spectral lines**.

Bohr proposed that each orbit can only contain a certain number of electrons. The arrangement of electrons in different atoms explained how they can form bonds in chemical reactions.

What did Bohr discover?

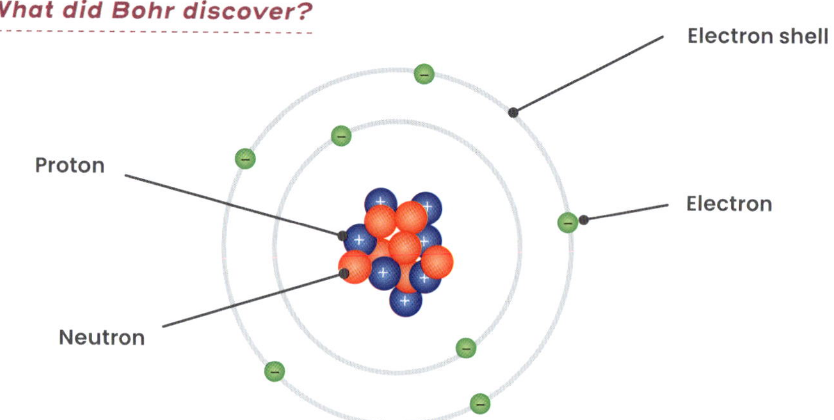

Figure 12.5 The Bohr model of the atom regards the atom as a small, dense nucleus surrounded by electrons in orbits of set size and energy.

6 Schrödinger took the Bohr model of the atom one step further

In 1926 an Austrian physicist named Erwin Schrödinger used mathematical equations to describe the probable location of an electron. Schrödinger worked out that electrons do not move in set paths (orbits) around the nucleus; instead, they move in clouds where their exact location is uncertain. Even though it is impossible to know the exact location of the electrons, Schrödinger's work shows that we are able to work out where they are likely to be found.

What did Schrödinger's work show?

Democritus

John Dalton

JJ Thomson

Ernest Rutherford

7 Chadwick discovered the neutron

James Chadwick was an English physicist who worked with Rutherford. In 1932, Chadwick bombarded beryllium atoms with alpha particles. He found that a new particle was ejected that had almost the same mass as the proton but no charge – the neutron. His finding revolutionised understanding of atomic structure and gained him a Nobel Prize in Physics in 1935.

During World War II, Chadwick was extensively involved in the Manhattan Project, a research project that studied nuclear fission reactions to produce nuclear weapons.

How did Chadwick discover the neutron?

8 The evolution of atomic theories and models

Table 12.1 Summary of theories and models about the atom

Date	Name	Discoveries or theories
450 BCE	Democritus	All substances are made of atoms.
1803	John Dalton	Atoms are solid spheres and cannot be divided. Atoms of an element are identical. Atoms of different elements combine to form new substances called compounds.
1904	J.J. Thomson	Atoms are composed of electrons. An atom is a sphere of positive charge (the 'pudding') with negatively charged electrons (the 'plums') scattered through it.
1910	Ernest Rutherford	Atoms are mostly empty space, with a tiny, dense, positively charged nucleus in the centre.
1913	Niels Bohr	Electrons move around the nucleus in orbits of fixed sizes and energies (like planets orbiting the Sun).
1926	Erwin Schrödinger	Electrons do not move in set paths around the nucleus, but in waves or clouds.
1932	James Chadwick	The atom contains a neutral subatomic particle – the neutron.

How has atomic theory changed over time?

Niels Bohr

Erwin Schrödinger

James Chadwick

CHECKPOINT 12.2 

1 Explain what atomic theory is.
2 List the concepts of the atomic theory proposed by Dalton.
3 According to Dalton, how do elements react to produce new elements?
4 Describe the atomic model proposed by Thomson.
5 Explain how a cathode ray tube works.
6 How did Rutherford discover the positively charged nucleus?
7 How was Rutherford's view of the atom similar to and different from that of Thomson?
8 How was the atomic model improved as a result of Bohr's discoveries?
9 Describe Bohr's evidence suggesting electrons orbit in shells.
10 Describe the experiment conducted by Chadwick.

CHALLENGE

11 Draw different models of atoms that have been developed to show how they have been refined over the time.

SKILLS CHECK

- I can define atomic theory.
- I can identify key scientists and outline their contribution to current atomic theory.

12.3 Radiation

At the end of this lesson I will be able to:

- **identify** that natural radioactivity arises from the decay of nuclei in atoms, releasing particles and energy.

KEY TERMS

atomic number
the number of protons in an atom

half-life
the time taken for half of a radioactive material to decay

ionise
move electrons from an atom or molecule

isotopes
atoms of the same element with the same number of protons but a different numbers of neutrons

penetrating power
how well radiation from radioactive materials can pass through matter

radioisotope
an isotope that emits radiation

LITERACY LINK

Complete this sentence: 'The term *isotope* is not the same as *radioisotope* because ...' (Hint: Use the key terms if you need to.)

NUMERACY LINK

The activity of a radioactive sample is measured as follows:

Time (min)	35	50	80	150
Activity (counts)	120	100	80	60

Graph the above data and draw a curve of best fit. Determine the half-life of the sample.

Some atoms are unstable because their nucleus contains too many neutrons or protons. The nucleus breaks down and releases energy. This energy is known as radiation and can be either alpha or beta particles or gamma rays. Ultimately, the atom is transformed into a new element with the release of the energy or particle.

1 Radioisotopes are unstable atoms that decay

Atoms of different elements have different numbers of protons in the nucleus. Atoms of the same element always have the same number of protons (same **atomic number**) but can have different numbers of neutrons.

Isotopes are atoms of the same element with the same atomic number but a different atomic mass. For example, carbon-12 has six protons and six neutrons. Its isotope carbon-14 also has six protons, but it has eight neutrons. An isotope of an element has the same number of protons, but a different number of neutrons.

Most atoms have stable nuclei, which contain enough neutrons to balance out the protons. These stable atoms do not break down. Atoms that are unstable have either too many protons or too many neutrons. **Radioisotopes** are radioactive isotopes of an element – unstable atoms that break down and release electromagnetic energy from their nucleus.

When a sample of radioisotope breaks down (or decays), it emits radiation. Over time, the level of radioactivity falls as the amount of radioisotope in the sample decreases. The **half-life** of a radioactive substance is the time it takes for a given amount of the substance to reduce by half. Half-lives range from a fraction of a second to thousands of years.

What are radioisotopes?

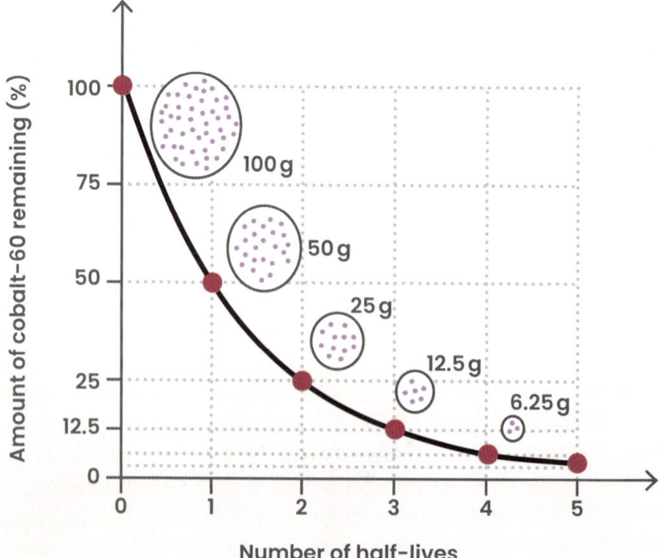

Figure 12.6 The decay curve for cobalt-60, which has a half-life of 5.27 years

❷ An unstable nucleus may emit α or β particles

When a nucleus has too many protons, the like charges of the protons repel each other. This causes the nucleus to break down and release a particle made up of two protons and two neutrons. This is called an alpha particle or alpha (α) radiation. A new element is also formed, one that has an atomic number two less than the original element.

When a nucleus has too many neutrons, a neutron breaks down into a proton and an electron. The proton is retained and the electron is emitted as beta (β) radiation. Because the element now has an additional proton, the atomic number increases by one, producing a new element.

Because they have a large charge, α particles **ionise** other atoms strongly. However, they have a low **penetrating power** because they lose energy each time they ionise an atom. Beta particles ionise atoms that they pass through, but not as strongly as α particles do; however, they are more likely to penetrate other atoms.

What is the difference between α and β radiation?

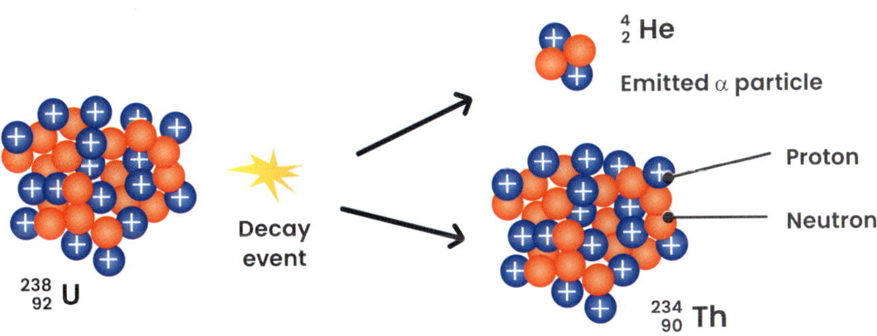

Figure 12.7 Uranium-238 has too many protons and is unstable. Therefore, it decays by emitting alpha radiation to become a new element, thorium, which is stable.

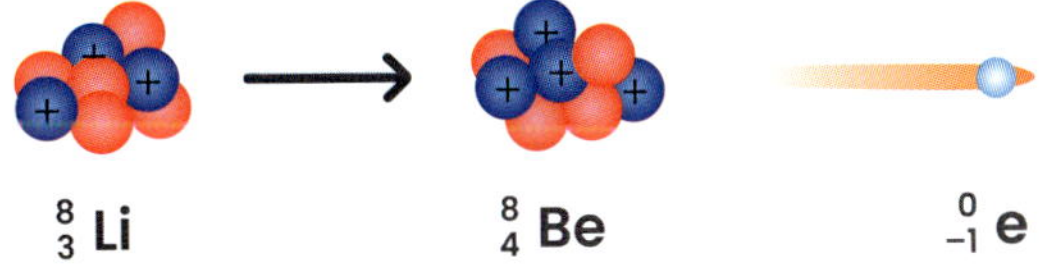

Figure 12.8 Lithium-8 is an unstable isotope of lithium. It decays by emitting beta radiation to form a new element, beryllium-8.

❸ Different types of radiation have different properties

Another type of radiation that is produced when a nucleus decays is gamma (γ) radiation. Gamma rays are high-energy radiation and are more penetrating than α or β radiation. Gamma rays are emitted when the protons and neutrons in an unstable nucleus rearrange and emit energy instead of a particle.

Gamma rays are waves, not particles. This means that they have no mass and no charge.

What is gamma radiation?

INVESTIGATION 12.3
Half-life coin experiment

CHECKPOINT 12.3 

1 Explain what radioactivity is.
2 Explain why some nuclei are unstable.
3 Describe why these radiation types occur.
  a alpha
  b beta
  c gamma
4 Explain the difference between nuclear decay by α and β radiation in terms of the new elements that are formed.

CHALLENGE

5 Strontium-90 has a half-life of 28.9 years. It emits β particles, which can be dangerous for human health. How many years must pass before its activity falls to one-eighth of its original value?
6 Research the carbon-14 radioisotope and how scientists have used it to estimate the ages of many of the mummified bodies that have been discovered in the past century.

SKILLS CHECK

- I can explain what radiation is and what happens during nuclear decay.
- I can identify the different particles and energy released when an atom's nucleus breaks down.

12.4 Nuclear energy

At the end of this lesson I will be able to:

- **evaluate** the benefits and problems associated with medical and industrial uses of nuclear energy.

KEY TERMS 

dosimeter
a device used to measure an absorbed dose of ionising radiation

ionising radiation
radiation, made up of particles, X-rays or gamma rays, that has sufficient energy to cause cancer

radiotherapy
a therapy that uses radiation to kill cancerous cells

LITERACY LINK 

Write a piece of persuasive writing either for or against the use of nuclear power.

NUMERACY LINK 

The amount of energy that can be generated from 1 kg of uranium is equivalent to 10 000 kg of mineral oil or 14 000 kg of coal. If the fission of 1 kg uranium produces 8×10^{10} kJ of energy, calculate how much energy is produced by 1 kg of mineral oil and 1 kg of coal.

Nuclear reactions release huge amounts of energy. This energy, which is in the form of radiation, can be used to generate electricity at nuclear power plants. In medicine, nuclear radiation can be used to treat and diagnose medical conditions.

There are many benefits associated with the use of nuclear energy. However, with these benefits there have also been the problems of nuclear disasters and nuclear waste.

1 Nuclear energy is used to detect and treat cancer

Radiation in medicine is used to diagnose medical problems and treat cancer. One major use is in medical imaging, where technologies such as PET scans use radiation to detect cancer sites. During a PET scan, diagnostic isotopes are injected in liquid form into the bloodstream. The scanner can then locate tumours and blood clots by detecting the radiation as it passes through body tissues.

The use of radiation to treat cancer is known as **radiotherapy**. Radiation can destroy the tumour and eliminate the cancer. The use of radiotherapy to destroy a tumour can also destroy the surrounding healthy cells. Although radiation may stop a tumour from growing, it can also cause side effects, such as nausea, swelling, skin irritation, ulcers, loss of hair and secondary cancers.

What are some uses of radiation in medicine?

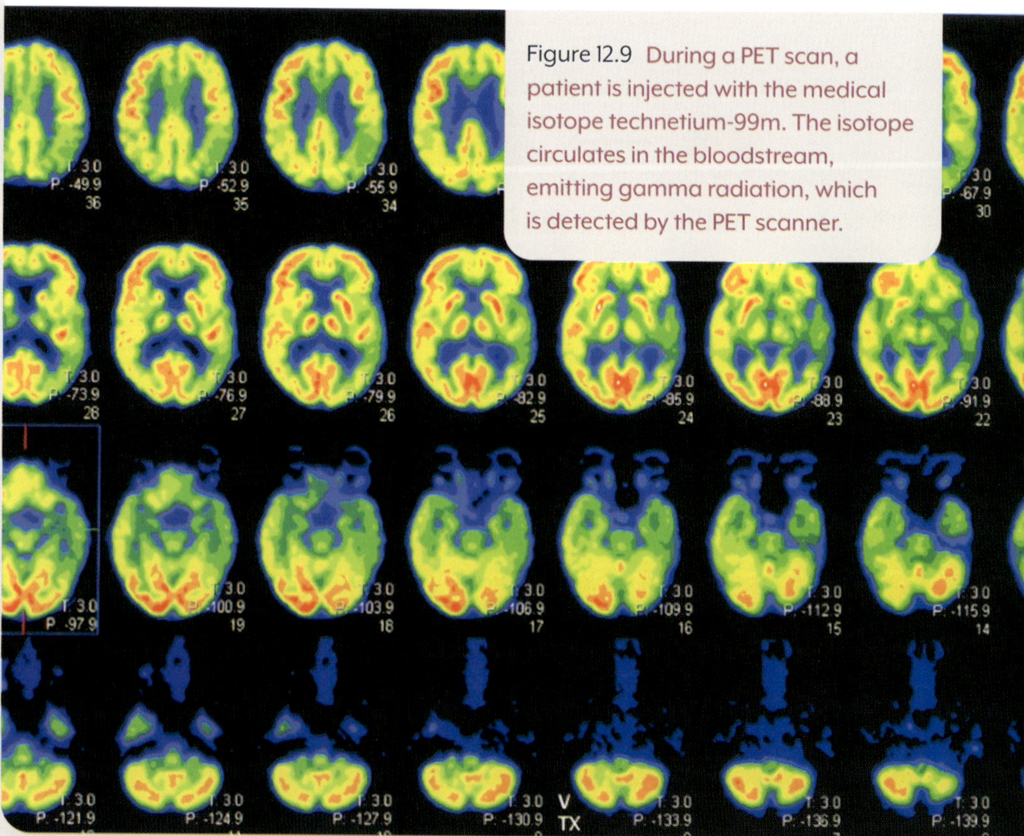

Figure 12.9 During a PET scan, a patient is injected with the medical isotope technetium-99m. The isotope circulates in the bloodstream, emitting gamma radiation, which is detected by the PET scanner.

❷ Nuclear energy has many industrial uses

Some common uses of radiation in industry are to sterilise food to extend its shelf life, check for cracks in underground pipes, and check the thickness of paper during manufacturing.

Nuclear energy can also be used to generate electricity instead of conventional methods that use fossil fuels. Nuclear power plants have low emissions of CO_2 and SO_2 gases and don't rely on coal, which is a non-renewable resource. Nuclear power uses much less fuel (in the form of uranium) than a coal-powered station does. Nuclear power plants produce far less background radiation than coal-powered plants and are far more effective at producing energy.

What are some uses of radiation in industry?

❸ Nuclear energy has safety concerns

Prolonged exposure to radiation damages living cells and can lead to diseases such as leukaemia and cancer. It can also damage DNA and lead to birth defects. In the past, scientists such as Marie Curie (1867–1934) have died as a result of exposure to dangerous radiation.

There are now guidelines that protect hospital staff and people who work in nuclear power stations. Workers wear special detectors called **dosimeters**, which monitor their exposure to radiation. They can also wear protective suits or work behind thick glass screens to shield them from radiation.

One of the main issues in the nuclear industry is nuclear waste. Nuclear waste contains radioactive elements such as plutonium, which has a half-life of 24 000 years. Workers handling such radioactive materials are exposed to dangerous **ionising radiation**, which causes cancer. Dangerous radioactive waste must be treated and then stored safely; for example, in concrete cylinders or by burying it deep underground.

In the past, there have been disasters at nuclear power plants. In 1986, a nuclear reactor at Chernobyl, Ukraine, exploded. This caused deaths and cancers in the surrounding area, and spread radioactive products as far as Europe. In 2011, a tsunami off the coast of Japan resulted in the meltdown of the Fukushima nuclear power plant, and release of radioactive contamination.

Other disadvantages of nuclear energy include the expense of constructing a safe nuclear power plant, which can operate for only 40 years and then has to be safely dismantled and removed.

What is the main safety concern when dealing with nuclear energy?

INVESTIGATION 12.4
Nuclear energy debate

CHECKPOINT 12.4 

1 Describe how a PET scan works.
2 What are some of the side effects of using radiotherapy?
3 Outline some advantages and disadvantages regarding the use of nuclear energy.
4 The use of nuclear energy has resulted in some large-scale catastrophes. Suggest why it continues to be used.
5 Outline some safety precautions that can be used when working with nuclear energy and radiation.
6 Write a list of pros and cons for the use of nuclear energy.
7 Describe the benefits and dangers of using radiation to treat cancer.

CHALLENGE

8 Research a disaster related to the use of nuclear energy. Prepare a short report and include how it happened and what effects it had.

SKILLS CHECK

- I can describe some benefits of nuclear energy in medicine and industry.
- I can describe some problems with the use of nuclear energy in medicine and industry.

CHAPTER SUMMARY

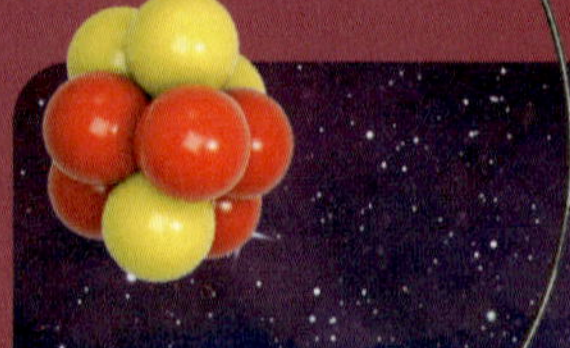

Matter is made up of atoms.

Matter

Pure substances

Impure substances (mixtures)
- Can be physically separated
- For example, air, milk, blood, salty water

Elements
- One type of atom
- Cannot be physically separated
- For example, gold (Au), nitrogen (N_2)

Compounds
- Two or more types of atoms
- Cannot be physically separated
- For example, salt (NaCl), carbon dioxide (CO_2)

Properties of subatomic particles

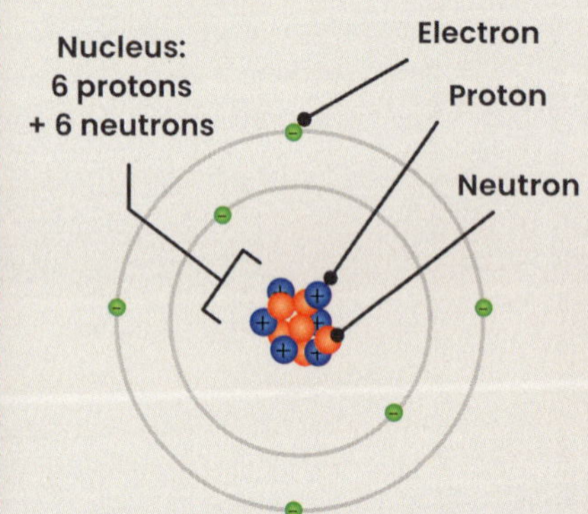

Nuclear energy

Pros:
- generates less emissions than CO_2
- can be used to treat cancer
- can extend the shelf life of food

Cons:
- risk of environmental disaster
- can lead to illness and disease
- disposal of nuclear waste

Properties of alpha, beta and gamma radiation

Radiation	Symbol	Composition	Charge	Relative penetrating power	Ionising power
Alpha	α	Alpha particles (proton–neutron pairs)	2+	Low	Strong
Beta	β	Beta particles (electrons)	1–	Medium	Medium
Gamma	γ	High-energy radiation (photons)	0	High	Very weak

Atomic theory – key thinkers

Democritus

John Dalton

JJ Thomson

Ernest Rutherford

Niels Bohr

Erwin Schrödinger

James Chadwick

FINAL CHALLENGE

1. Draw an arrow between the scientist and the description of their contribution.

Democritus	Proposed that atoms are positively charged spheres with electrons embedded like plums in a plum pudding
Bohr	Proposed that substances consist of tiny particles called atoms
Thomson	Proposed the atomic theory that all matter is made up of atoms
Dalton	Discovered that electrons orbit the nucleus in energy levels
Rutherford	Discovered the existence of neutrons in the nucleus
Chadwick	Proposed that the atom is mostly empty space, with a dense, positively charged nucleus in the centre

2. Describe the three parts of the atom.
3. What holds the parts of an atom together?
4. Identify four concepts of the atomic theory proposed by John Dalton.

5. Describe how Rutherford proved that the atom is mostly empty space.
6. Describe the two situations when an isotope would become unstable.
7. Compare the three types of radiation in terms of penetrating ability and ionising power. (Hint: Refer to Table 12.1, page 181.)

8. Create a table that summarises the main points under the headings of 'Uses of radiation in medicine' and 'Problems of radiation in medicine'.
9. Using your understanding of nuclear power to generate electricity, propose reasons for moving away from using fossil fuels towards using nuclear power.

10. Research the consequences of the nuclear disasters at Chernobyl, Ukraine, in 1986 and Fukushima, Japan, in 2011.
11. Nuclear power plants produce nuclear waste. Make some suggestions as to how to best address this problem.

13 The periodic table

The periodic table is a way of sorting elements into groups based on their physical and chemical properties.

The left side of the periodic table contains some very reactive elements, such as hydrogen, caesium and potassium. Hydrogen can be used to make bombs that are even more powerful than atomic bombs. Pure cesium and potassium have to be stored in oil so they don't come in contact with any other element. Otherwise, they quickly catch fire and explode.

As you move towards the centre of the periodic table, the elements tend to become more stable. However, as you move past the centre and to the right, they become increasingly unstable again. Fluorine, for example, is the most reactive non-metal element in the whole table. The last group, the noble gases, is the least reactive group in the whole table.

1 LEARNING LINKS

Brainstorm as much as you can remember about these topics.

96
Cm
Curium
247

Atomic number
Symbol
Atomic mass

An element can be identified by its atomic number and atomic mass.

Most metals are good conductors of heat and electricity.

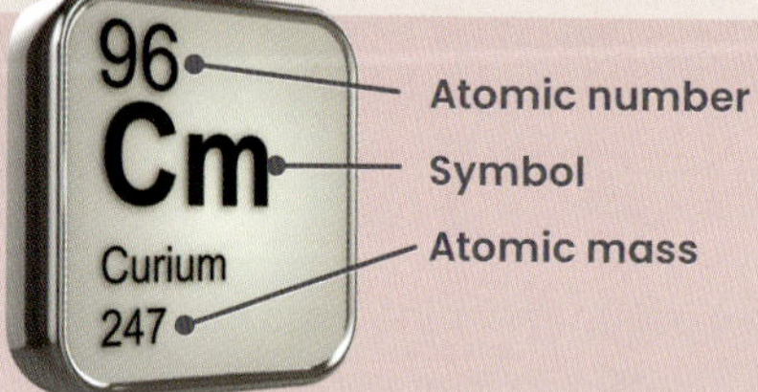

Non-metals have a wide range of properties.

2 SEE–KNOW–WONDER

List three things you can **see**, three things you **know** and three things you **wonder** about this image.

Serious threat in the next 100 years
Rising threat from increased use
Limited availability, future risk to supply
Plentiful Supply
Synthetic
From conflict minerals
Elements used in a smart phone

3 CRITICAL + CREATIVE THINKING

The reverse: List five things that are NOT made of elements.

Different uses: Think of as many different uses for helium as possible.

The alphabet: Think of a word about the periodic table for each letter of the alphabet.

4 THE MYSTERY OF YTTERBY!

Strangely, one of the most important places in the history of chemistry is a small village called Ytterby on a tiny Swedish island. The village had a quarry that was used for mining a mineral called feldspar. In this single quarry, seven previously undiscovered rare earth elements were found. Four of them were named after Ytterby itself – ytterbium, terbium, erbium and yttrium – and there was also holmium, thulium and gadolinium. Almost 6% of all of the elements ever discovered were found on an island you could walk around in a few hours.

13.1 Elements and their atoms

At the end of this lesson I will be able to:

- **identify** the atom as the smallest unit of an element, which can be represented by a symbol
- **distinguish** between the atoms of some common elements by comparing information about the numbers of protons, neutrons and electrons.

KEY TERMS 

atom
the smallest particle of matter, made up of electrons orbiting a nucleus of protons and neutrons

chemical symbol
one or two letters that represent an element

element
a pure substance made up of one type of atom

LITERACY LINK 

Construct the longest possible word using element symbols. (You can use each symbol only once.)

NUMERACY LINK 

One proton has a positive charge and one electron has a negative charge. What is the overall charge of an atom if it has:

a 3 protons and 2 electrons?
b 16 protons and 18 electrons?

All matter is made of different elements. These elements are made up of atoms, which we cannot see. Some elements are stable and exist on their own. Other elements are very unstable and tend to bond with other elements. Scientists have identified 118 different elements. Most of these are naturally occurring elements, and the rest have been made by scientists.

1 Every element is unique

Elements are the basic building blocks of all matter. An **element** is a pure substance made up of only one type of **atom** – the smallest unit of an element. For example, oxygen is made up of only oxygen atoms.

Each element is assigned a unique **chemical symbol** using one or two letters. The first letter is always written as a capital letter and, when there is one, the second letter is always a lower-case letter. This ensures that scientists from different countries, who speak different languages, still use the same symbols in their work. The names of elements have different origins. Some elements were named according to their properties, some were named after scientists, and other were named after places.

The symbols of many elements are the first letter of their English names; for example, O for oxygen and C for carbon. Other symbols are made up of two letters from their English name; for example, Cl for chlorine and Mg for magnesium. You will find these symbols easy to identify.

Other symbols are not as easy to identify because they come from the Latin, Greek or German names of the elements. These elements were discovered centuries ago. For example, the symbol for:

- gold (Au) comes from the Latin name *aurum*
- iron (Fe) comes from the Latin name *ferrum*
- sodium (Na) comes from the Latin name *natrium*
- mercury (Hg) comes from the Greek word *hydrargyros*
- tungsten (W) comes from the German word *wolfram*.

How are elements assigned a chemical symbol?

Figure 13.1 The element 'curium' is named after physicist and chemist Marie Curie.

❷ Different elements have different numbers of protons

Each element has a unique number of protons. For example, every atom with 6 protons is carbon, while every atom with 7 protons is nitrogen.

However, the same element can have different numbers of neutrons and electrons. Carbon-12 is the most common form of carbon, and it contains 6 protons and 6 neutrons. Carbon-14 is rarer and contains 6 protons and 8 neutrons. Every single carbon atom will always have 6 protons – if it didn't, it wouldn't be carbon. But the number of neutrons within those atoms can change.

Similarly, atoms of the same element can have different numbers of electrons. A neutral atom has the same number of protons and electrons, whereas a charged atom (called an ion) has different numbers of protons and electrons. An oxygen atom has 8 protons, so a neutral oxygen atom has 8 electrons. However, an oxygen ion has 8 protons and 6 electrons.

Table 13.1 shows the first ten elements, as well as the number of protons, the average number of neutrons and the number of electrons in a neutral atom.

What are elements?

Table 13.1 Composition of the most common forms of the first ten elements

Element name	Symbol	Protons	Neutrons	Electrons
Hydrogen	H	1	0	1
Helium	He	2	2	2
Lithium	Li	3	4	3
Beryllium	Be	4	5	4
Boron	B	5	6	5
Carbon	C	6	6	6
Nitrogen	N	7	7	7
Oxygen	O	8	8	8
Fluorine	F	9	10	9
Neon	Ne	10	10	10

❸ The periodic table lists all of the elements

All of the 118 elements that science has discovered are shown on the periodic table.

The table is more than just a 1–118 list, though. Its structure groups together elements with similar properties. It also shows us how many protons each element contains.

The full periodic table is shown on the next two pages.

How many elements have been identified to date?

CHECKPOINT 13.1 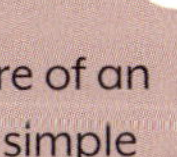

1 Describe the structure of an atom and provide a simple diagram.

2 What is the symbol for:
   a sodium?
   b silver?
   c sulfur?

3 Explain why elements are given a chemical symbol and describe how these symbols might be chosen.

4 The number of protons, not the number of neutrons or electrons, is what makes an element unique. Explain why.

CHALLENGE

5 Carry out research to identify the most unstable element on the periodic table. Describe the element and include how many protons, neutrons and electrons it has.

SKILLS CHECK

- I can give an example of the number of protons, electrons and neutrons in at least one element.
- I can explain what an element is.

13.1 continued...

13.1 continued...

Elements and their atoms

Figure 13.2 The modern periodic table

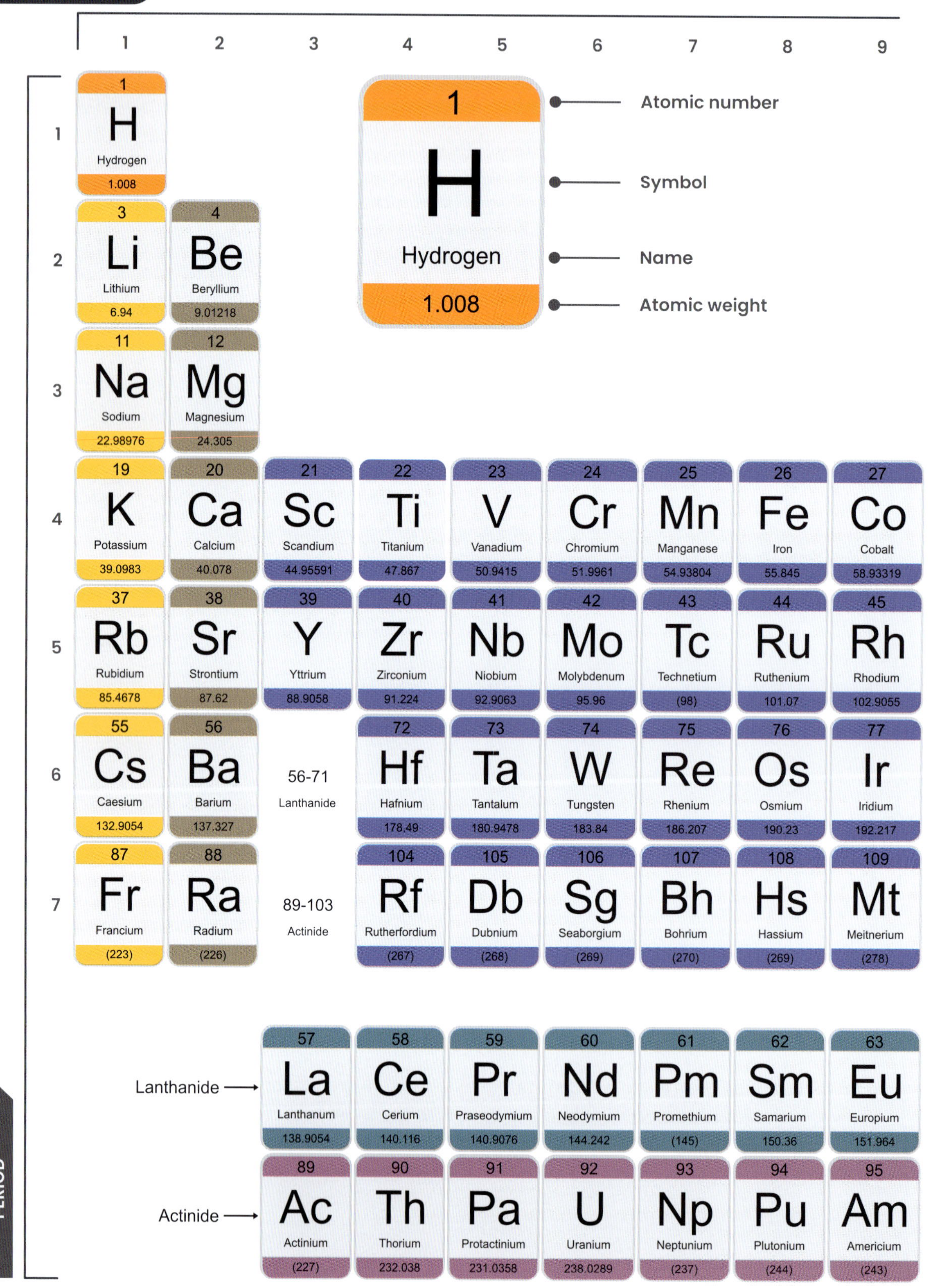

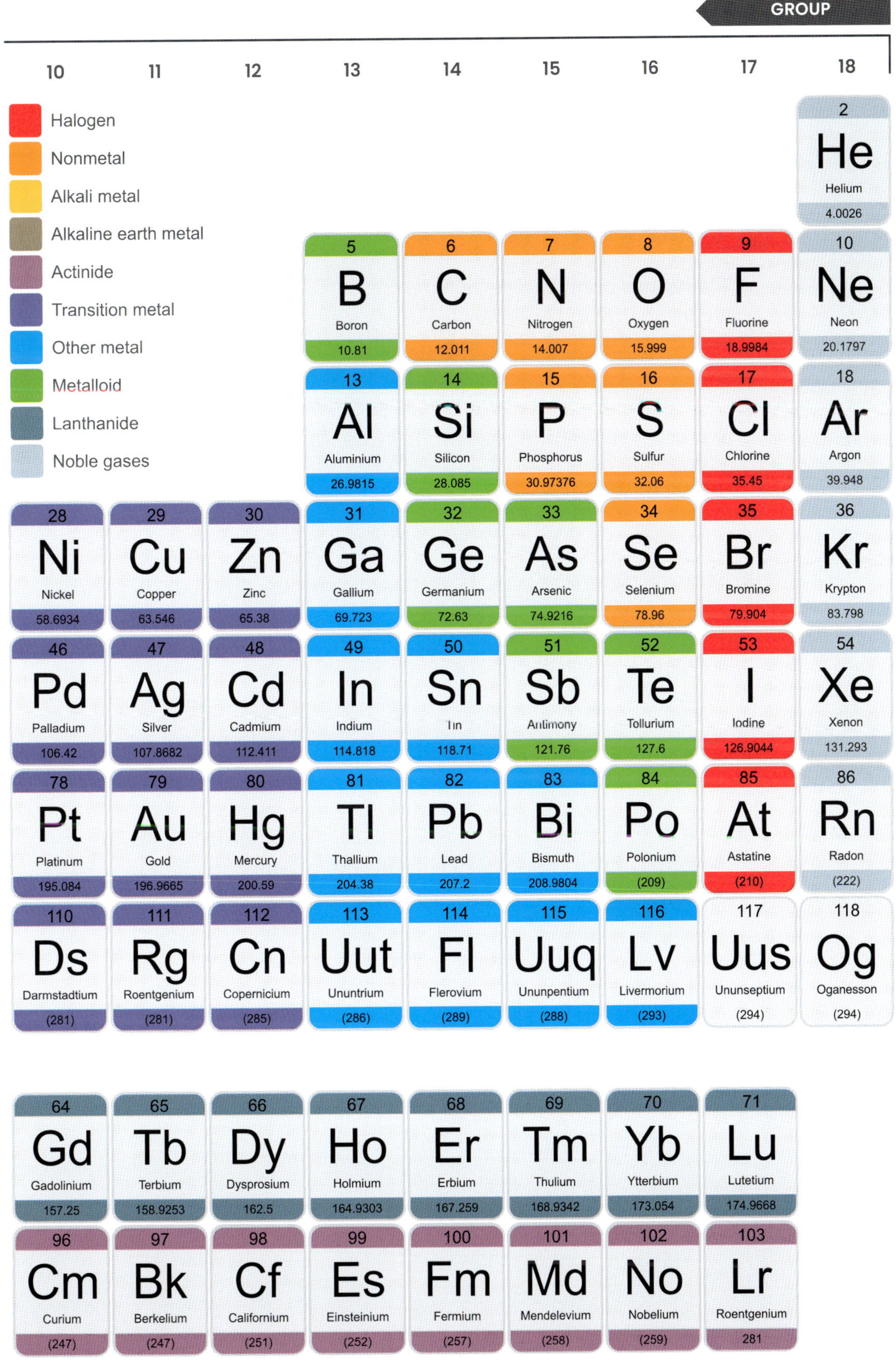
GROUP
10
11
12
13
14
15
16
17
18
Halogen
Nonmetal
Alkali metal
Alkaline earth metal
Actinide
Transition metal
Other metal
Metalloid
Lanthanide
Noble gases
2 He Helium 4.0026
5 B Boron 10.81
6 C Carbon 12.011
7 N Nitrogen 14.007
8 O Oxygen 15.999
9 F Fluorine 18.9984
10 Ne Neon 20.1797
13 Al Aluminium 26.9815
14 Si Silicon 28.085
15 P Phosphorus 30.97376
16 S Sulfur 32.06
17 Cl Chlorine 35.45
18 Ar Argon 39.948
28 Ni Nickel 58.6934
29 Cu Copper 63.546
30 Zn Zinc 65.38
31 Ga Gallium 69.723
32 Ge Germanium 72.63
33 As Arsenic 74.9216
34 Se Selenium 78.96
35 Br Bromine 79.904
36 Kr Krypton 83.798
46 Pd Palladium 106.42
47 Ag Silver 107.8682
48 Cd Cadmium 112.411
49 In Indium 114.818
50 Sn Tin 118.71
51 Sb Antimony 121.76
52 Te Tollurium 127.6
53 I Iodine 126.9044
54 Xe Xenon 131.293
78 Pt Platinum 195.084
79 Au Gold 196.9665
80 Hg Mercury 200.59
81 Tl Thallium 204.38
82 Pb Lead 207.2
83 Bi Bismuth 208.9804
84 Po Polonium (209)
85 At Astatine (210)
86 Rn Radon (222)
110 Ds Darmstadtium (281)
111 Rg Roentgenium (281)
112 Cn Copernicium (285)
113 Uut Ununtrium (286)
114 Fl Flerovium (289)
115 Uuq Ununpentium (288)
116 Lv Livermorium (293)
117 Uus Ununseptium (294)
118 Og Oganesson (294)
64 Gd Gadolinium 157.25
65 Tb Terbium 158.9253
66 Dy Dysprosium 162.5
67 Ho Holmium 164.9303
68 Er Erbium 167.259
69 Tm Thulium 168.9342
70 Yb Ytterbium 173.054
71 Lu Lutetium 174.9668
96 Cm Curium (247)
97 Bk Berkelium (247)
98 Cf Californium (251)
99 Es Einsteinium (252)
100 Fm Fermium (257)
101 Md Mendelevium (258)
102 No Nobelium (259)
103 Lr Roentgenium 281

13.2 Organising elements

At the end of this lesson I will be able to:

- **describe** the organisation of elements in the periodic table by their atomic number.

KEY TERMS 

atomic number
the number of protons in an atom

group
a column in the periodic table

mass number
the sum of the protons and neutrons in an atom

period
a row in the periodic table

LITERACY LINK 

Create a mind map, using the key words, and at least two additional words of your choice.

NUMERACY LINK 

The atomic number of potassium is 19 and its mass number is 39. Determine the number of protons, electrons and neutrons in a potassium atom.

The atomic number of bromine is 35 and its mass number is 80. Determine the number of protons, electrons and neutrons in a bromine atom.

A wide variety of substances exist in nature. These substances are made up of elements, which have different atoms. The periodic table is a way of classifying these elements. In the periodic table, elements are organised by their atomic number and grouped according to their properties. When the periodic table was being developed, scientists left gaps for elements that were still unknown at the time. They could predict the properties of the undiscovered elements by studying the properties of the surrounding elements.

1 Atomic number is the number of protons

The **atomic number** is the number of protons that each atom of the element has in its nucleus. The elements on the periodic table are arranged by increasing atomic number. In the periodic table, atomic number is usually written above the element's symbol.

For example, every atom of silicon (Si) has 14 protons, so the atomic number of silicon is 14. Similarly, every atom of oganesson (Og) has 118 protons so the atomic number of oganesson is 118.

How are elements organised in the periodic table?

2 Mass number is the number of protons and neutrons

The number of protons and neutrons in the nucleus is known as the **mass number**.

Mass number = number of protons + number of neutrons

The periodic table doesn't list the mass number of elements. Instead, it shows the atomic mass, which is the the average number of protons and neutrons for all natural isotopes of an element. This is always a decimal number.

How is mass number calculated?

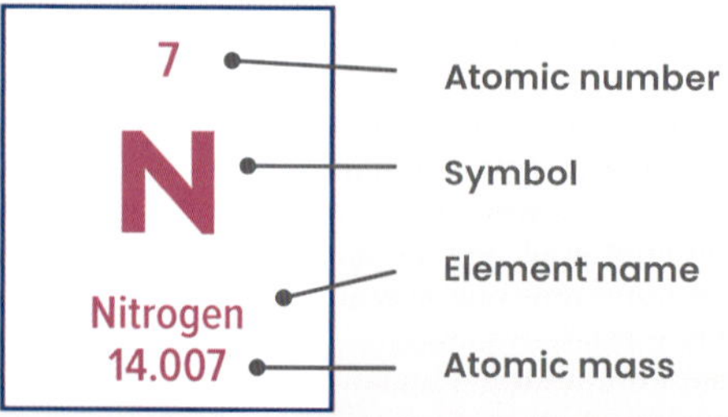

Figure 13.3 The periodic table includes information for every known element, including the element's name and symbol, atomic mass and atomic number.

3 Elements in the same group share some properties

In the periodic table, elements are arranged into 18 **groups**, the vertical columns based on their physical and chemical properties. Groups indicate the number of valence (outermost) shell electrons – the number of the group is related to the number of valence electrons and chemical properties of the elements within it.

Group 1 metals are called the alkali metals and have one valence electron. They are very reactive and are soft, shiny, malleable and conduct heat and electricity. Group 2 metals are the alkaline earth metals and have two electrons in their valence shell. They have similar metallic properties to group 1 metals, except they are less reactive.

The middle of the periodic table contains the transition metals, which are all solids except mercury, which is liquid. To the right of the transition metals are the semimetals (or metalloids), such as boron, silicon, germanium, arsenic, antimony and tellurium. Their properties are a blend of metal and non-metal properties.

The non-metals are found on the right-hand side of the periodic table. Group 17 is known as the halogens, the most reactive non-metal group. Halogens have seven electrons in their valence shell and are all gases except bromine, which is liquid at room temperature, and iodine and astatine, which are solids.

Group 18 is known as the noble gases. They are also known as the inert gases because they are unreactive.

The horizontal rows are called **periods**, and they are numbered 1 to 7. Elements in the same period have the same number of electron shells, or orbits in which electrons can be found. Potassium (K) and bromine (Br) have very different properties, but they both have four electron shells and are found in period 4.

Two parts of the periodic table are broken out into separate rows. These are the lanthanides (57–70) and the actinides (89–102). These elements are rare and have very similar properties, so they're broken out into their own rows so that they don't disrupt the arrangement of the table.

What does an element's group and period tell us?

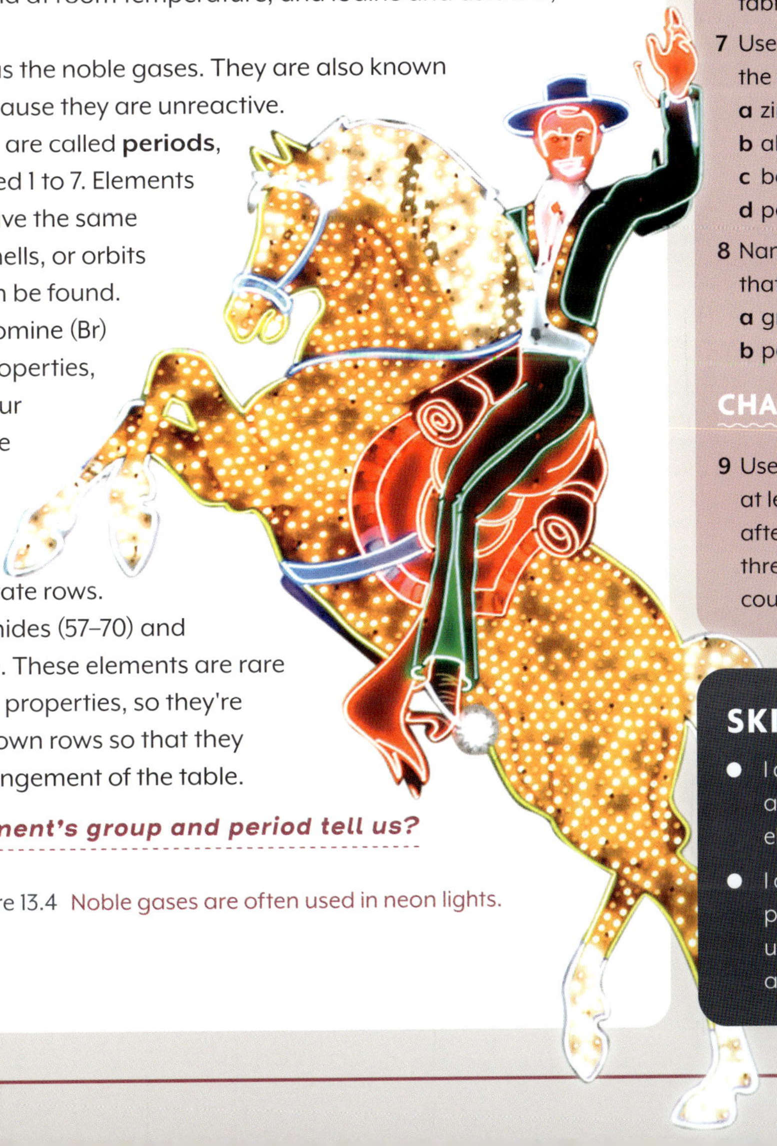

Figure 13.4 Noble gases are often used in neon lights.

CHECKPOINT 13.2 

1 Recall how many elements are in the periodic table.

2 Name and describe groups 1 and 2 of the periodic table.

3 Describe the elements found in the middle of the periodic table.

4 Recall what type of elements are found on the right-hand side of the periodic table.

5 Identify which subatomic particle is linked to the atomic number.

6 Explain what the groups and periods are in the periodic table.

7 Use the periodic table to find the atomic number of:
- a zinc
- b aluminium
- c boron
- d potassium.

8 Name at least three elements that are in the same:
- a group as potassium
- b period as calcium.

CHALLENGE

9 Use the periodic table to find at least three elements named after scientists and at least three elements named after countries.

SKILLS CHECK

- I can identify what the atomic number of an element is.
- I can describe how the periodic table is organised, using terms such as *groups* and *periods*.

13.3 Properties of elements

At the end of this lesson I will be able to:

- **relate** the properties of some common elements to their position in the periodic table
- **predict**, using the periodic table, the properties of some common elements.

KEY TERMS 

ductile
can be drawn out into a wire

malleable
can be bent

metalloid
an element with properties similar to both metals and non-metals

LITERACY LINK 

Create a mind map using the key words and at least two additional words of your choice.

NUMERACY LINK 

Determine whether the following elements are solids, liquids or gases at room temperature (25 °C) based on their boiling and melting points:

a BP = 440 °C, MP = 115 °C.
b BP = –196 °C, MP = –210 °C.
c BP = 59 °C, MP = –7 °C.

The elements in the periodic table are divided into three main groups – metals, non-metals and metalloids – based on their physical and chemical properties. By exploring properties of elements, scientists have been able to use them in many ground-breaking discoveries.

1 Metals are malleable and ductile

Metallic elements are found on the left-hand side of the periodic table. Metals are good conductors of heat and electricity. They are shiny, **malleable** (can be bent) and **ductile** (can be drawn out into a wire).

The elements in group 1 of the periodic table are lithium (Li), sodium (Na), potassium (K), rubidium (Rb), caesium (Cs) and francium (Fr). They are known as alkali metals. Alkali metals are shiny and soft and are the most reactive metals. They react with oxygen, water and acids in varying degrees. These metals become more reactive as you go down the group.

The elements in group 2 of the periodic table are beryllium (Be), magnesium (Mg), calcium (Ca), strontium (Sr), barium (Ba) and radium (Ra). They are known as the alkaline earth metals. Elements in group 2 are shiny, soft, reactive metals. Radium is also radioactive.

Where are metals located in the periodic table?

2 Transition metals have useful and unique properties

In the middle of the periodic table, the elements in groups 3–12 are called the transition metals. Transition metals have a range of physical and chemical properties. They can form many different compounds with other elements. Many of these compounds are brightly coloured.

Figure 13.5 Metals are shiny, malleable, ductile and good conductors of heat and electricity.

Some transition metals are reactive. For example, iron (Fe) forms rust with oxygen and water. Others are good conductors of electricity and heat. For example, copper is used extensively in electrical equipment, such as wiring and motors. Copper (Cu) is also used in roofing and plumbing and has many important industrial uses. Other transition metals are not good conductors or very reactive. For example, although gold (Au) is a good conductor, it's one of the least reactive metals. Gold has many uses, including in jewellery.

Where are transition metals located in the periodic table?

INVESTIGATION 13.3A
Physical properties of elements

INVESTIGATION 13.3B
Chemical properties of elements

3 Non-metals include the halogens and noble gases

On the right-hand side of the periodic table are the non-metals. The two groups of non-metals are the elements of group 17, the halogens, and group 18, the noble gases.

The halogens are the most reactive group of non-metals. The halogens are fluorine (F), chlorine (Cl), bromine (Br), iodine (I) and astatine (At). Halogens react with metals to form ionic compounds. They also react with themselves to form covalent compounds.

The six noble gases in group 18 are helium (He), neon (Ne), argon (Ar), krypton (Kr), xenon (Xe) and radon (Rn). The noble gases are also known as the inert gases because they are very unreactive.

Hydrogen is also a non-metal but it is often put into group 1 with the metals because it has one electron in its valence shell, like the elements of group 1.

Where are non-metals located in the periodic table?

Figure 13.6 Non-metals are dull and brittle, and don't conduct electricity or heat.

CHECKPOINT 13.3

1 Identify the names of groups 1, 2, 17 and 18.

2 a Where on the periodic table are metals located?
b Identify some properties of metals.

3 Identify the reactivity of metals.

4 Why is hydrogen, a non-metal, placed in group 1 with metals?

5 a Where on the periodic table are non-metals located?
b Identify some properties of non-metals.

6 Draw an outline of the periodic table showing where you would find the noble gases, alkali metals, alkaline earth metals, halogens and transition metals.

CHALLENGE

7 Choose an element. Research when the element was discovered, who discovered it, how it is found in nature and its properties and uses.

4 Metalloids are semimetals

On the periodic table, the metals and the non-metals are separated by a group of six elements – boron (B), silicon (Si), germanium (Ge), arsenic (As), antimony (Sb) and tellurium (Te). Polonium (Po) and astatine (At) are also sometimes included in this group.

These elements have a combination of the properties of both metals and non-metals and so are known as **metalloids**. Metalloids are usually brittle, sometimes shiny solids. Most metalloids have important applications in industry. For example, silicon is a semiconductor of electricity – it conducts electricity under certain conditions. For this reason, it is used in electronics.

Where are metalloids located in the periodic table?

SKILLS CHECK

- I can identify properties of both metals and non-metals.
- I can relate properties to where elements are located on the periodic table.
- I can predict the location of an element on the periodic table from its properties.

13.4 Development of the periodic table

At the end of this lesson I will be able to:

- **outline** some examples to show how creativity, logical reasoning and scientific evidence available at the time contributed to the development of the modern periodic table.

KEY TERMS 

chemical property
a property of a substance that is observed during a chemical reaction

Dalton's atomic theory
the theory that states that all matter is made up of tiny particles

physical property
a property of a substance that can be measured, such as colour, melting and boiling points, density and hardness

LITERACY LINK 

Write an A4 page biography about one of the scientists from this section: Dalton, Döbereiner, Mendeleev or Moseley.

NUMERACY LINK 

Construct a timeline showing the development of the periodic table including names of scientists and their contributions. Ensure the timeline is scaled correctly.

The periodic table developed over thousands of years and is the work of many scientists. The Ancient Greek philosopher Aristotle believed that matter consisted of four elements: water, fire, earth and air. Alchemists developed knowledge of many new elements.

With continued research and scientific evidence, scientists started to group elements according to their physical and chemical similarities. This grouping was important in shaping the current periodic table.

1 Dalton produced an early periodic table of 20 elements

English scientist John Dalton proposed his **atomic theory** in 1808. This theory states that all matter is made of tiny particles called atoms. Dalton produced a table of the mass numbers of 20 elements. He used his own symbols (which were quite difficult to remember) for the elements, and many of the atomic masses he used have been found to be incorrect.

Figure 13.7 Dalton produced an early table of 20 elements.

What did Dalton produce?

2 Döbereiner grouped elements into triads

Scientists continued to expand Dalton's table as they discovered new elements. In 1829, German chemist Johann Döbereiner saw a relationship between the properties of the 55 elements known at the time. He observed trends in the **physical** and **chemical properties** of some elements, known as 'triads'. This grouping of elements was the beginning of the periodic table.

What contribution did Döbereiner make to the periodic table?

3 Newlands arranged elements by mass number

In 1864, English chemist John Newlands was the first person to arrange the 60 known elements in order of increasing mass number. Newlands is known for his law of octaves. When he arranged elements in seven groups of eight, every eighth element seemed to have similar properties. However, some elements did not fit the pattern; for example, Newlands put the metal iron into the same group as the non-metals oxygen and sulfur.

Figure 13.8 Newlands arranged elements in order of increasing mass number.

What was Newland's contribution to the periodic table?

4 Mendeleev arranged elements into groups and left gaps

In 1869, Russian chemist Dmitri Mendeleev arranged the known elements in order of mass number, putting the known 'families' into vertical columns.

He also left gaps in the periodic table, for elements that were not yet discovered. Mendeleev predicted that the unknown elements would have similar chemical properties to those of their 'family'. Later, when the elements were discovered, their properties closely matched Mendeleev's predictions.

Figure 13.9 Mendeleev arranged elements by atomic mass into vertical columns, or 'families', and predicted the properties of unknown elements.

In 1869, German chemist Lothar Meyer constructed a similar table to that of Mendeleev by comparing physical properties of elements.

What was Mendeleev's contribution to the periodic table?

5 Moseley arranged elements by atomic number

In 1913, English chemist Henry Moseley suggested that it was the atomic number, not the mass number, that was related to the physical and chemical properties of elements. He refined the previous periodic table and came up with a more accurate version with fewer elements missing.

What was Moseley's contribution to the periodic table?

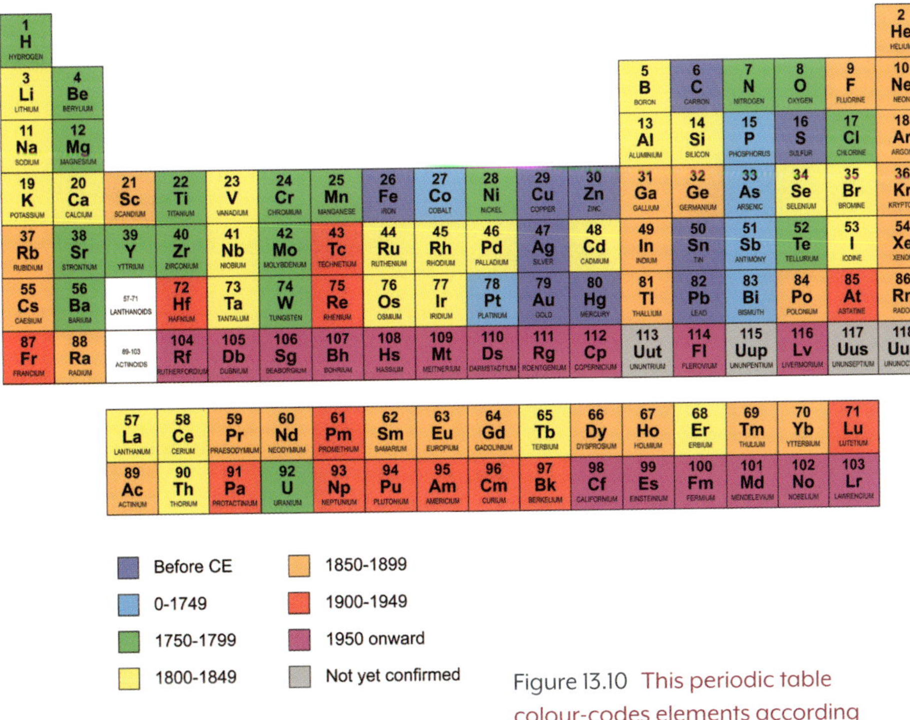

Figure 13.10 This periodic table colour-codes elements according to when they were discovered.

CHECKPOINT 13.4 

1 Identify some differences between the early periodic table and the modern periodic table.

2 How did Moseley's suggestion change Mendeleev's periodic table?

3 Describe the law of octaves proposed by Newlands.

4 Over time, more and more elements have been discovered. Suggest some reasons for this.

5 Is there a pattern in the classification of the elements on the periodic table, from the early table made by Dalton to the current one? Explain your response.

6 Research Mendeleev's periodic table. Compare Mendeleev's original periodic table with the current periodic table. What are the similarities and differences?

CHALLENGE

7 Draw up a table with three columns headed 'Period of discovery', 'Elements in order of discovery' and 'Total number of elements'. Research all the elements in order of their discovery and complete your table. The periods of discovery could be Ancients, 1600s, 1700s, 1800s, 1900s and 2000s.

SKILLS CHECK

- I can identify the main contributions of scientists in the development of the periodic table.
- I can outline the scientific processes used to develop the periodic table.

CHAPTER SUMMARY

The **periodic table** has evolved over many years and groups together elements based on their physical and chemical similarities.

- All matter is made up of **elements**.
- An element is made up of **identical atoms**.
- Every element has a **unique** symbol.

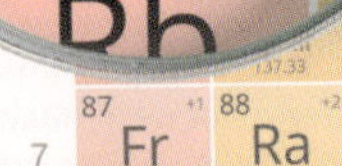

Dalton proposed the atomic theory and created a table of 20 elements and their atomic numbers and masses.

Döbereiner identified the relationships between the 55 known elements at the time, and grouped elements based on their common traits.

Newlands organised the 60 known elements at the time in order of increasing mass number. However, the rows of his table showed too many elemental differences between the columns.

Mendeleev arranged known elements in order of atomic mass into vertical columns, or 'families'. This allowed Mendeleev to predict the properties of unknown elements.

Moseley refined the periodic table in 1913, suggesting that an element's physical and chemical properties were related to the atomic number, not the atomic mass.

Most element symbols are easy to identify (e.g. C for carbon). Some symbols are not as obvious because they originate from the Latin, Greek or German name of the element (e.g. Fe for iron comes from the Latin word *Ferrum*.)

The **atomic number** is the number of protons in the nucleus of an atom.

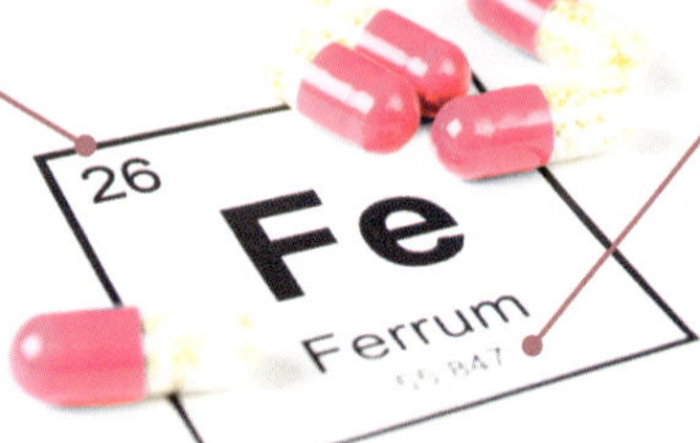

The **mass number** is the number of protons and neutrons in the nucleus of an atom.

Metals:
- are good conductors of heat and electricity
- are shiny, malleable and ductile
- react with oxygen, water and acids (to certain degrees).

Metalloids have properties of both metals and non-metals.

Non-metals:
- some react with metals, or themselves (halogens)
- some are not reactive (noble gases)
- do not conduct electricity and heat
- are dull and brittle.

★ FINAL CHALLENGE ★

1 Identify the symbols for the elements chromium, curium, strontium and antimony.

2 Identify the number of protons, neutron and electrons for the elements calcium, aluminium and xenon.

LEVEL 1

50XP

LEVEL UP!

3 Use the periodic table to determine the information to complete this table for elements 11–20.

Element	Symbol	Atomic number	Mass number	Number of protons	Number of neutrons	Number of electrons

4 Describe some physical and chemical properties that distinguish metals from non-metals.

5 This list shows scientists and their contributions to the discovery of the atom. Draw an arrow between the scientist and their contribution.

Mendeleev	Produced a table showing the symbols and atomic masses of 20 elements
Döbcreiner	Arranged the 60 known elements in order of increasing atomic mass
Dalton	Arranged the known elements in order of mass number, putting the known 'families' into vertical columns
Newlands	Grouped elements belonging to the same 'family' according to their physical and chemical properties
Moseley	Constructed a similar table to that of Mendeleev by comparing physical properties of elements with atomic mass

6 Create a table and summarise the main points under the headings of 'Properties and uses of common metals' and 'Properties and uses of common non-metals'.

7 Predict which group the following elements belong to, based on the characteristics described:

a Element A: shiny, ductile and very reactive

b Element B: good conductor of heat and electricity

c Element C: brittle, shiny, solid.

14 Chemical reactions

Have you ever heard the saying 'your eyes are bigger than your stomach'? Well, the world's largest burger costs $8000 and weighs a whopping 813 kilograms. That's a lot of burger! Sometimes we eat things that are much bigger than our stomachs, so how is that even possible? The answer is chemical reactions. These reactions can be fast or slow, and are involved in our digestion and the processes we use to obtain energy. Chemical reactions for digestion begin straight away, in our mouths, starting to break the food down before it even gets to our stomachs. In our stomach, strong acids break down the food even more, so we can eat more of that burger!

1 LEARNING LINKS

Brainstorm as much as you can remember about these topics.

Many important chemical reactions in non-living systems involve the transfer of energy.

Many important chemical reactions in living systems involve the transfer of energy (e.g. respiration and acid–base reactions in digestion).

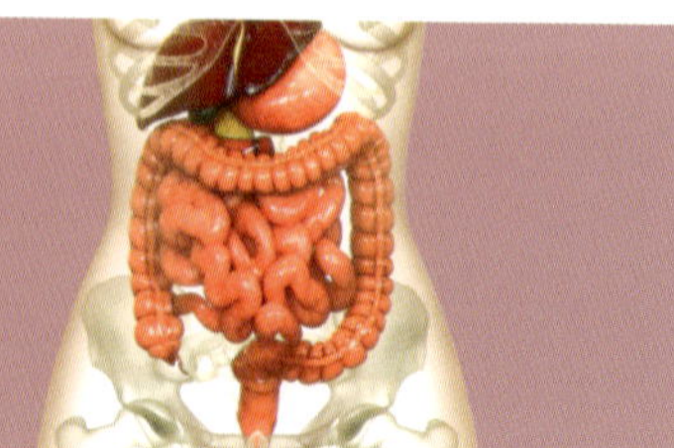

Many organs of the digestive system secrete different substances.

2 SEE–KNOW–WONDER

List three things you can **see**, three things you **know** and three things you **wonder** about this image.

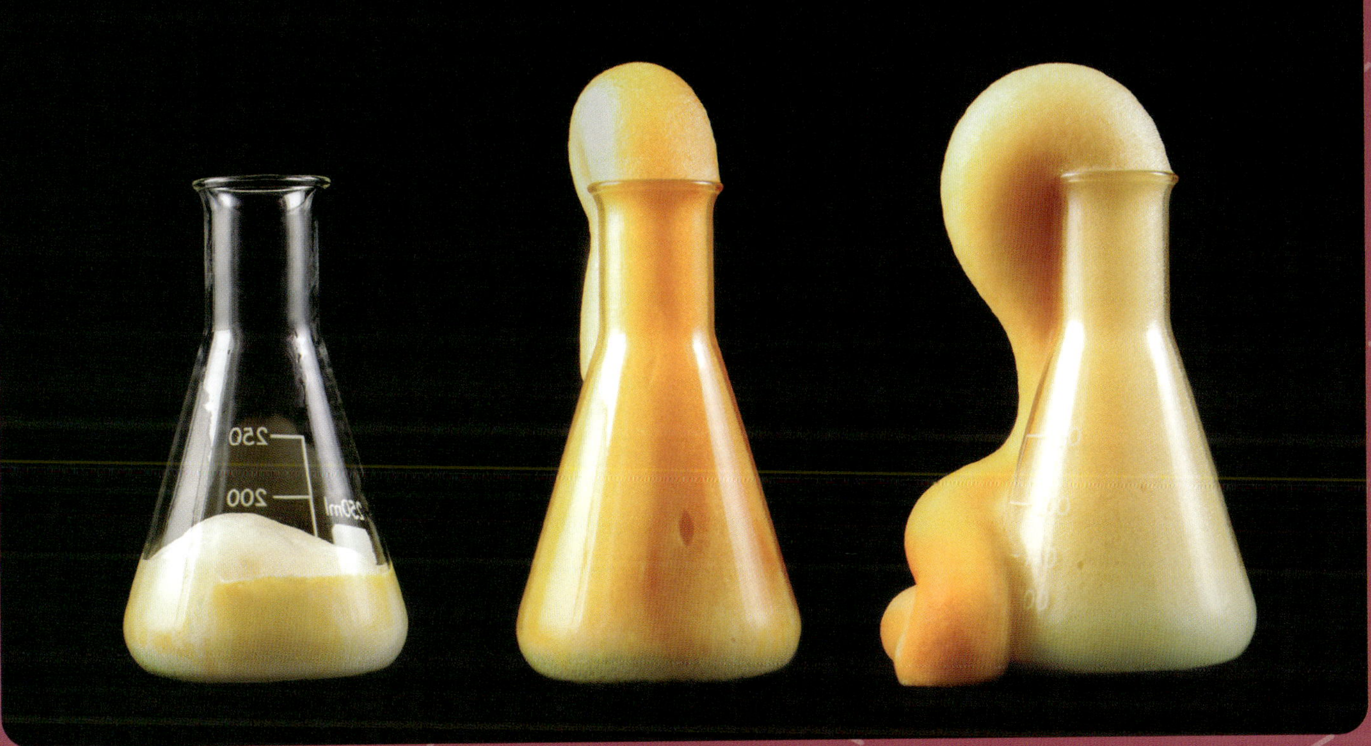

3 CRITICAL + CREATIVE THINKING

The reverse: List things that will never be involved in a chemical reaction.

The brainstorm: Brainstorm a list of ways to prevent cars from rusting.

The different uses: How many different uses can you think of for a very strong acid?

4 THE GREENEST!

The Statue of Liberty, one of the most iconic landmarks in the world, is found in New York. The statue is famous for its green colour – however, it was originally brown! The Statue of Liberty is made of copper, and over the years the copper in the statue reacted with the oxygen in the air to form a coating of copper oxide. It is this copper oxide that gives the statue its familiar green colour. This is an example of a chemical reaction, one that happened very slowly and in front of our very eyes.

14.1 Atoms and matter

At the end of this lesson I will be able to:

- **recall** that all matter is made up of atoms and has mass.

KEY TERMS 

atom
the smallest possible particle of an element

compound
a substance made up of two or more types of atoms bonded together in fixed ratios

element
a pure substance made up of only one type of atom

isotopes
atoms of the same element with the same number of protons but different numbers of neutrons

mass
the amount of matter in an object

matter
physical substance that has mass and takes up space

LITERACY LINK 

Design and write a numbered method for an investigation that aims to prove that all matter has mass. (Hint: You can use a balance.)

NUMERACY LINK 

Consider the formula for sulfuric acid, H_2SO_4. List the elements, their ratios and the total number of atoms in the compound.

Everything in the universe is made up of matter. Matter is anything living or non-living that occupies space and has mass. It can be in the form of a solid, liquid, gas or even plasma. You, water, books, tables, the Sun, Earth, cars and trees are all made up of matter.

Matter is made of atoms, and those atoms determine a material's physical and chemical properties. An atom is the smallest basic unit of matter, so small that you cannot see it even with very strong optical microscopes. So, in order to understand the world around you, you need to visualise what an atom is and how it behaves.

1 Matter is made up of different types of atoms

Everything is made up of a combination of different types of **atoms**. Atoms of identical size, mass, and similar properties are called elements. These are listed in the periodic table of elements.

Atoms consist of a positive nucleus containing protons and neutrons surrounded by electrons orbiting the nucleus. The mass of the atom is concentrated inside its nucleus and is the sum of the mass of the protons and neutrons.

What are the main components of an atom?

Figure 14.1 Matter is made up of atoms. Atoms contain protons and neutrons inside the nucleus and electrons in energy levels outside the nucleus.

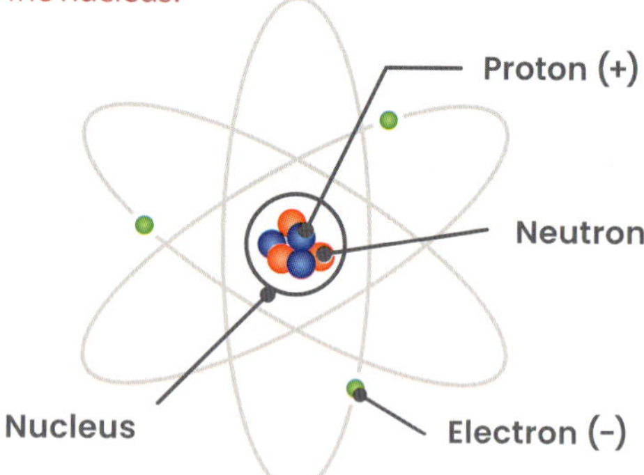

2 Pure substances can be elements or compounds

Elements are pure substances that contain the same type of atoms. For example, the metal sodium contains only sodium atoms, each with 11 protons, 11 electrons and 12 neutrons. All of the known elements are listed in the periodic table of elements.

Most of the elements are formed naturally; however, some elements, such as technetium, californium and neptunium, have been made by scientists and only exist for a short time. These elements are made in a nuclear reactor or particle accelerator.

A **compound** is a chemical substance made up of atoms of two or more elements held together in fixed ratios by chemical bonds. Some examples of compounds are water (H_2O), carbon dioxide (CO_2), glucose ($C_6H_{12}O_6$) and salt (NaCl).

What is a compound?

Figure 14.2 New elements can be made in particle accelerators.

3 Matter has mass

Mass is the amount of **matter** in an object. A small object such as a copper bell might have a lot of mass. A large object such as a balloon filled with helium might have very little mass.

It's important to remember that mass is not weight. Mass is a measure of the matter in an object, while weight is a measure of gravity's pull on an object. An object's mass is constant; you can only change its mass by changing the object, such as breaking off a piece of it.

What is the difference between mass and weight?

4 Isotopes are different forms of the same element

Some elements exist in different forms called **isotopes**. Isotopes have different numbers of neutrons in the nuclei of their atoms. This means that isotopes have different mass numbers and some different properties. For example, many isotopes are radioactive. However, all isotopes of an element have the same atomic number and occur in the same place on the periodic table.

The isotopes of hydrogen (Figure 14.3) have one proton but different numbers of neutrons. They also have different names.

What is an isotope?

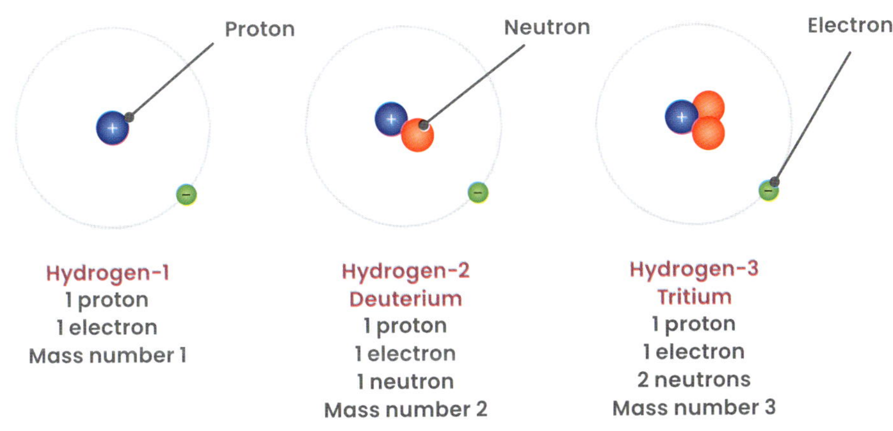

Figure 14.3 The isotopes of hydrogen

CHECKPOINT 14.1 

1 Explain what matter is.
2 Describe the difference between an element and a compound.
3 Identify two elements and two compounds.
4 Identify the different parts of an atom.
5 Draw the atomic structure of an atom and label the nucleus, protons, electron and neutrons.
6 What is an isotope, and how do isotopes of the same element differ?
7 Identify the isotopes of hydrogen.
8 State the number of protons, electrons and neutrons in the carbon isotopes C-12, C-13 and C-14.

CHALLENGE

9 Research to find out why scientists made technetium.

SKILLS CHECK

- I can explain what matter is.
- I can describe the structure of an atom.
- I can explain what an isotope is.

14.2 Bonds between atoms

At the end of this lesson I will be able to:

- **recall** that all matter is made up of atoms and has mass
- **classify** compounds into groups based on their bonding type.

KEY TERMS

covalent bond
a bond in which two atoms share one or more pairs of electrons

ion
a charged atom

ionic bond
a bond in which some atoms gain and some lose electrons, becoming ions

lattice
an interlaced structure or pattern

metallic bond
a bond in which free electrons move around metal ions

molecule
the smallest unit of a covalent compound that can take part in a chemical reaction

valency
an element's power to combine with other atoms

LITERACY LINK

Explain the meanings of the terms *molecule* and *compound* and use examples to show the difference between them.

All matter is made up of atoms, but atoms are incredibly small particles. Any piece of matter that you can see, even something as small as a grain of sand, contains millions of atoms.

In order for atoms to stick together to make a compound, they need to form bonds with other atoms. These bonds are formed by the atoms' electrons, and there are three different types of bonds that can form.

1 Covalent compounds are bound together by shared electrons

Electrons move around the nucleus of atoms in different orbits or 'shells'. The electrons in the outermost shell, called the **valence** shell, move around to bond atoms together.

When two or more non-metallic atoms bond, they share electrons in their valence shells. The shared electrons orbit the nuclei of all the bonded atoms, producing a **molecule**. This kind of bond is called a **covalent bond**, and compounds formed this way are called covalent compounds.

While a few non-metallic elements, such as helium, naturally occur as single atoms, most non-metallic elements appear in nature as molecules containing two or more atoms.

Table 14.1 Some common molecules and their formulas

Name	Formula
Ammonia	NH_3
Carbon dioxide	CO_2
Chlorine	Cl_2
Glucose	$C_6H_{12}O_6$
Hydrogen	H_2
Ozone	O_3
Water	H_2O

Pure substances

Elements

Compounds

Atomic

Molecular

Molecular

Ionic

Atoms of an element, e.g. helium (He)

Molecules of an element, e.g. oxygen (O_2)

Molecules of a compound, e.g. water (H_2O)

Lattice of a compound, e.g. salt (NaCl)

Figure 14.4 Classification of elements, molecules and compounds.

A molecule is the smallest unit of a covalent compound that can take part in a chemical reaction. Some examples of molecules are water, oxygen, nitrogen and glucose. Some more molecules and their formulas are listed in Table 14.1.

How are covalent compounds formed?

2 Ionic compounds are bound together by electron loss and gain

Some atoms gain or lose electrons when they bond, creating **ions**. Ions are still atoms, but they are atoms with an electrostatic charge. When atoms lose electrons, they form positively charged ions; when they gain electrons, they form negatively charged ions.

When a metal atom bonds with a non-metal atom, the metal loses electrons and the non-metal gains them. This creates an ionic compound, such as sodium chloride (NaCl), made up of positive and negative ions. Rather than individual molecules, the ions arrange themselves into a **lattice**. The **ionic bonds** in an ionic compound are very strong, because of the electrostatic attraction between the positive and negative ions.

When atoms lose electrons, they form positively charged ions; when they gain electrons, they form negatively charged ions. Ions are shown by writing a + or – sign at the upper-right of the atomic symbol, such as Na^+ and Cl^-. Some ions are formed by gaining or losing more than one electron. We indicate this by writing a number in front of the + or –, such as O^{2-} or Hg^{2+}.

How are ionic compounds formed?

3 Metal atoms are bound together by free electrons

In metallic bonding, all of the metal atoms lose electrons, rather than gain or share them. These lost electrons form a 'sea' of free electrons that surrounds the positive metal ions. The metal ions are closely packed together and are held in place by their attraction to the sea of electrons.

Metallic bonds tend to be weaker than ionic bonds, but the large numbers of free electrons make metallic compounds good conductors of heat and electricity.

How are metals formed?

Figure 14.5 Free electrons move easily around metal ions in a metal.

CHECKPOINT 14.2

1 Explain the difference between elements and compounds.
2 Explain what a compound is.
3 Name three compounds.
4 Draw a table that shows three molecules found in air and write the formula of each next to them.
5 How are ions formed?
6 Glucose has the formula $C_6H_{12}O_6$. How many atoms of carbon, hydrogen and oxygen are there in one molecule of glucose?
7 How are covalent bonds formed?
8 Which are stronger – ionic bonds or covalent bonds?
9 Do metals form positive or negative ions when they lose electrons?

CHALLENGE

10 Compare and contrast ionic, metallic and covalent bonds.

SKILLS CHECK

- I can explain what a compound is.
- I can classify compounds into groups based on common chemical characteristics.

14.3 Identifying chemical compounds

At the end of this lesson I will be able to:

- **identify** a range of compounds using their names and chemical formulas.

KEY TERMS 

chemical formula
chemical symbols showing the ratio of elements to one another

polyatomic ion
two or more ions bonded together and acting as a single charged unit

prefix
a word or number placed before another word

LITERACY LINK 

Create a poster to use as a reference when naming chemical compounds.

NUMERACY LINK 

If the formula for a particular ionic compound is X_2O_3. Determine the charge (valency) of X.

A chemical formula is a shorthand way of writing the name of an element or compound. It tells you the number and type of atoms in a compound. You can then use these formulas to write chemical equations that model reactions.

1 There are rules for naming covalent compounds

While many compounds have simple, everyday names – water, vinegar, bleach – they also have scientific names. These two-word names are based on their **chemical formula**.

To make sure formulas can be understood in every language, scientists follow four rules when naming compounds.

1 Determine whether the compound is ionic or covalent. If it contains a metal, it is ionic. If it contains two or more non-metals, it is covalent.
2 If it is covalent, name the element that comes first in the formula.
3 Name the second element, making sure it ends with 'ide'; for example, carbon fluoride (CF), hydrogen chloride (HCl).
4 If there is more than one of each type of atom in the formula, add a **prefix** to the element's name. Read the first element normally, but add the prefix before the second element; for example, phosphorus pentoxide (P_2O_5), carbon dioxide (CO_2). For example, the chemical formula for water is H_2O. Following our naming rules, (1) we can see that it contains no metals and so is covalent, (2) the hydrogen comes first, (3) the oxygen becomes oxide, and (4) there are two hydrogens, and so these need the prefix 'di'. So, water's scientific name is dihydrogen monoxide.

Table 14.2 Number of atoms and the prefixes to use

Number of atoms	1	2	3	4	5	6	7	8
Prefix	mono	di	tri	tetra	penta	hexa	hepta	octa

Why do scientists follow rules when naming compounds?

Figure 14.6 Water, vinegar, baking soda and bleach are the common names for, respectively, dihydrogen monoxide, ethanoic acid, sodium bicarbonate and sodium hypochlorite.

❷ Ionic compounds require additional rules

Although the basic rules for naming compounds are simple, it can become more complicated when naming ionic compounds.

The first change is that the metal atom in the compound is always named first. Some ionic compounds are made up of more than two elements. In this case, there is a group of elements bonded together into a **polyatomic ion**. When this happens, the metal is named first, then the polyatomic ion, which has its own name; for example, sodium sulfate (Na_2SO_4), copper hydroxide ($Cu(OH)_2$). (These names don't always follow the 'ide' rule.)

Table 14.3 Common polyatomic ions

Name	Formula
Ammonium	NH_4^+
Carbonate	CO_3^{2-}
Hydroxide	OH^-
Nitrate	NO_3^-
Nitrite	NO_2^-
Phosphate	PO_4^{3-}
Sulfate	SO_4^{2-}
Hydrogen sulfate	HSO_4^-
Sulfite	SO_3^{2-}

To determine the chemical formula for an ionic compound, you can use the swap-and-drop method.

Step 1 Write the ions with the positive ion on the left and the negative ion on the right.

Step 2 Swap the charge number on the left-hand side to become a subscript on the right-hand side.

Step 3 Swap the charge number on the right-hand side to become a subscript on the left-hand side.

Step 4 Simplify, using brackets and the lowest common factor.

For example, calcium carbonate is a compound in which calcium (Ca^{2+}) is bonded with carbonate (CO_3^{2-}). To determine the chemical formula of calcium carbonate, using the swap and drop method:

Step 1
Write the ions:

Ca^{2+} and CO_3^{2-}

Step 2
Swap left-hand charge:

Ca and $(CO_3^{2-})_2$

Step 3
Swap right-hand charge:

Ca_2 $(CO_3)_2$

Step 4
Simplify:

$CaCO_3$

Some metals may have different ionic forms. Add roman numerals after the name of the metal to show the ionic form. For example, iron can exist as Fe^{2+} and Fe^{3+}, which form different compounds – iron(II) chloride is $FeCl_2$, while iron(III) chloride is $FeCl_3$.

What is a polyatomic ion?

Figure 14.7 Iron can have different ionic forms: iron(II) and iron(III). Both are soluble in water but they form different coloured solutions.

CHECKPOINT 14.3 

1 Explain why chemical formulas are used instead of chemical names.

2 When naming a compound, which element should you write first?

3 Name these compounds.
- a FeS
- b KBr
- c $PbSO_4$
- d $CuCO_3$
- e $Ca(OH)_2$

4 Write the chemical formulas for these compounds.
- a potassium nitrate
- b barium sulfate
- c sodium oxide
- d ammonium chloride

5 Explain what a polyatomic ion is.

CHALLENGE

6 Research where acetylsalicylic acid, calcium hydroxide, calcium carbonate and monosodium glutamate are found in your home.

SKILLS CHECK

- I can identify a range of compounds using their names and chemical formulas.
- I can use rules to name chemical compounds.

14.4 Writing chemical equations

At the end of this lesson I will be able to:

- **construct** word equations from observations and written descriptions of a range of chemical reactions
- **deduce** that new substances are formed during chemical reactions by rearranging atoms rather than creating or destroying them.

KEY TERMS 

coefficient
a number placed before the chemical in a formula or chemical equation

conservation of mass
mass cannot be created or destroyed

product
a substance formed in a chemical reaction

reactant
a substance that starts a chemical reaction

LITERACY LINK 

Create a flow chart to demonstrate the steps you would use to balance a chemical reaction.

NUMERACY LINK 

Refer to the equation
$N_2 + 3H_2 \rightarrow 2NH_3$

If there are 3 molecules of N_2, how many molecules of H_2 are required, and how many molecules of NH_3 are produced?

One of the unbreakable laws of the universe is that matter cannot be created or destroyed; this is the law of **conservation of mass**. One of the effects of this law is that the amount of matter at the start of a chemical reaction is always the same as the amount of mass left at the end. The atoms and molecules may be different, but the total mass never changes.

An important application of this law is that it lets us model a chemical reaction by writing a chemical equation, which is very similar to an equation in maths. A chemical equation shows what atoms are involved before the reaction and how they recombine afterwards.

1 Start by listing the reactants and the products

The first step in writing a chemical equation is to list all of the substances involved. The **reactants** are the substances that you have before the reaction takes place, while the **products** are the substances you will have after the reaction has finished.

Write the reactants on the left side of an arrow, and the products on the right side, using plus signs to separate each material on each side. Although we use formulas in the equation, it's often easiest to start by writing the names of the materials. This is known as a word equation. You can then write the formulas underneath the words.

A simple chemical reaction is the one in which nitrogen and hydrogen combine to form ammonia, which is an important component of fertilisers. This takes place at high temperatures. The chemical equation for this reaction would start off looking like this:

nitrogen + hydrogen → ammonia

$$N_2 + H_2 \rightarrow NH_3$$

What is the difference between a reactant and a product?

Figure 14.8 Producing ammonia fertiliser from nitrogen in the air allows farmers to grow more crops.

❷ Balance the number of atoms on each side

The law of conservation of mass says that matter cannot be created or destroyed. Whenever it looks as though matter has been destroyed, it has actually been lost in the form of gas.

In a chemical reaction, the atoms of the reactants rearrange to form the molecules of the products. Because no new atoms can be created or destroyed, this means that the number of atoms of each element must be the same on both sides of the equation. If the numbers do not match, you need to balance the equation by adding more molecules to one or both sides. Do this by adding **coefficients** (numbers) in front of the reactants or products to show that there are more molecules of that substance.

Consider the ammonia formation equation:

$$N_2 + H_2 \rightarrow NH_3$$

Step 1:

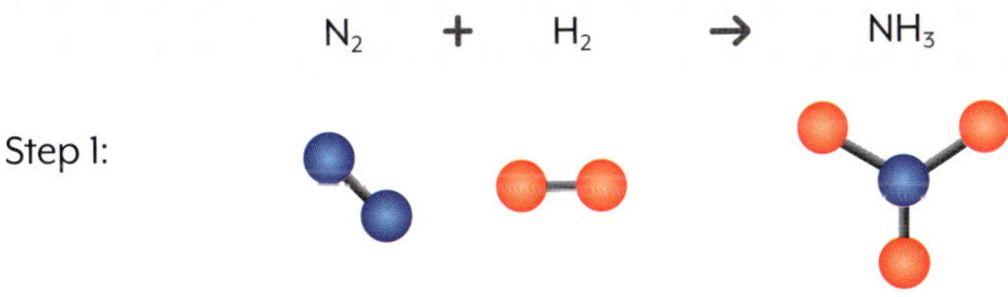

This equation is unbalanced because there is not the same number of nitrogen atoms and hydrogen atoms on each side.

First, balance the number of nitrogen atoms by adding a 2 in front of the NH_3 on the right-hand side:

$$N_2 + H_2 \rightarrow 2NH_3$$

Step 2:

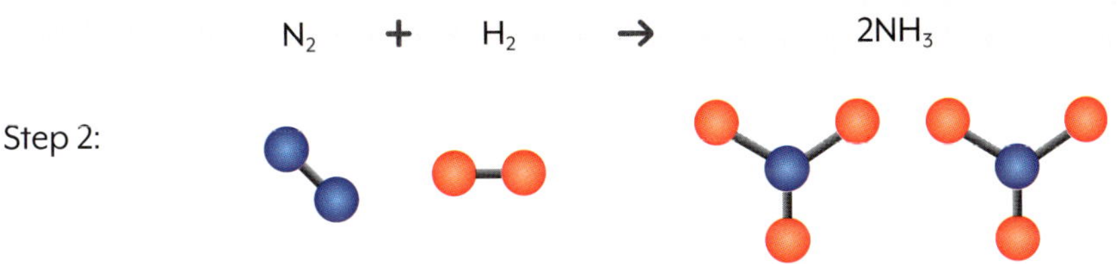

There are now two nitrogen atoms on each side of the equation. However, there are now two hydrogen atoms on the left-hand side and $2 \times 3 = 6$ hydrogen atoms on the right-hand side. To balance it, put a 3 in front of the H_2 on the left-hand side:

$$N_2 + 3H_2 \rightarrow 2NH_3$$

Step 3:

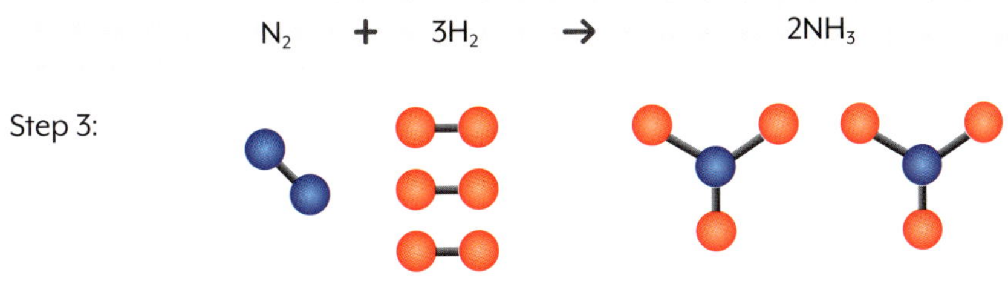

The equation is now balanced, with the same number of atoms on each side. The equation also shows us that one nitrogen molecule and three hydrogen molecules react to form two ammonia molecules.

Why do we need to balance the atoms on each side of a chemical equation?

CHECKPOINT 14.4 

1 Explain what the law of conservation of mass is and how it applies to chemical reactions.

2 Explain what the reactants and products are in a chemical reaction.

3 Consider this chemical reaction

$Fe_2O_3 + C \rightarrow Fe + CO_2$

Identify the:

a products

b reactants

c number of iron atoms in each of the reactants and products

d number of oxygen atoms in each of the reactants and products.

4 Balance these equations:

a $Fe_2O_3 + C \rightarrow Fe + CO_2$

b $P_4 + O_2 \rightarrow P_2O_5$

c $NaBr + Cl_2 \rightarrow NaCl + Br_2$

CHALLENGE

5 Research and identify a situation where matter is converted into energy during a chemical reaction.

SKILLS CHECK

- I can balance chemical equations by making sure that the number and types of atoms are the same on both sides of the arrow.
- I can construct word equations from observations and written descriptions of a range of chemical reactions.
- I can explain the law of conservation of mass and how it applies to chemical reactions.

14.5 Acids and bases

At the end of this lesson I will be able to:

- **classify** compounds into groups based on common chemical characteristics.

KEY TERMS 

caustic
able to burn or corrode organic tissue through chemical action

concentration
the amount of a substance in a volume of solution

corrosive
highly reactive and damaging or destructive to another substance

neutralise
to make something chemically neutral

pH
a figure expressing the acidity or alkalinity of a solution

LITERACY LINK 

Create a Venn diagram to compare and contrast characteristics of acids and bases.

NUMERACY LINK 

The pH scale goes from 1 to 14, where each number differs by a factor of 10. For example, pH 2 is ten times more acidic than pH 3.

a How much more acidic is pH 5 than 7?

b Compare the acidity levels of pH 9 and 6.

Chemical compounds can be grouped according to their common characteristics. Two of the most common groups are acids and bases. They are used in cleaning products, swimming pools and kitchens. Many of the foods we eat are either acidic or basic, and acids and bases help digest food.

Figure 14.9 Weak acids and bases occur naturally in some foods.

❶ Acids produce large amounts of hydrogen ions

An acid is a **corrosive** chemical substance that produces hydrogen ions (H^+) when mixed with water. The hydrogen ions can react with the other substances to produce salts (ionic substances) and other substances such as water and hydrogen gas.

The higher the **concentration** of hydrogen ions produced by an acid, the higher its acidity. Strong acids are very dangerous, especially when they contact skin and eyes. That's why you must always wear protective clothing and eyewear when working with acids in the lab.

Weak acids are much safer to use, and are important in our diet. Citrus fruits such as oranges and lemons contain a weak acid, called citric acid, which contributes to their sour flavour. Soft drinks also contain weak acids, as does coffee. There's even some very, very weak acid in milk.

What happens when an acid is mixed with water?

❷ Bases produce large amounts of hydroxide ions

A base is a substance that reacts with an acid to **neutralise** it. Bases that dissolve in water are called alkalis. When mixed with water, bases produce hydroxide ions (OH^-), which are one atom of oxygen bonded to one atom of hydrogen, with a negative charge.

Although bases are not acidic, they are **caustic**, and can be just as dangerous as acids. Household cleaning products are strong bases. Bases such as sodium hydroxide and ammonia react with oils and fats, which is why they are used in many household cleaners. Weak bases are found in toothpaste, conditioners, antacid tablets and baking powder.

There are three types of bases.

- Metal hydroxides (e.g. potassium hydroxide) contain metals bonded with hydroxide (OH^-)
- Metal oxides (e.g. zinc oxide) contain metals bonded with oxide (O_2^-)
- Metal carbonates (e.g. copper carbonate) contain metals bonded with carbonate (CO_3^{2-}).

What are alkalis?

INVESTIGATION 14.5A
Action of acids and bases – cleaning coins

INVESTIGATION 14.5B
Acid or base?

INVESTIGATION 14.5C
The effect of indicators on acids and bases

Figure 14.10 This symbol is used on containers to let people know that the chemical inside is corrosive and should be handled with care.

3 The pH scale measures acidity

The acidity of a solution is measured on a scale called the **pH** scale. Acids have a low pH, while bases have a high pH.

A solution with a pH of 7 is neutral, while a solution with a pH below 7 is an acid and a solution with a pH above 7 is a base.

You can measure pH by adding an indicator to a solution and matching its colour to a chart or using a pH meter. Indicators are substances that change to different colours depending on whether they're mixed with an acid or a base. One of the most common indicators is litmus, which turns red when mixed with acids and blue when mixed with bases.

What is the pH of a neutral substance?

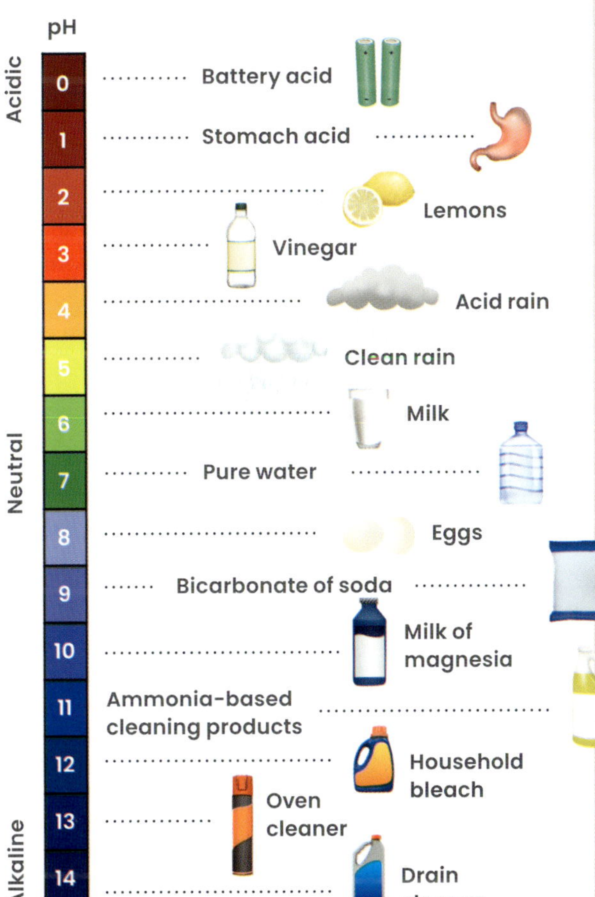

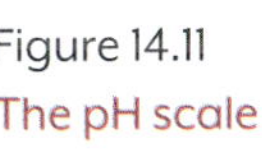

Figure 14.11
The pH scale

CHECKPOINT 14.5

1 List three properties of acids.
2 Identify three acids.
3 List two properties of bases.
4 Identify three bases.
5 What is the pH scale?
6 Identify the pH of water.
7 Copy and complete these sentences.
   Acids have a pH range of ________. Bases have a pH range of ________.
8 Explain why strong acids and bases can be dangerous.

CHALLENGE

9 Research the chemical behaviour of indicators in acids and bases and identify the strongest acid on Earth. Give some examples of how corrosive it is.

SKILLS CHECK

- I can explain what acids and bases are and give examples of each.
- I can describe the pH scale, including the range for acids and bases.

14.6 Acid reactions

At the end of this lesson I will be able to:

- **investigate** a range of important chemical reactions that occur in non-living systems and involve energy transfer, including neutralisation and the reaction of acids.

KEY TERMS 

carbonate
a substance containing the elements carbon and oxygen

neutralisation reaction
a reaction involving an acid and a base to produce water and a salt

strong acid
an acid that ionises completely in water

weak acid
an acid that partially ionises (loses or gains electrons) in water

LITERACY LINK 

Write a short creative story that includes acids, bases and neutralisation.

NUMERACY LINK 

Magnesium was added to hydrochloric acid at different temperatures and the time of reaction recorded:

Temp (°C)	0	15	25	40
Time to react (s)	36	21	15	14

Construct a line graph for this data.

Acids can be dangerous substances because they are corrosive. Some metals are very reactive when exposed to acids, producing hydrogen gas and metallic salts. But when an acid and a base react, they neutralise each other, making the products safer.

1 Acids and bases neutralise each other

In a **neutralisation reaction**, an acid and a base react to form a salt and water. This happens because acids are a source of hydrogen ions (H^+), while bases are a source of hydroxide ions (OH^-).

In an acid–base neutralisation reaction, the H^+ from the acid and the OH^- from the base combine to form water, which is neutral and has a pH of 7:

$$\text{Hydrogen} + \text{hydroxide} \rightarrow \text{water}$$
$$H^+ + OH^- \rightarrow H_2O$$

The other parts of the acid and base combine to produce a salt.

The reaction of a **strong acid** with a strong base results in a neutral solution with a pH of 7 and a neutral salt. For example, hydrochloric acid reacts with sodium hydroxide to form water and the neutral salt sodium chloride (NaCl):

$$\text{Hydrochloric acid} + \text{sodium hydroxide} \rightarrow \text{water} + \text{salt}$$
$$HCl + NaOH \rightarrow H_2O + NaCl$$

The reaction of a strong acid with a weak base produces a solution with pH < 7, containing water and an acidic salt. The reaction of a strong base with a **weak acid** produces a solution with pH > 7, containing water and a basic salt.

What is a neutralisation reaction?

2 Acids react with metals to produce salts and hydrogen

Acids react with metals. Some metals are more reactive than others. You can see this in the reactivity series of metals in Table 14.4.

The metals at the top of the reactivity series, such as potassium and sodium, react violently, while those at the bottom react very little. This is one reason why gold and silver are used to make jewellery, rather than iron and zinc. Gold and silver are chemically very unreactive, and they keep their shiny surface even when exposed to acids and oils in the air or on your skin.

Table 14.4 The reactivity of different metals with acids

Metal	Reactivity
Potassium (K)	Most reactive
Sodium (Na)	
Calcium (Ca)	
Magnesium (Mg)	
Aluminium (Al)	
Zinc (Zn)	
Iron (Fe)	
Tin (Sn)	
Lead (Pb)	
Copper (Cu)	
Silver (Ag)	
Gold (Au)	Least reactive

Acids react with metals to form salt and hydrogen gas. More reactive metals react faster, which you can see by how fast the hydrogen gas bubbles are released.

Acid + metal → salt + hydrogen

The salt formed depends on the acid the metal reacts with. For example, magnesium reacts with hydrochloric acid to produce magnesium chloride and hydrogen gas:

Hydrochloric acid	+	magnesium	→	magnesium chloride	+	hydrogen
2HCl	+	Mg	→	$MgCl_2$	+	H_2

Which metals are more reactive than magnesium?

3 Acids react strongly with metal carbonates

A metal **carbonate** is a compound containing metal, carbon and oxygen. Acids react with metal carbonates to form salts, carbon dioxide and water.

Acid + metal carbonate → salt + carbon dioxide + water

For example, calcium carbonate (a compound used to settle an upset stomach) reacts with hydrochloric acid to produce calcium chloride, carbon dioxide gas and water:

Hydrochloric acid	+	calcium carbonate	→	calcium chloride	+	carbon dioxide	+	water
2HCl	+	$CaCO_3$	→	$CaCl_2$	+	CO_2	+	H_2O

Carbonates give off the gas carbon dioxide when they react with acids. This is why people burp when they take calcium carbonate to settle an upset stomach. The carbonate reacts with the stomach acids to produce carbon dioxide, which fills up the stomach and has to be released by burping.

To test if the gas released in a reaction is carbon dioxide, bubble the gas through limewater. Carbon dioxide turns limewater milky or cloudy.

How can we test if a gas is carbon dioxide?

Figure 14.12 Calcium carbonate reacts with hydrochloric acid to produce calcium chloride and carbon dioxide, which you can see as bubbles.

INVESTIGATION 14.6A
Reactions of acids with metals

INVESTIGATION 14.6B
Reactions of acids with carbonate

CHECKPOINT 14.6

1 Write the general neutralisation equation.
2 Identify the salt formed when sulfuric acid reacts with zinc.
3 Describe what happens when an acid reacts with a metal.
4 a Which metal is more reactive – magnesium or aluminium?
  b How could you show your answer to part a was true?
5 Identify the gas formed when a metal carbonate reacts with an acid.
6 Identify the salt formed when nitric acid reacts with magnesium carbonate.
7 Explain how calcium carbonate can help an upset stomach.

CHALLENGE

8 Design an experiment to see if increasing the concentration of hydrochloric acid affects the rate of its reaction with zinc metal. The concentrations of hydrochloric acid given are 0.2, 1.0 and 1.5 mol/L. In your answer, include the controlled, independent and dependent variables.

SKILLS CHECK

- I can describe what happens when an acid reacts with a base.

14.7 Combustion reactions

At the end of this lesson I will be able to:

- **investigate** a range of types of important chemical reactions that occur in non-living systems and involve energy transfer, including combustion.

KEY TERMS

combustion
a reaction that involves burning in the presence of oxygen to release heat

hydrocarbon
a compound made up of only hydrogen and carbon

oxidation
a reaction taking place in the presence of oxygen

soot
a black form of carbon formed by incomplete combustion

transform
change from one form to another

LITERACY LINK 

Summarise this section into a postcard addressed to your teacher.

NUMERACY LINK 

Balance the following complete combustion reactions:

$CH_4 + O_2 \rightarrow CO_2 + H_2O$

$C_4H_8 + O_2 \rightarrow CO_2 + H_2O$

Energy is essential for life. Biological reactions and chemical reactions involve energy transfer. Some chemical reactions need energy to start them and others release energy in the form of heat, light and sound. Combustion, the reaction of acids with metals and neutralisation reactions are some reactions that release energy.

Figure 14.13 Energy transformation occurs when a match is struck.

1 Combustion reactions involve oxygen

An **oxidation** reaction is a reaction that occurs in the presence of oxygen. There are two types of oxidation reactions – combustion and corrosion. A **combustion** reaction takes place between a compound and oxygen to produce heat and a new product. The most obvious example of a combustion reaction is the burning of wood in a fire; another example is the respiration that happens within your cells.

All of the products formed in combustion reactions are oxides. When a metal is burnt in oxygen, a metal oxide is formed and energy is released. For example, burning magnesium in oxygen produces heat and light energy and a white solid product called magnesium oxide.

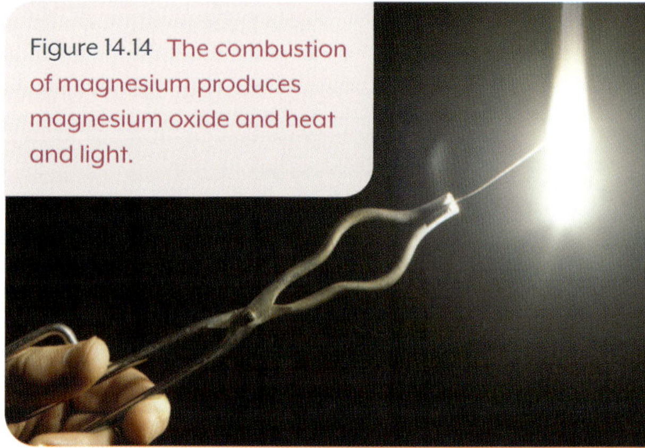

Figure 14.14 The combustion of magnesium produces magnesium oxide and heat and light.

Magnesium + oxygen → magnesium oxide

When a hydrocarbon such as natural gas or methane is burnt, the products are usually carbon dioxide and water. This is because **hydrocarbons** only contain carbon and hydrogen. Methane undergoes a combustion reaction in the presence of oxygen to form carbon dioxide and water:

Methane + oxygen → carbon dioxide + water

We use combustion reactions to provide energy for our daily activities. Power plants burn coal to provide us with electricity. The combustion of petrol in cars provides the energy to make them move. In a combustion reaction, most of the energy is transformed from chemical energy to heat energy.

What is an example of a combustion reaction?

INVESTIGATION 14.7A
Combustion of fuels

INVESTIGATION 14.7B
Complete and incomplete combustion reactions

❷ Incomplete combustion reactions produce carbon monoxide and carbon

If there is plentiful oxygen, complete combustion of hydrocarbons occurs. But if there is a limited supply of oxygen, incomplete combustion may occur. Instead of producing carbon dioxide and water, incomplete combustion produces carbon monoxide and water or carbon (**soot**) and water.

For example, the complete combustion reaction of propane (C_3H_8) requires five molecules of oxygen for every molecule of propane:

Propane	+	oxygen	→	carbon dioxide	+	water
C_3H_8	+	$5O_2$	→	$3CO_2$	+	$4H_2O$

If there are only 3.5 oxygen molecules for every molecule of propane, incomplete combustion will produce carbon monoxide and water:

Propane	+	oxygen	→	carbon monoxide	+	water
C_3H_8	+	$3.5O_2$	→	$3CO$	+	$4H_2O$

If there are only two oxygen molecules for every molecule of propane, the reaction will produce carbon and water:

Propane	+	oxygen	→	carbon	+	water
C_3H_8	+	$2O_2$	→	$3C$	+	$4H_2O$

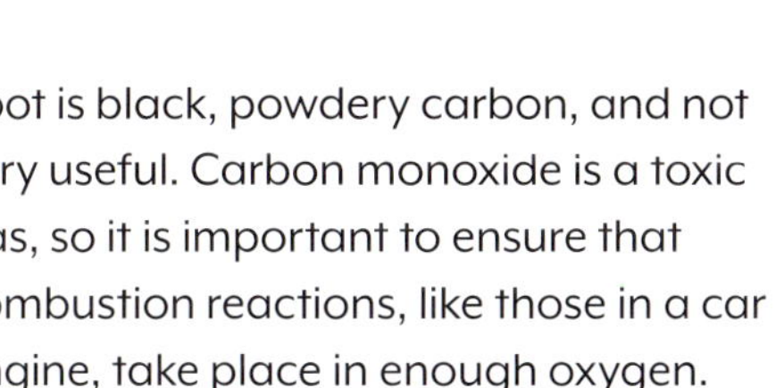

Soot is black, powdery carbon, and not very useful. Carbon monoxide is a toxic gas, so it is important to ensure that combustion reactions, like those in a car engine, take place in enough oxygen.

What is the difference between combustion and incomplete combustion?

Figure 14.15 The combustion of rocket fuel powers this shuttle into space.

CHECKPOINT 14.7

1 Explain what combustion is.
2 Why is combustion an important chemical reaction?
3 When a metal is burnt in oxygen, what is formed?
4 Identify the products of a complete combustion reaction.
5 When does incomplete combustion occur?
6 Identify the possible products of an incomplete combustion reaction.
7 What is needed for a combustion reaction to start?
8 What is the energy transformation taking place in a combustion reaction?

CHALLENGE

9 Research and provide a summary of the major combustion reactions currently used on Earth and what they are used for.

SKILLS CHECK

- I can explain what combustion reactions are.
- I can give two examples of combustion reactions.

14.8 Corrosion and decomposition

At the end of this lesson I will be able to:

- **investigate** a range of types of important chemical reactions that occur in non-living systems and involve energy transfer, including corrosion and decomposition.

KEY TERMS 

alloy
a mixture of two or more metals

degradation
the breaking down of a substance

LITERACY LINK 

Create a pamphlet aimed at boat owners showing the different ways to prevent their boats from rusting.

NUMERACY LINK 

A scientist analysed an 8.5 g piece of metal and recorded that there was 3.4 g of iron oxide (rust) on it. Calculate the percentage of the metal that had rusted.

Figure 14.16 The iron in this old car has corroded to produce orange-brown rust.

Corrosion is a natural process. It is the gradual **degradation** of metals by chemical reactions with substances in their environment. When metals react with oxygen, compounds form on the surface of metals. The oxygen can be in the air, water or salt water. Most metals corrode, some faster than others. Corrosion affects the properties of a metal structure, such as its strength and appearance.

1 Iron corrodes to create rust

Rusting is the corrosion of iron. The scientific name for rust is iron oxide (Fe_2O_3), and it forms so easily that pure iron is rarely found in nature. Rust forms as a flaky red-brown solid on iron structures. For rusting of iron to occur, you need water and oxygen:

$$\text{Iron} + \text{oxygen} \rightarrow \text{iron oxide}$$

$$4Fe + 3O_2 \rightarrow 2Fe_2O_3$$

Rust causes a lot of damage to buildings, cars and ships because it does not form a protective layer on the metal's surface. Water can get through to the metal underneath, leading to further corrosion. Little bits of rust flake off, leaving the rest of the metal exposed to oxygen. Eventually all of the metal corrodes.

Corrosion and combustion reactions are very similar. They are both oxidation reactions, involving oxygen, and they both give off heat. However, combustion is a fast reaction that gives off a lot of heat, while corrosion may take years and gives off very small amounts of heat.

How is corrosion different from combustion?

2 Corrosion can be prevented in various ways

To prevent corrosion, you need a barrier between the metal surface and the water and oxygen. Some ways to prevent corrosion are:

- painting the metal
- coating the metal with plastic
- coating iron with a protective layer of zinc – galvanising
- coating the metal with another metal – electroplating
- attaching the metal to a more reactive metal, which will be 'sacrificed', leaving the less reactive metal intact – sacrificial protection
- forming an **alloy**, which is a substance made up of two or more metals that is often stronger and more resistant to corrosion. For example, stainless steel is an alloy of iron and chromium or nickel. Stainless steel is slower to corrode than iron, which is why it is used to make pots, pans and utensils
- use passivating metals, which are metals that form inactive surface layers that prevent further corrosion. For example, aluminium reacts with oxygen in the air to form a protective coating of aluminium oxide (Al_2O_3), which acts as a barrier to prevent rust.

What are three ways in which iron can be prevented from rusting?

INVESTIGATION 14.8A
Corrosion of iron

INVESTIGATION 14.8B
Preventing corrosion

3 Decomposition reactions break down substances

In decomposition reactions, one substance breaks down into two or more simpler substances. Most decomposition reactions require energy to get them started.

Thermal decomposition is started by heat energy. This is an easy reaction to see in the laboratory, if you heat a reactive substance over a Bunsen burner. For example, if you heat copper carbonate, it decomposes into copper oxide and carbon dioxide gas:

Copper carbonate	→	copper oxide	+	carbon dioxide
$CuCO_3$	→	CuO	+	CO_2

The carbon dioxide released can be tested by bubbling it through limewater. The limewater should go milky in the presence of carbon dioxide.

Figure 14.17 Heating a metal carbonate causes it to decompose and produce the metal oxide and carbon dioxide gas.

Electrical decomposition is better known as electrolysis. It is usually performed by passing an electrical current through a liquid reactant. For example, if you apply electricity to water, it decomposes into hydrogen and oxygen:

Water	→	hydrogen	+	oxygen
$2H_2O$	→	$2H_2$	+	O_2

Photochemical decomposition is triggered by light energy. This can often be a very slow reaction, so we don't notice it much in our everyday lives. For example, when silver chloride is exposed to light, it slowly decomposes into silver and chlorine ions:

Silver chloride	→	silver ions	+	chlorine ions
$AgCl$	→	Ag^+	+	Cl^-

What is corrosion?

CHECKPOINT 14.8

1 What is corrosion?
2 When is corrosion also known as rusting?
3 Identify the chemical name and formula of rust.
4 What is necessary for rusting to occur?
5 List two methods of preventing corrosion.
6 List three types of decomposition reaction.
7 Give an example for each type of decomposition reaction you listed in question 6 and write the word equation for each.
8 Explain what a passivating metal is and give an example of one.
9 Explain what happens when water undergoes electrolysis.

CHALLENGE

10 You are given four items – a gold coin, an iron nail, a piece of copper plate and some aluminium foil. Which do you think will corrode faster and why? Write a method to test your hypothesis.

SKILLS CHECK

- I can explain what a decomposition reaction is and give some examples of decomposition reactions.

14.9 Precipitation reactions

At the end of this lesson I will be able to:

- **investigate** a range of types of important chemical reactions that occur in non-living systems and involve energy transfer, including precipitation.

A precipitation reaction occurs when two salt solutions are mixed, and two products are formed – an insoluble solid and a soluble salt. The insoluble solid may be **suspended** for a while but eventually settles to the bottom of the container. Precipitation reactions can be used to identify the presence of certain compounds or ions.

Figure 14.18 Solutions are clear liquids. When a precipitate forms, a solid drops out of solution.

KEY TERMS 

aqueous
dissolved in water to create a solution

dissociate
to split apart into ions

precipitate
an insoluble product

spectator ion
an ion that does not take part in the reaction

suspended
distributed throughout a fluid

LITERACY LINK 

Identify some other words you could use instead of *soluble*, *precipitate* and *reaction*.

NUMERACY LINK 

A student repeated a precipitation reaction four times and measured the quantity of lead iodide that had formed. She recorded 0.234 g, 0.247 g, 0.238 g and 0.241 g. Calculate the average mass of the lead iodide for this experiment. Express your answer in mg.

1 Only ionic compounds form precipitates

When ionic compounds are placed in water, the water molecules pull them apart and they **dissociate** into ions. The ions are then free to form new compounds with other ions. This is why only ionic compounds form **precipitates**.

Consider what happens when solutions of potassium iodide and lead nitrate are mixed together. The potassium iodide solution is made up of positive potassium ions (K^+) and negative iodide ions (I^-). The lead nitrate solution is made up of positive lead ions (Pb^{2+}) and negative nitrate ions (NO_3^-). Once these ions are mixed up, a precipitation reaction occurs, and the insoluble lead and iodide ions combine to form a solid. This produces a solution of potassium nitrate, with a precipitate of insoluble yellow lead iodide floating within it (Figure 14.19).

Figure 14.19 When a solution of potassium iodide is added to a solution of lead nitrate, a yellow precipitate of lead iodide is formed.

The potassium and the polyatomic nitrate ion do not take part in the reaction. They are called **spectator ions**, because they are not involved in the reaction. They are shown in the complete ionic equation for the reaction but not in the net ionic equation:

Potassium iodide	+	lead nitrate	→	lead iodide	+	potassium nitrate
2KI	+	$Pb(NO_3)_2$	→	PbI_2	+	$2KNO_3$

Complete ionic equation: $2K^+ + 2I^- + Pb^{2+} + 2NO_3^- \rightarrow PbI_2 + 2K^+ + 2NO_3^-$

Net ionic equation: $Pb_2^+ + 2I^- \rightarrow PbI_2$

What are precipitation reactions?

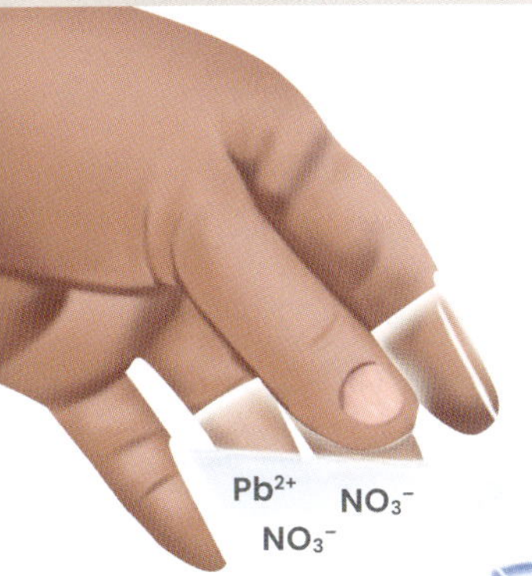

A solution of lead nitrate, which is made up of lead ions and nitrate ions in water, is added to a solution of potassium iodide, which is made up of potassium ions and iodide ions in water.

Lead iodide forms as a bright yellow precipitate because it is insoluble in water. The potassium and nitrate ions stay in solution.

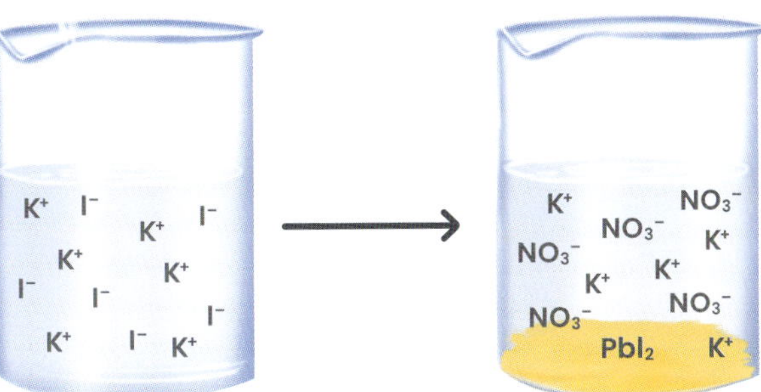

Figure 14.20 Precipitation reaction of potassium iodide with lead nitrate

INVESTIGATION 14.9
Precipitation reactions

❷ Solubility rules help to identify precipitates

If a reaction produces a precipitate, you may want to identify it. You can use solubility rules to help identify a precipitate. The solubility rules are a list of rules for ionic compounds that tell you whether certain ions will form precipitates with other ions or stay in solution. Table 14.5 is a list of solubility rules. As well as these rules, all group 1 salts and all ammonium (NH_4^+) salts are soluble and do not form precipitates.

For example, the rules tell you that in the previous reaction of potassium iodide with lead nitrate, all nitrates are soluble but lead iodide is insoluble. This helps identify the yellow precipitate as lead iodide. Similarly, you can see that sodium chloride (table salt) is soluble but silver chloride is not soluble.

Table 14.5 The solubility rules

Ions	Solubility
Nitrates (NO_3^-)	All soluble
Chlorides (Cl^-)	All soluble except for AgCl, $PbCl_2$ and $HgCl_2$
Iodides (I^-)	All soluble except for AgI, PbI_2 and HgI_2
Sulfates (SO_4^{2-})	All soluble except for $PbSO_4$, $CaSO_4$ and $BaSO_4$
Carbonates (CO_3^{2-})	All insoluble except for group 1 carbonates (e.g. Na_2CO_3, K_2CO_3) and $(NH_4)_2CO_3$
Hydroxides (OH^-)	All insoluble except for group 1 hydroxides (e.g. NaOH, KOH), $Ca(OH)_2$, $Ba(OH)_2$ and NH_4OH

What is an example of a solubility rule?

CHECKPOINT 14.9 

1 What is a precipitation reaction?
2 How can you tell if a precipitation reaction occurs?
3 How would you describe potassium iodide and lead nitrate solutions?
4 What are spectator ions in a precipitation reaction?
5 What does the net ionic equation of a precipitation reaction contain?
6 True or false? All nitrates and solutions of group 1 compounds are soluble.

CHALLENGE

7 Write the net ionic equations for the following reactions. You may use the solubility rules to help you.
a $NaCl + AgNO_3 \rightarrow NaNO_3 + AgCl$
b $CaCl_2 + Na_2CO_3 \rightarrow CaCO_3 + 2NaCl$
c $2NaOH + CuSO_4 \rightarrow Na_2SO_4 + Cu(OH)_2$
d $BaCl_2 + K_2SO_4 \rightarrow BaSO_4 + 2KCl$
e $FeI_2 + 2KOH \rightarrow 2KI + Fe(OH)_2$

SKILLS CHECK

- I can explain what a precipitate is and how they are formed.
- I can use the solubility table to predict which compounds form a precipitate.

14.10 Chemical reactions in living systems

At the end of this lesson I will be able to:

- **identify** some examples of important chemical reactions that occur in living systems and involve energy transfer, including respiration and reactions involving acids such as those that occur during digestion.

KEY TERMS 

aerobic
in the presence of oxygen

anaerobic
in the absence of oxygen

carbohydrase
a type of enzyme that digests carbohydrates

catalyst
substance that speeds up a chemical reaction

lipase
a type of enzyme that digests lipids (fats)

protease
a type of enzyme that digests proteins

NUMERACY LINK

Describe and explain the trend of this graph.

Rate of respiration

Oxygen concentration

Respiration is one of the most important processes that occur in all living things because it provides organisms with the energy needed to carry out life processes. Without a constant supply of this energy, we wouldn't be able to do anything! Most living things require oxygen to respire, so it's vital that plants and animal cells receive oxygen.

Figure 14.21 During intense exercise, your muscles carry out anaerobic respiration.

1 Respiration produces energy

Respiration is the process of using glucose to obtain energy, and is a type of combustion reaction. There are two types of respiration – **aerobic** (requiring oxygen) and **anaerobic** (not requiring oxygen).

Aerobic respiration happens when there is plenty of oxygen present. Glucose (from food) reacts with oxygen to produce carbon dioxide, water and energy:

$$\text{Glucose} + \text{oxygen} \rightarrow \text{carbon dioxide} + \text{water} + \text{energy}$$

$$C_6H_{12}O_6 + 6O_2 \rightarrow 6CO_2 + 6H_2O + \text{energy}$$

Anaerobic respiration happens where there is little or no oxygen present. During vigorous exercise, your heart and lungs can't get oxygen to your muscles fast enough to keep up with the demands of aerobic respiration, so the muscles carry out anaerobic respiration. Anaerobic respiration produces lactic acid, which can cause your muscles to cramp and get tired.

The energy produced from respiration is stored in a chemical called adenosine triphosphate, or ATP. This energy is used for:

- muscular action
- growth and repair of cells
- allowing chemical reactions to take place
- keeping your body temperature constant
- sending messages along nerves.

What is the difference between aerobic and anaerobic respiration?

2 Digestion involves a series of chemical reactions

During digestion, large food molecules are broken down into smaller molecules that the body can use. The smaller molecules, such as glucose, are absorbed through the walls of the small intestine into the bloodstream. Chemical digestion occurs in your mouth, stomach and small intestine.

Most chemical reactions in the human body need enzymes to work. Enzymes are biological **catalysts** that speed up the rate of chemical reactions. They reduce the amount of energy needed to start a reaction. Enzymes work best at a specific pH and normal body temperature. Stomach enzymes work best at pH 4–5. This pH is maintained by the presence of hydrochloric acid in the stomach.

Chemical digestion involves a number of enzymes. The three main types of enzyme are **carbohydrases**, **proteases** and **lipases**. Their reactions are summarised in Table 14.6.

Table 14.6 The three main types of enzymes involved in digestion

Enzyme type	Reaction	Example
Carbohydrases	Break down carbohydrate molecules	In the mouth, amylase breaks down carbohydrates into maltose, which is broken down to glucose in the small intestine.
Proteases	Break down protein molecules	In the stomach, pepsin breaks down proteins into peptides and some amino acids.
Lipases	Break down lipid (fat) molecules	In the small intestine, lipase breaks down complex fat molecules into fatty acids and glycerol molecules.

What role do enzymes play in the human body?

Figure 14.22 The action of enzymes on protein, starch and fats

A protein molecule is made up of many different amino acids.

Protease breaks down protein molecules.

Amino acids

A starch molecule is a carbohydrate made up of many glucose molecules.

Carbohydrase breaks down carbohydrate molecules.

Glucose

A fat molecule is made up of fatty acids and glycerol molecules.

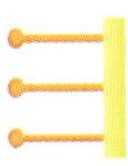

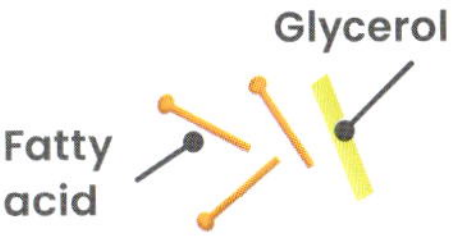

Fatty acid

Glycerol

CHECKPOINT 14.10 

1 Describe respiration.
2 Explain the importance of respiration.
3 Where does respiration take place?
4 What chemical reaction happens during respiration?
5 How is the energy produced during respiration used?
6 Describe the process of digestion.
7 What are enzymes?
8 Explain how enzymes help in the digestive process.
9 What is the purpose of hydrochloric acid in your stomach?

CHALLENGE

10 Write a short report detailing the process of digestion, including examples of chemical reactions that take place during digestion.

SKILLS CHECK

- I can describe some of the chemical reactions that occur during digestion.
- I can explain what respiration is and give the equation of cellular respiration.

CHAPTER SUMMARY

All matter is made up of **atoms** which have **mass**.

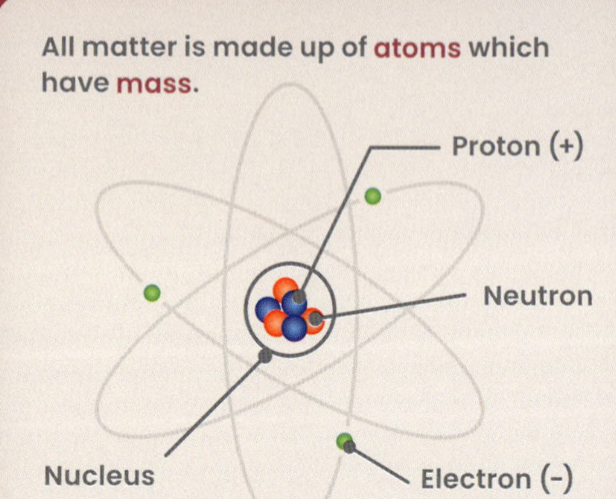

A **chemical equation** shows what atoms are involved before the reaction and how they recombine afterwards.

Chemical formulas:

- are a shorthand way of writing the name of an element or compound
- tell you the number and type of atoms in a compound
- can be used to write equations that model reactions.

The **pH scale** measures acidity.

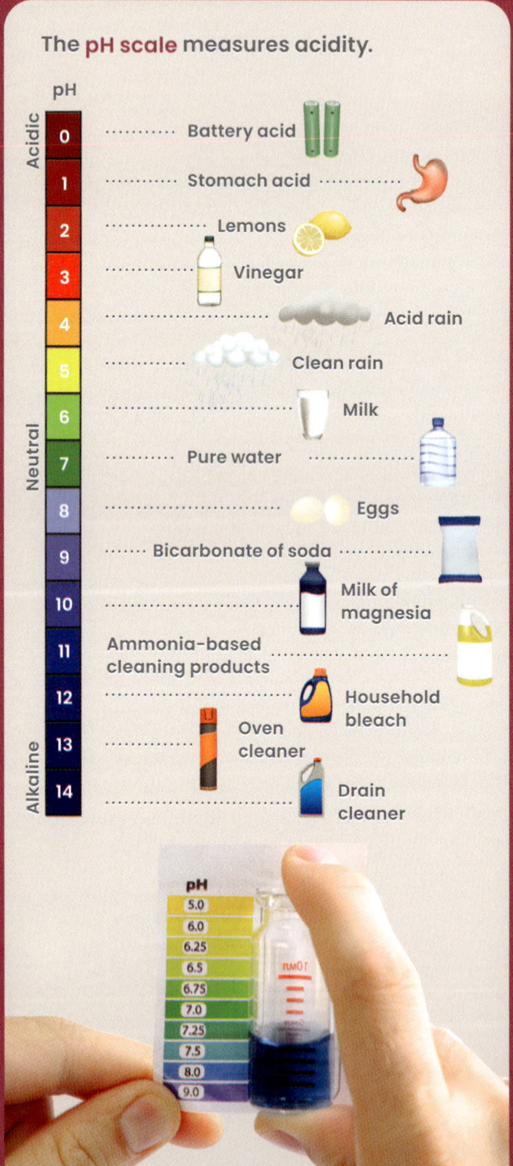

Atoms bond to form elements, covalent compounds or ionic compounds.

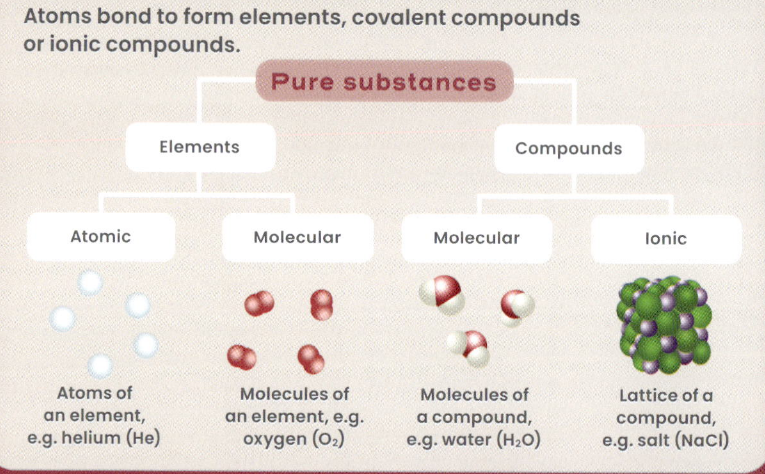

Two common groups of chemical compounds are **acids** and **bases**.

- **Corrosion** is a natural process that can be prevented.
- **Precipitation** reactions are formed by mixing salt solutions.
- **Combustion** reactions involve a compound and oxygen.
- **Incomplete combustions** are a result of limited oxygen supply.

FINAL CHALLENGE

1. Explain the difference between an element and a compound.
2. Draw and annotate the pH scale.
3. Suggest why corrosion occurs and how to prevent it.

4. Provide a diagram of an atom, and label with the following labels: protons, neutrons and electrons.
5. Identify the products and reactants in the following chemical reaction:

   $CH_4 + 2O_2 \rightarrow CO_2 + 2H_2O$

6. How many carbon, hydrogen and oxygen atoms can be found in the products and reactants of the reaction in question 5?

7. Balance the following chemical reactions:
   a $Fe + O_2 \rightarrow Fe_3O_4$
   b $Mg + HCl \rightarrow MgCl_2 + H_2$
   c $S + O_2 \rightarrow SO$
8. Explain how compounds are named.
9. What are precipitates and how are they formed?

10. Explain what covalent and ionic bonds have in common, and how they differ.
11. Describe what a combustion reaction is and what the products of combustion are.
12. Explain how neutralisation reactions work and give an example using a chemical equation.

13. Explain what an isotope is.
14. Give an example of an important chemical reaction that takes place in the body. Describe the reaction and suggest why it is so important.

15 Rates of reactions

Have you ever wondered why some reactions are very slow (for example, the spoiling of milk and corrosion) while others are fast (for example, the combustion reactions that happen in bushfires)? The rate of a reaction is a measure of how quickly reactants are used up and products are formed. If this happens fast, then the rate of reaction is high. If a reaction takes a long time to complete, then the reaction rate is low.

1 LEARNING LINKS

Brainstorm as much as you can remember about these topics.

Explosions lead to many different energy transformations.

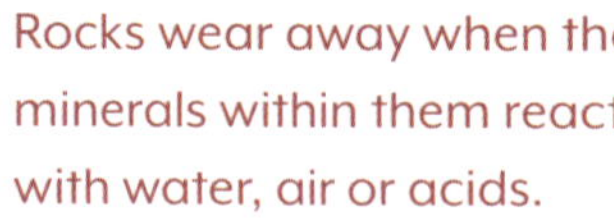

Rocks wear away when the minerals within them react with water, air or acids.

A precipitation reaction produces a product that is not soluble in water – a precipitate. In most precipitation reactions, the products are ionic compounds.

2 SEE–KNOW–WONDER

List three things you can **see**, three things you **know** and three things you **wonder** about this image.

3 CRITICAL + CREATIVE THINKING

What if ... combustion reactions (burning things) was banned in Australia?

The variations: How many different ways can you use chemical reactions every day?

The combination: List two features of a catalyst and two features of a submarine. Now add those attributes together to create a new object.

4 DISASTER!

Huge chemical disasters are rare, but they still happen all over the world. Perhaps the worst such disaster was a major gas leak at a pesticide plant in Bhopal, India, in 1984. Almost 10 000 people died as a result of poisoning from the leak, and more than 500 000 people were injured. The site of the leak is still contaminated, and people living in Bhopal still become ill from exposure to toxic chemicals.

15.1 Exothermic and endothermic reactions

At the end of this lesson I will be able to:

- **identify** that chemical reactions involve energy transfer and can be exothermic or endothermic.

KEY TERMS 

electrolysis
a decomposition reaction using electricity

endothermic
a reaction that absorbs energy in the form of heat

exothermic
a reaction that releases energy in the form of heat

LITERACY LINK 

Summarise this section into a tweet of exactly 280 characters.

NUMERACY LINK 

A sample of sodium hydroxide was dissolved in a test tube of water at 19 °C. The temperature increased to 28 °C. A sample of potassium nitrate was dissolved in a test tube of water at 20 °C. The final temperature was 11 °C. Determine the temperature change for each and state which reaction was endothermic and which was exothermic.

Chemical reactions involve more than just elements and compounds – they also involve energy. This energy usually takes the form of heat. Some reactions release heat, such as burning a candle. Others require heat to be added, such as cooking an egg. Without energy being transferred, chemical reactions simply would not happen.

1 Exothermic reactions release heat, while endothermic reactions absorb heat

In **exothermic** reactions, the reactants have more energy than the products. This extra energy is released as heat. Exothermic reactions include:

- respiration
- combustion reactions
- corrosion
- neutralisation reactions
- burning magnesium.

In **endothermic** reactions, the products have more energy than the reactants. This energy has to be added or absorbed, mainly in the form of heat. Endothermic reactions include:

- photosynthesis
- bread baking
- an egg cooking
- **electrolysis** of water
- thermal decomposition reactions.

What are two differences between exothermic and endothermic reactions?

Figure 15.1 Fuel being burnt in a rocket launch is an example of an exothermic reaction.

2 Bonds break and form during chemical reactions

Why do chemical reactions involve energy? It's because of the bonds between the atoms within the molecules. These bonds, which hold molecules together, contain chemical potential energy. During a chemical reaction, bonds in the reactants break, and new bonds form to make the products.

In all chemical reactions, energy is needed to break the bonds of the reactants, and energy is released when new bonds form in the products. For example, methane reacts with oxygen to form carbon dioxide and water:

$$CH_4 + 2O_2 \rightarrow CO_2 + 2H_2O$$

Methane + oxygen → carbon dioxide + water

As bonds within methane and oxygen break, new bonds form between carbon, hydrogen and oxygen to form carbon dioxide and water.

Why is energy needed in chemical reactions?

3 Chemical reactions produce or absorb heat

To see whether a reaction is exothermic or endothermic, we need to look at the overall energy of the reaction – whether energy was required to make the reaction proceed or whether energy was released.

In an endothermic reaction, the energy required to break the bonds in the reactants is greater than the energy released when the products are formed. This means that the reaction takes in more energy overall, so the temperature decreases around the reaction container or surroundings. Endothermic reactions give rise to cold surroundings.

Figure 15.2
In a cold pack, ammonium nitrate reacts with water and absorbs energy from the surroundings in an endothermic reaction.

Chemical equations for endothermic reactions can be written like this:

Reactants + ENERGY → products

An exothermic reaction occurs when the energy used to break the bonds in the reactants is less than the energy released when new bonds are made in the products. This extra energy is given off as heat and the temperature of the surroundings increases. Exothermic reactions give rise to hot surroundings.

Chemical equations for exothermic reactions can be written like this:

Reactants → products + ENERGY

Why is energy released in an exothermic reaction?

INVESTIGATION 15.1
Exothermic and endothermic reactions

CHECKPOINT 15.1

1 How is an endothermic reaction different from an exothermic reaction?
2 Give two examples of endothermic reactions and two examples of exothermic reactions.
3 Compare the energy required to break and form bonds in both endothermic and exothermic reactions.
4 Which of the following reactions are exothermic and which are endothermic? Give reasons for your choice.
 a baking biscuits
 b burning a piece of paper
 c lighting a candle
 d using a cold pack on an injured leg
5 Bonds break during chemical reactions. Explain why.

CHALLENGE

6 Research how endothermic and exothermic reactions are used in everyday life. Give at least three examples of each.

SKILLS CHECK

- I can explain why chemical reactions involve energy transfer.
- I can explain what endothermic and exothermic reactions are.

15.2 Combustion and respiration

At the end of this lesson I will be able to:

- **compare** combustion and respiration as types of chemical reactions that release energy but occur at different rates.

KEY TERMS

combustion
an exothermic reaction requiring the presence of oxygen

hydrocarbon
a compound made up of carbon and hydrogen

respiration
exothermic reaction within cells that releases energy for the body's use

spontaneous
not planned

LITERACY LINK

Write a speech aimed at students in your class, explaining the environmental consequences of the combustion of coal for electricity. Explain the science behind combustion and some ideas for combating the problem.

NUMERACY LINK

Australia's electricity is generated from a number of sources: black coal (47%), brown coal (15%), natural gas (21%), crude oil (2%) and renewables (15%). Construct a pie chart for this data.

Respiration and combustion are both exothermic reactions, so they release energy. You can observe a combustion reaction in a fire – it gives off heat, light and sound energy. Respiration occurs in cells and involves 'burning' food in order to turn it into a form of energy that cells can use. Both combustion and respiration are exothermic chemical reactions that release energy, but they happen at very different rates.

1 There are different types of combustion reactions

All **combustion** reactions are exothermic reactions, releasing heat. They also all require fuel and oxygen. However, there are important differences between the various kinds of combustion reactions.

Complete combustion happens when a fuel, such as a **hydrocarbon**, burns in plentiful amounts of oxygen to release carbon dioxide and water.

Hydrocarbon + oxygen → carbon dioxide + water + energy

Incomplete combustion happens when a fuel burns in a limited amount of oxygen, and the reaction does not go to completion. The products are water and carbon monoxide, or sometimes carbon in the form of soot. You can see incomplete combustion happening when a car or truck gives off a lot of black smoke from its exhaust. By varying the amount of oxygen used, you can change the product so that it is less sooty and toxic.

Hydrocarbon + limited oxygen → carbon monoxide + water + energy

or Hydrocarbon + limited oxygen → carbon + water + energy

Spontaneous combustion is when a substance bursts into flames without external energy being applied. It happens when the internal temperature of the substance increases because of other chemical reactions. Once the temperature reaches a point at which the substance would normally catch fire, it can trigger combustion. Spontaneous combustion has been observed in coal mines, forests, large compost heaps and hay bales.

What is the difference between complete and incomplete combustion?

Figure 15.3 Spontaneous combustion can be very dangerous.

2 Combustion reactions power the world

Can you imagine a life without combustion reactions? Our whole world depends on combustion. Some combustion reactions are obvious, such as fires, but others are subtler.

Cars burn petrol and diesel so that they can transport us from one place to another. We use gas ovens and stoves for cooking. Burning fossil fuels such as coal provides us with electricity to light and warm our homes.

However, combustion reactions can be dangerous. Rapid combustion happens in an explosion, such as fireworks. Rapid combustion can cause a lot of damage. Slow combustion, such as burning a piece of wood in the fireplace, is much safer. The log takes a long time to burn, but keeps on giving out heat energy to keep us warm.

Name three everyday combustion reactions.

Figure 15.4 Rapid combustion gives out energy fast and is over very quickly.

Figure 15.5 Slow combustion produces energy over a long period.

3 Respiration is a form of combustion

Respiration is a natural process that is constantly happening in all living cells. It uses glucose produced during digestion, as well as the oxygen we breathe in, and produces carbon dioxide, water and energy.

Respiration is a biochemical reaction, and all chemical reactions require some energy to start. The energy required to start a respiration reaction is produced from previous respiration reactions and stored in a chemical called ATP (adenosine triphosphate).

Respiration is a form of slow combustion. Another example of slow combustion is corrosion, which can take years.

What are the products of respiration?

CHECKPOINT 15.2 

1 Define:
   a *respiration*
   b *combustion.*
2 Identify the vital ingredient required for combustion to occur.
3 Explain what complete combustion means.
4 True or false? Combustion and respiration are both exothermic.
5 Give an example of spontaneous combustion.
6 Give three reasons why combustion reactions are important to us.
7 What are the waste products of combustion? Are they toxic?
8 Where does the energy needed to start respiration come from?
9 Draw a table to show the similarities between respiration and other types of combustion.

CHALLENGE

10 Research cellular respiration. Write a half-page report, using the following prompts: Where does it take place? Which organisms and types of cells use cellular respiration? What is the chemical equation for respiration?

SKILLS CHECK

- I can explain what combustion reactions are.
- I can explain what respiration is.
- I can compare combustion and respiration reactions, including the rate of each.

15.3 Temperature and concentration

At the end of this lesson I will be able to:

- **describe** the effects of factors such as temperature and concentration on the rate of some common chemical reactions.

KEY TERMS 

concentration
the number of particles in a given volume

kinetic energy
energy of motion

LITERACY LINK 

Re-write the first paragraph in this section to be more formal and scientific.

NUMERACY LINK 

Observe the following data:

Concentration of reactant (g/L)	Time of reaction (s)
0	0
2	8
4	14
6	20
8	24
10	24

Draw a line graph and describe the pattern of results.

Reactions can be fast, such as an explosion or firing a missile. Reactions can also be slow, such as a ship rusting or rocks weathering, which take many years to happen. The rate of a chemical reaction is how fast the reaction can be completed; that is, the speed at which reactants are converted into products. Many factors influence the rate of reaction, including temperature, concentration, the presence of a catalyst and the surface area involved.

Figure 15.6 (a) Wood burning is a fast chemical reaction, which may be completed in a few hours. (b) Iron rusting is a slow chemical reaction, which can take many years.

1 Many factors can affect reaction rate

Every chemical reaction makes products from the reactants at a certain rate, but this rate is not fixed. Many different factors affect reaction rate, including:

- concentration
- temperature
- pressure
- surface area
- catalysts.

It's possible to adjust many of these factors. In some cases, you might want to reduce the reaction rate to make the reaction safer to handle. Or you might want to increase the rate of a reaction so that products are made more quickly. This is very common in the chemical industry, where it is important to reduce costs and maximise profits.

What are three factors that can affect reaction rate?

2 Raising temperature increases reaction rate

Molecules must collide in order to react. Increasing the temperature of a system increases the average **kinetic energy** of the particles in the system. As the average kinetic energy increases, the particles move faster, so they collide more frequently and possess greater energy when they collide.

The reaction rate of nearly all reactions increases with increasing temperature. This means that the hotter the reactants, the faster the reaction will be. Likewise, the cooler the reactants, the slower the reaction.

We change temperature to adjust reaction rates every day. You boil water before making a cup of tea because the tea dissolves faster through hot water. If you make a cup of tea from room-temperature water, it will take much longer. Similarly, we keep food in the refrigerator to slow down the chemical reactions that cause it to spoil over time.

Figure 15.7 The reactions in glow sticks give off light. When a glow stick is placed in warm water, it glows more brightly because the reactions happen faster at higher temperatures.

How does changing the temperature affect reaction rates?

3 Concentrated substances react more quickly

The **concentration** of a substance is a measure of how many particles are within the volume of a liquid solution or a gas's container. Increasing the concentration of the reactants increases the number of particles within a given volume. The more crowded the particles are, the more collisions occur. This increase in the frequency of collisions leads to an increase in the rate of the reaction.

Another way to adjust concentration for a gas is by changing the pressure. The pressure of a system increases when the volume of the reactant vessel decreases. This means that there are now the same number of particles in a smaller volume, leading to more collisions between particles.

Why does increasing the concentration of reactants increase the rate of reaction?

Figure 15.8 Concentrated sulfuric acid reacts dramatically with sucrose (table sugar).

INVESTIGATION 15.3A
The effect of concentration on the rate of a reaction

INVESTIGATION 15.3B
The effect of temperature on the rate of a reaction

CHECKPOINT 15.3

1 Explain why temperature can increase the rate of a chemical reaction.
2 List some factors that affect the rate of chemical reactions.
3 Outline why collisions between the reacting particles are important for a chemical reaction to take place.
4 Will increasing the concentration of one of the reactants increase the rate of the reaction? Explain your answer.
5 Is it always a good thing to increase the rate of a reaction? Explain your answer.
6 Explain how the rate of a gaseous reaction is affected when the volume of the container decreases.

CHALLENGE

7 Research why reaction rate needs to be controlled in industry. What compromises are made regarding temperature and concentration, and why?

SKILLS CHECK

- I can describe the effect of temperature on the rate of some common chemical reactions.
- I can describe the effect of concentration on the rate of some common chemical reactions.

15.4 Surface area and catalysts

At the end of this lesson I will be able to:

- **describe** the effects of factors such as surface area and catalysts on the rate of some common chemical reactions.

KEY TERMS 

activation energy
the energy required to start a reaction

catalyst
substance that increases the rate of a chemical reaction without being used up

enzyme
a biological catalyst that increases the rate of reactions in cells

surface area
the area of the outermost layer of an object

LITERACY LINK 

Write a paragraph that uses these words: catalyst, activation energy, enzymes and rate of reaction.

NUMERACY LINK 

Determine the surface area of a cube with 1 cm sides.

The cube is cut up into 8 equal pieces. Calculate the surface area of the new cubes. Does the large cube or the smaller cubes together have the larger surface area?

Temperature and concentration are relatively easy factors to control. We adjust these factors every day in our kitchens. Other factors are harder to control, and may require special equipment or materials. These factors are often adjusted in industrial or manufacturing processes, although we can adjust them in smaller ways in our own lives.

1 Increasing surface area increases reaction rate

Surface area is the exposed area of a solid substance. If you increase the surface area of a substance, there are more particles at the surface that can react with another substance, so the reaction rate increases.

Consider the reaction of zinc metal with an acid (Figure 15.9). If the zinc is present as a large cube, the total surface area of the cube is the sum of all the faces of the cube. If you break the zinc cube into many smaller pieces, the total surface area greatly increases, and more area is exposed to react with the acid. If you crush zinc into powder, its surface area increases even further. This form of zinc reacts the fastest.

An everyday example is that small pieces of food cook faster. These pieces of food have a large surface area, compared to their volume, exposed to heat. Medicinal tablets are another example. When you take tablets or pills, different tablets dissolve at different rates depending on their surface areas.

Why does increasing the surface area increase the rate of a chemical reaction?

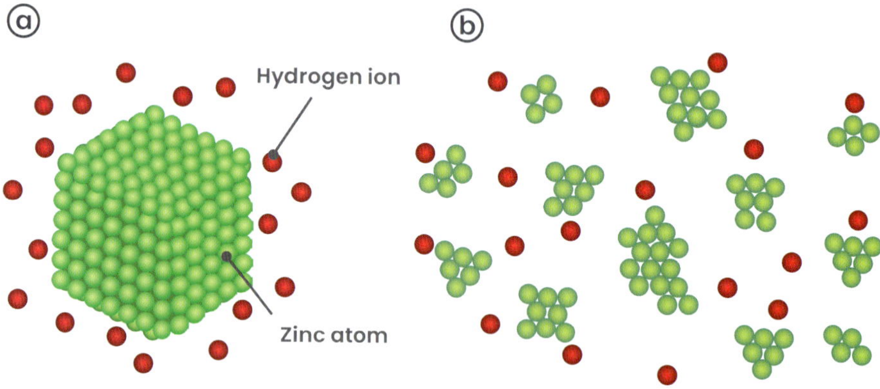

Figure 15.9(a) Hydrogen ions from the acid can collide only with the outer layer of zinc atoms in the large cube of zinc. The reaction rate is slow. (b) When the zinc block is broken into smaller pieces, more surface is exposed for the hydrogen ions to collide with and the rate of the reaction increases.

❷ Catalysts increase the rate of reaction

A **catalyst** is a substance that increases the rate of a chemical reaction without being used up. It enables the reaction to occur at a lower temperature.

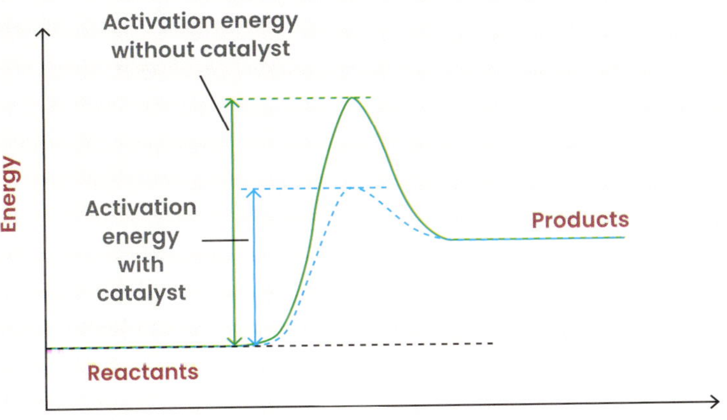

Figure 15.10 The chemical reaction that is catalysed requires less activation energy.

Catalysts lower the **activation energy** of a reaction, which is the energy required to start a reaction. Catalysts are specific, which means that each catalyst increases the rate of some reactions but not others.

Catalysts are important in industry because they reduce reaction times and enable chemical processes to take place at lower temperatures and pressures. So they reduce the cost of production of valuable products.

Enzymes are biological catalysts. They are proteins that help speed up chemical reactions in the body.

Here are some common uses of catalysts.

- Catalytic converters in cars contain platinum, which catalyses the conversion of toxic carbon monoxide into carbon dioxide.
- Biological catalysts in detergents remove protein stains, such as egg yolk, that are otherwise hard to dislodge.
- Catalysts in the paper industry break down paper pulp to produce smooth paper for magazines.
- Catalysts turn milk into yoghurt and petroleum into plastic milk jugs, CDs and bicycle helmets.

Why are catalysts important in our lives?

Figure 15.11 Saliva contains enzymes that start the chemical reactions of digestion.

INVESTIGATION 15.4A
The effect of a catalyst on the rate of a reaction

INVESTIGATION 15.4B
The effect of surface area on the rate of a reaction

CHECKPOINT 15.4 

1 Explain how surface area can influence the rate of chemical reactions.

2 Explain what activation energy is.

3 Explain the significance of the number of collisions to the rate of a chemical reaction.

4 Explain what a catalyst is and give an example of a catalyst.

5 Explain how enzymes are a type of catalyst.

6 True or false? A crushed substance has a large surface area.

CHALLENGE

7 Research catalysts in industry. Identify a catalyst and state where it is used. How does this catalyst help improve the rate of this reaction? Write the word and formula equations for the reactions that take place. What would have happened if this catalyst hadn't been used?

SKILLS CHECK

- I can describe the effect of increasing surface area on the rate of some common chemical reactions.
- I can describe the effect of a catalyst on the rate of some common chemical reactions.

15.5 Making decisions about science

At the end of this lesson I will be able to:

- **analyse** how social, ethical and environmental considerations can influence decisions about scientific research related to the development and production of new materials.

KEY TERMS 

carbon capture
the process of trapping carbon dioxide at its emission source so that it does not enter the atmosphere

implication
likely outcomes or consequences of an action

LITERACY LINK 

Explain these terms: social, ethical and environmental.

NUMERACY LINK 

For every 1000 kg of cement that is produced, 900 kg of CO_2 is emitted into the environment. Express this as a percentage.

How much CO_2 would be emitted from the production of 5 tonnes of cement?

There are many different fields of scientific research and technological development. All around the globe, scientists and engineers are creating new materials, technologies and inventions, many of which have the promise of improving the world in some way.

However, just because a new technology is invented, or a new material is developed, it doesn't mean we should use it right away. Scientists have to consider many different factors during their research. Let's consider one example.

1 Carbon capture helps reduce carbon dioxide emissions

One field of materials science that has received a lot of attention in the 21st century is **carbon capture**. Carbon dioxide (CO_2) is a vital gas for plants, but human technology creates far more of it than the environment can process. Excess CO_2 pollutes the atmosphere and is a major contributor to climate change, so we need to find ways to reduce CO_2 levels around the world.

Carbon capture is a common way of dealing with CO_2. It is an engineering process that compresses CO_2 into a liquid form, which can then be stored underground.

How might carbon capture be useful?

Figure 15.12 The major sources of carbon dioxide emissions are burning fossil fuels for electricity, heat and transport.

❷ There are possible alternatives to carbon capture

While carbon capture helps reduce the amount of CO_2 pollution in the atmosphere, it's not a perfect solution. It's an expensive process that uses a lot of energy, and there's the risk of leakage from the storage sites. Because of these issues, scientists are also researching other ways to deal with carbon dioxide emissions.

In 2019, scientists from RMIT in Melbourne developed a low-cost technique for converting atmospheric CO_2 into solid particles of carbon. This process uses liquid metal as a catalyst, and adds energy in the form of an electric current. The chemical reaction causes the CO_2 to condense into flakes of carbon at room temperature.

Another useful material could be Ferrock, an alternative to concrete created by American scientist David Stone. Concrete is used in buildings throughout the world, but the process of making it emits large amounts of CO_2. Ferrock contains steel dust and silica (ground glass), and the process of making it actually uses up CO_2 instead of creating it.

Figure 15.13 Atmospheric carbon dioxide could be converted back into coal.

Why are alternatives needed for carbon capture?

❸ Many factors influence decisions about scientific research

Scientists need to consider the **implications** of new technologies and materials being put into widespread use. There are many different types of considerations.

- Social considerations: what impact could this have upon society and the way people or organisations behave?
- Ethical considerations: is it morally right to use this new material?
- Environmental considerations: what impact could this new technology have on the environment?

Consider the solid carbon made from atmospheric CO_2. One of the potential uses for reformed carbon is as fuel – but that means that the captured CO_2 would be released into the atmosphere again. What impact would this have on the environment – would it help or actually make things worse?

Considerations such as this mean that scientific research must be done carefully. Scientists, organisations and governments need to think through the implications of every new technology.

Name three types of considerations that influence scientific research.

CHECKPOINT 15.5 

1 Why do chemists have an important role to play in the manufacture of new materials?

2 What do you think the statement 'scientists turn carbon dioxide into coal' means?

3 Explain how gaseous carbon dioxide is converted into liquid carbon dioxide.

4 What are the two main ingredients of Ferrock?

5 Explain why Ferrock might have less environmental impact than concrete.

6 Describe three important concerns to consider when conducting scientific research.

7 If scientists do turn carbon dioxide back into coal on a large scale, how could this help the environment? How could it hurt it?

CHALLENGE

8 Research the uses of fuel cells. What are the social, ethical and environmental implications of using fuel cells instead of combustion engines?

SKILLS CHECK

- I can discuss how social, ethical and environmental considerations influence decisions about scientific research.

15.6 Careers in chemistry

At the end of this lesson I will be able to:

- **describe** examples to show where advances in science and/or emerging science and technologies significantly affect people's lives, including generating new career opportunities in areas of chemical science such as biochemistry and industrial chemistry.

Industrial chemistry is the use of chemical processes on an industrial scale, usually for producing materials. The scientists who work in this area are often referred to as industrial chemists. Biochemistry is a related field, because many biological items – medicines, cosmetics, even food – are also produced and developed on an industrial scale. The growth of materials science means that many new career opportunities are appearing in these scientific areas.

KEY TERMS 

biochemist
a scientist who studies the chemical processes in living organisms

industrial chemist
a scientist who converts raw materials to products on a commercial scale

1 Chemists work in a broad range of industries

We often use the word 'chemist' to mean someone who works in a pharmacy, but it really means a scientist who specialises in chemistry.

The diverse research done by chemistry experts includes the discovery of new medicines and vaccines, forensic analysis for criminal cases, improving understanding of environmental issues and developing new chemical products and materials such as cosmetics, paints, plastics, food and drink. Researchers are often required to create and develop new chemical engineering techniques.

Industrial chemists work in chemical plants and laboratories across many industries, producing building materials and paints, paper, petroleum, plastics and advanced materials. They can spend a lot of time in analytical laboratories, and in the plants themselves, but may also work in offices when they do theoretical research.

What are three fields of research for industrial chemists?

LITERACY LINK 

Write a job advertisement for a biochemist or an industrial chemist. Make sure to include what the job involves, what tasks they will be required to undertake and what qualifications are required.

NUMERACY LINK 

A chemist analyses the zinc oxide (ZnO) content of three brands of sunscreen. Brand A contains 30 g ZnO in 200 g sunscreen, Brand B contains 150 mg in 75 g and Brand C contains 7.7 g in 0.045 kg. Which brand of sunscreen contains the most ZnO?

Figure 15.14 Industrial chemists work in chemical plants and laboratories in many different industries.

❷ Chemical engineers focus on solving problems

While industrial chemists often focus on research, chemical engineers tend to be more hands-on in their work. They apply scientific knowledge and techniques to problems.

Chemical engineers work in a number of areas of industry, including oil and gas, energy, water treatment, plastics, toiletries, pharmaceuticals and food and drink. While processes differ within each of these areas, chemical engineers are directly involved in the design, development and manufacture of chemical products and materials.

Daily tasks for chemical engineers include ensuring the efficiency and safety of chemical processes, adapting the chemical make-up of products to meet environmental or economic needs, and applying new technologies to improve existing processes.

What are three sectors in which chemical engineers work?

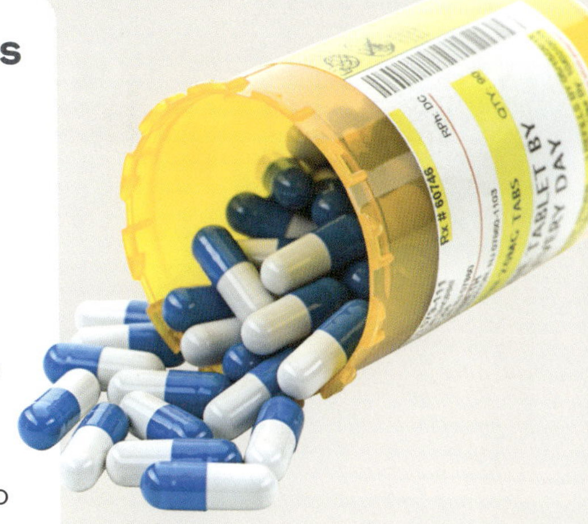

Figure 15.15 Dosed medicine is one of the most important chemical engineering inventions of the modern era.

❸ Biochemists study the chemistry of living things

Biochemists study the chemical processes in living organisms to understand how certain chemical reactions happen in tissues, and record the effects of medicines. Their aim is to improve our quality of life by understanding living organisms at the molecular level. This includes running laboratory experiments to develop effective medicines and collecting cell samples from animals and plants in order to understand how genetic traits are carried.

Biochemists often work in research roles in the pharmaceutical and biotechnology industries, in food technology, toxicology and vaccine production. Many teach at universities in addition to researching and collaborating with their peers.

What are three fields of research for biochemists?

Figure 15.16 Biochemists study organisms at the molecular level.

CHECKPOINT 15.6 

1 What are some areas of research chemists are involved in?
2 Explain what an industrial chemist does.
3 Which other professionals do industrial chemists work alongside?
4 Which areas or sectors do chemical engineers work in?
5 Outline the aim of biochemists.
6 Describe the variety of work biochemists do.

CHALLENGE

7 Research five other careers in science and write a half-page report of your findings.
8 Carry out research to find out what subjects students should study in senior high school if they wish to go to university to study for a career in science.

SKILLS CHECK

- I can describe the role of an industrial chemist.
- I can describe the role of a biochemist.

CHAPTER SUMMARY

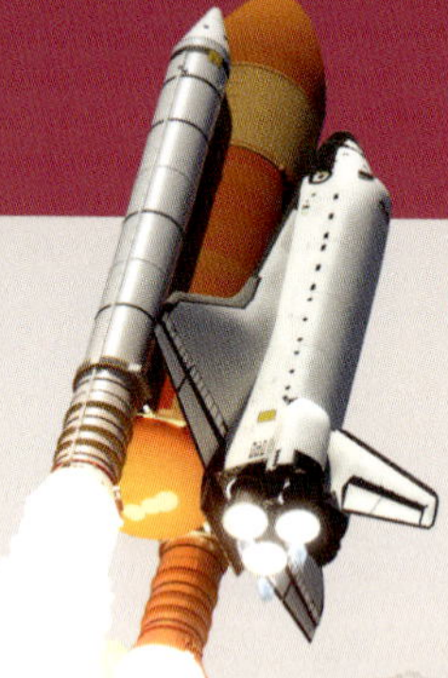

Exothermic reactions release heat.

Combustion and respiration are exothermic reactions that release energy.

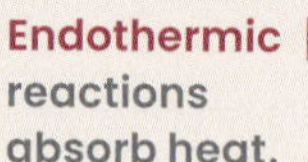

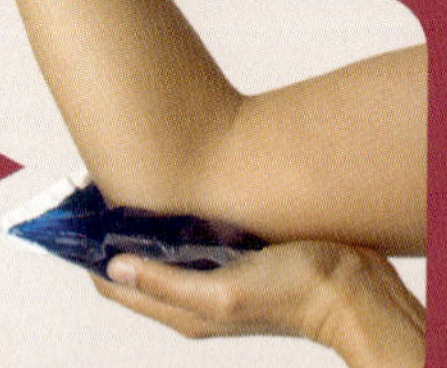

Endothermic reactions absorb heat.

Chemical reactions involve elements, compounds and energy.

Factors affecting reactions rates include:

- concentration
- temperature
- pressure
- surface area
- catalysts.

A **chemist** is a scientist who specialises in chemistry.

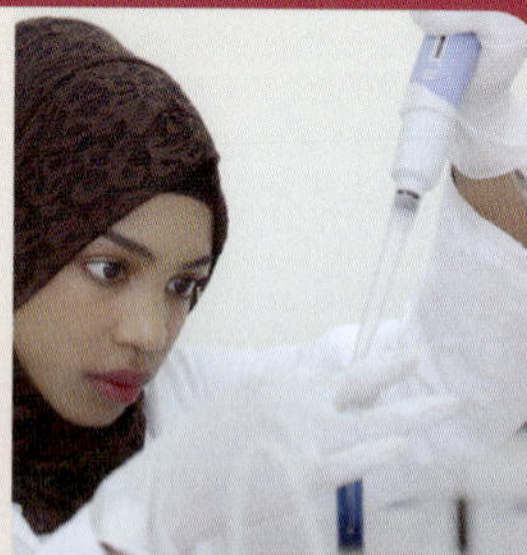

Industrial chemists often focus on research.

Chemical engineers apply hands-on knowledge to problems.

Biochemists study the chemistry of living things.

Increasing surface area increases reaction rate.

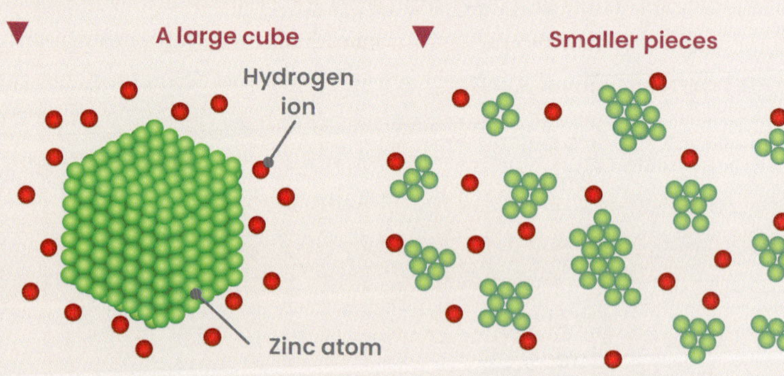

A large cube:

- Smaller surface area
- Lower reaction rate

Smaller pieces:

- Larger surface area
- Higher reaction rate

Scientists have to consider many different factors in their research.

Social considerations: what impact could this have on society?

Environmental considerations: what impact could this have on the environment?

Ethical considerations: is it morally right?

FINAL CHALLENGE

1. Explain the difference between endothermic and exothermic reactions.
2. Give an example of an exothermic reaction and provide the word equation for it.
3. Describe two similarities between combustion and respiration.

LEVEL 1

50XP

LEVEL UP!

4. Give an example of incomplete combustion. What are its products?
5. Where does respiration occur and why is it important?
6. Name some factors that increase the rate of reaction.

LEVEL 2

100XP

7. Explain what bond breaking and bond formation mean, using the example below in your response.

   $CH_4 + 2O_2 \rightarrow CO_2 + 2H_2O$

8. Explain what activation energy is and why it must be supplied to start almost all reactions.
9. Use a diagram to describe what happens in a chemical reaction when there is an increase in temperature.

10. Explain, in terms of particles, how increasing the concentration of a reactant increases the rate of reaction.
11. Increasing the surface area of a reactant increases the rate of reaction. Explain why.
12. Describe the role of catalysts in increasing the rate of a chemical reaction.

13. Create an A2 or A1 sized poster that details different careers in chemical sciences. Make sure to include 'highlights' of each particular career.

INVESTIGATIONS

PHYSICAL WORLD

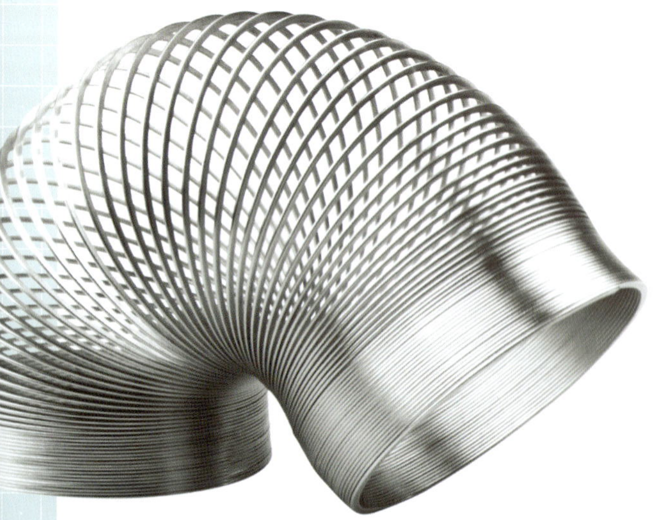

EARTH AND SPACE

LIVING WORLD

CHEMICAL WORLD

Investigation 1.1

Convection currents

AIM

To investigate the movement of convection currents

MATERIALS

- potassium permanganate crystals
- hot water
- 500 mL beaker
- tweezers
- drinking straw

METHOD

1 Copy the results table into your notebook, adding a title.

PART 1: HOT CONVECTION

2 Fill the beaker to the 375 mL level with hot water. Place the straw in the beaker.

3 With tweezers, drop a crystal of potassium permanganate into the straw located in the centre of the beaker. Make sure that the crystal drops to the bottom of the container.

4 Carefully observe and draw the motion of the purple stain.

PART 2: COLD CONVECTION

5 Fill the beaker to the 375 mL level with cold water. Place the straw in the beaker.

6 With tweezers, drop a crystal of potassium permanganate into the straw located in the centre of the beaker. Make sure that the crystal drops to the bottom of the container.

7 Carefully observe and draw the motion of the purple stain.

DISCUSSION

1 Explain how thermal energy moves in the water.

2 Refer to the diagrams in your results table to discuss the differences between hot and cold convection currents.

3 Identify the direction of a cold current.

HANDLE THE HOT BEAKER WITH CARE. DISPOSE OF WASTES APPROPRIATELY. POTASSIUM PERMANGANATE IS TOXIC AND LEAVES STAINS, SO AVOID CONTACT WITH IT.

RESULTS TABLE I1.1

	Hot convection	Cold convection
Observations		
Sketch		

Investigation 1.2

Waves in a slinky

AIM

To investigate the movement of waves in a slinky and to calculate the frequency of waves

MATERIALS

- slinky
- stopwatch
- masking tape
- metre ruler

METHOD

1. Copy the results table into your notebook, adding a title.
2. With a partner, stretch the slinky along the ground and mark the length with masking tape.
3. Measure this distance with the ruler.
4. Produce small transverse (up and down) waves. Use the stopwatch and count how many waves are produced in 10 seconds. This is your frequency in hertz (Hz).

$$\frac{\text{number of waves}}{\text{time}}$$

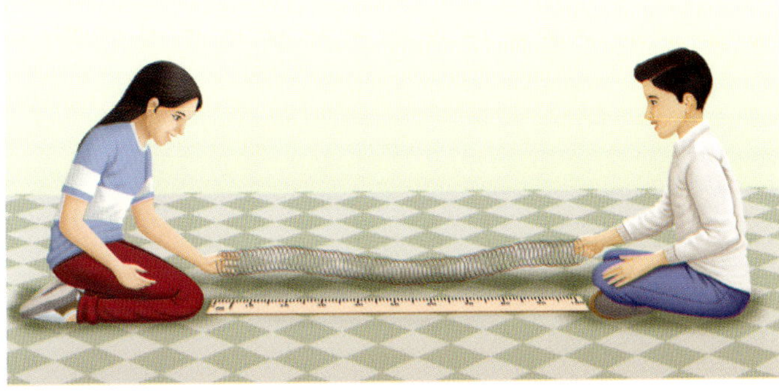

5. Produce large transverse (up and down) waves. Use the stopwatch and count how many waves are produced in 10 seconds.
6. Produce small longitudinal waves (back-and-forth). Use the stopwatch and count how many waves are produced in 10 seconds.
7. Produce large longitudinal waves (back-and-forth). Use the stopwatch and count how many waves are produced in 10 seconds.
8. Repeat steps 4–7 twice and average your three results for each.
9. Record all your measurements in your results table.

CONCLUSION

The two waves investigated are ______________ and ______________. The ______________ wave produced by the slinky has an up-and-down motion and the ______________ wave produced by the slinky has a back-and-forth motion. The average frequency calculated for the longitudinal wave is ______________ per second (Hz) and the average frequency calculated for the transverse wave is ______________ waves per second (Hz).

DISCUSSION

1. List the independent, dependent and controlled variables.
2. Which wave is faster?
3. Draw and label the two types of waves investigated.
4. Explain the difference between longitudinal and transverse waves.
5. Describe how frequency is calculated.

RESULTS TABLE I1.2

Wave	Small waves (number of waves per second) (Hz)				Large waves (number of waves per second) (Hz)			
	Trial 1	Trial 2	Trial 3	Average	Trial 1	Trial 2	Trial 3	Average
Longitudinal								
Transverse								

Investigation 1.5A

The law of reflection

AIM

To investigate the law of reflection

MATERIALS

- plane mirror
- concave and convex mirrors
- light box and power supply
- single slit plate
- protractor
- pencil
- sheet of white paper

METHOD

1 Copy the results table into your notebook, adding a title.
2 Set up your equipment, using the plane (flat) mirror, as shown in the diagram. Trace the mirror and the incident and reflected rays of light onto the paper.
3 Measure the angles that the incident ray (incoming ray) and reflected ray make to the normal. Record your results in the results table.
4 Repeat your experiment using at least three different angles of incidence.
5 Predict what will happen when rays of light strike a curved mirror. Will they be spread out (diverged) or brought together (converged)? Test your predictions with a concave mirror and a convex mirror.

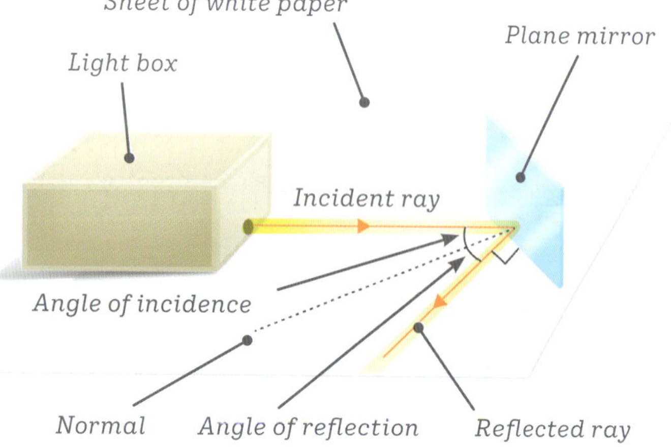

RESULTS

TABLE I1.5A

Ray	Angle of incidence (degrees)	Angle of reflection (degrees)
1		
2		
3		

Concave mirror diagram

Convex mirror diagram

DISCUSSION

1 Briefly describe the law of reflection.
2 Do your results support the law of reflection?
3 Compare what happened when light hit a concave mirror and a convex mirror.

CONCLUSION

Copy and complete.
The results show that: (*respond to the aim*).

Investigation 1.5B

Refraction

AIM

To observe refraction of light using different lenses

MATERIALS

- rectangular prism
- concave and convex lenses
- light box and power supply
- single slit plate
- pencil
- sheet of white paper

METHOD

1. Copy the results table into your notebook, adding a title.
2. Set up your equipment as shown in the diagram.
3. Direct a single ray of light to the centre of the prism.
4. Use the pencil to trace the path of the ray as it is refracted through the lens and out of the prism.
5. Remove the prism and use the protractor to measure the angle of incidence and angle of refraction. Record your results.
6. Repeat steps 3–5 using three angles of light from a light box with a concave lens and then with a convex lens. Trace the path of rays using your pencil.

RESULTS TABLE I1.5B

Ray	Angle of incidence (degrees)	Angle of refraction (degrees)
Light entering glass from air		
Light entering air from glass		

Concave lens diagram

Convex lens diagram

DISCUSSION

1. Compare the angle of light before it enters the rectangular prism and after it exits. Why is there a difference?
2. Compare what happened when light hit a concave lens and a convex lens.

CONCLUSION

Copy and complete.

The results show that: (*respond to the aim*).

Investigation 2.1

Acceleration changes due to mass

AIM

To investigate the effect of increasing mass on an object's acceleration

MATERIALS

- dynamics trolley
- mass carrier
- 2 × 100g masses
- 2 × 200 g masses
- 500 g mass
- bench mounted pulley
- fishing line
- stopwatch
- masking tape

METHOD

1 Copy the results table into your notebook, adding a title.
2 On a desk or benchtop, place two strips of masking tape about 1 metre apart. These strips represent the start and finish lines. Make sure there is at least a trolley's length of benchtop before and after each strip of tape.

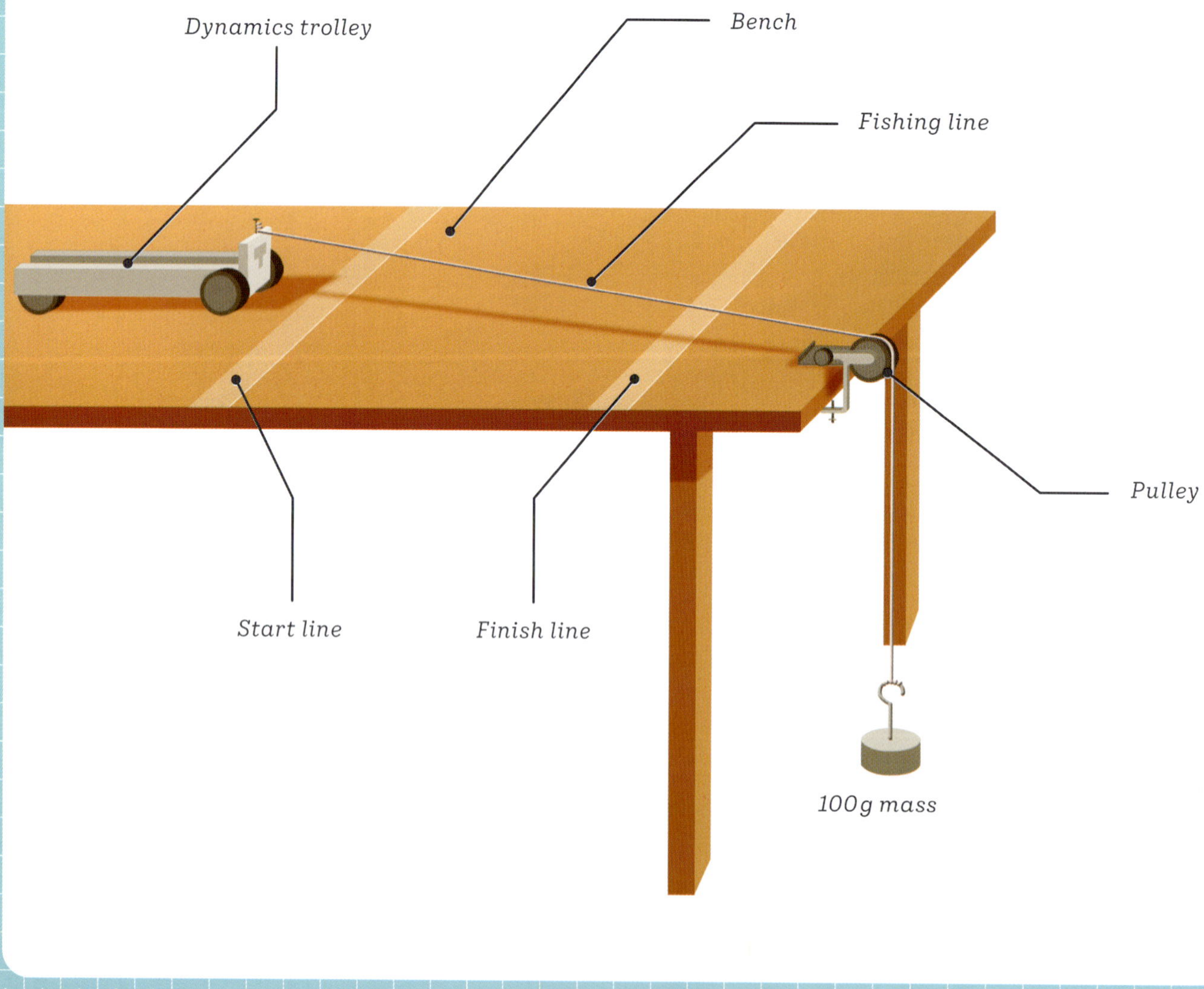

Dynamics trolley
Bench
Fishing line
Pulley
Start line
Finish line
100g mass

RESULTS **TABLE I2.1**

Mass on trolley (g)	Time (s) for trolley to travel from start line to finish line			
	Trial 1	Trial 2	Trial 3	Average
100				
200				
300				
400				
500				

3 Attach a 100 g mass to one end of the fishing line. Thread the other end over the pulley and attach it to the dynamics trolley, as shown in the diagram.
4 Release the mass so that it drops straight down to the ground. Use the stopwatch to time how long the trolley takes to travel from the start line to the finish line. Make sure to catch the trolley before it falls off the table. Perform three trials and record the times in your results table.
5 Add a 100 g mass to the trolley and repeat step 4.
6 Repeat step 4 with masses of 300, 400 and 500 g on the trolley. Record all your data.

DISCUSSION

1 Do your results follow a pattern? If so, describe what this is.
2 How do your results confirm Newton's second law?
3 What errors may have occurred in this experiment, and how could you overcome them next time?
4 Which part of the apparatus provides the force that accelerates the trolley?
5 Are there any other forces acting on the trolley? If so, what are they?

CONCLUSION

Copy and complete.

The results show that: (*respond to the aim*).

Investigation 2.2

Ticker timers

AIM

To investigate the relationship between speed, distance and time, using a ticker timer machine.

MATERIALS

- ticker timer
- 1 m strip of ticker tape
- 50 g mass
- sticky tape
- scissors
- power-pack
- 2 × electrical leads

METHOD

1 Copy the results table into your notebook, adding a title and rows as needed.
2 Use the sticky tape to attach the end of the ticker tape to the 50 g mass.
3 Thread the other end of the ticker tape through the ticker timer.
4 Start the ticker timer and drop (don't throw!) the 50 g mass to the ground.
5 Disconnect the tape from the mass and look at the tape. There should be a lot of ink dots on the tape. If the dots are difficult to see, repeat steps 2–4 with another strip of tape.
6 Draw a line through the first clear dot, and then every fifth dot after that. This represents a time of 0.1 seconds.
7 Measure the distance between each marked dot. This represents how far the mass travelled in each 0.1 second interval. Record the distance travelled in each interval in your results table. Note: You will need extra rows because you will have more than three intervals.
8 Find the speed of each interval in cm/s by dividing the distance travelled by the time taken (0.1 seconds). Enter the speed of each into the table.

DISCUSSION

1 What do you notice about the speed of each interval? What does this tell you about the speed of the mass through its journey?
2 What errors were present in the experiment?
3 What does this experiment indicate about the speed of a falling object?
4 If you performed this experiment using a high-speed video camera instead of a ticker timer, do you think you would record the same speeds? Why or why not?

CONCLUSION

Copy and complete.

The results show that: (*respond to the aim*).

RESULTS TABLE I2.2

Interval	Time of interval (s)	Distance travelled (cm)	Average speed (cm/s)
1	0.1		
2	0.1		
3	0.1		

Investigation 2.3

Vehicles and pedestrians

AIM

To investigate the speed travelled by vehicles and pedestrians

MATERIALS

- stop watch
- trundle wheel
- coloured markers or cones

METHOD

PART A

RECORDING THE SPEED OF VEHICLES

1 Copy the results table into your notebook, adding rows as needed.

2 Use a trundle wheel to measure out 30 m along the footpath beside the road. Place cones or markers at each end of the 30 m space.

3 Position one student at the start and one at the end of the 30 m space.

4 One student uses a stop watch to time how long it takes for a car to travel from the starting marker to the finishing marker. Record this time in the results table.

5 Repeat step 4 for a total of ten cars. If any pedestrians or cyclists travel the length of the 30 m space, record their time as well in order to compare it later to the speed the cars are travelling.

PART B

RECORDING THE SPEED OF PEDESTRIANS

6 Use the trundle wheel to measure out 10 m along the footpath beside the road. Place cones or markers at each end of the 10 m space.

7 Position one student at the start and one at the end of the 10 m space.

8 One student uses a stop watch to time how long it takes for Student 1 to walk the 10 m space. Record this time in the results table.

9 Repeat step 8, but instead of walking normally, record the speed for walking backwards, hopping, running and an activity of the student's choosing.

10 Students switch places and collect data for Student 2.

QUESTIONS

1 What is the formula used to calculate speed?

2 Consider the speed limit of the road you were on. Were any vehicles over the speed limit?

3 Which activity had the fastest student speed? Which had the lowest?

4 What errors may have led to inaccurate results during this investigation? How could you improve the method to ensure better accuracy?

RESULTS

TABLE I2.3

Vehicle	Distance	Time	Activity	Distance	Time
1	30 m		Student 1 walking	10 m	
2	30 m		Student 2 walking	10 m	
3	30 m		Student 1 walking backwards	10 m	
4	30 m		Student 2 walking backwards	10 m	
5	30 m		Student 1 hopping	10 m	
6	30 m		Student 2 hopping	10 m	
7	30 m		Student 1 running	10 m	
8	30 m		Student 2 running	10 m	
9	30 m		Student 1 student's choice	10 m	
10	30 m		Student 2 student's choice	10 m	

Investigation 2.4

Balloon rockets

AIM

To demonstrate Newton's third law, using balloon rockets

MATERIALS

- balloon
- fishing line
- sticky tape
- large drinking straw

METHOD

1 Inflate the balloon to about the size of a basketball. Hold the mouth of the balloon closed, but don't tie it off.
2 Attach the straw to the balloon as shown in the diagram. Thread the fishing line through the straw.
3 Ask a fellow student to hold one end of the fishing line while you hold the other, keeping it taut. Make sure the balloon is near one end of the fishing line so that most of the line is out the front of the balloon rocket.
4 Let go of the mouth of the balloon. Record your observations.

QUESTIONS

1 Describe the motion of the balloon relative to the air escaping from it.
2 How did the motion of the balloon demonstrate Newton's third law of motion? (Hint: In which direction did the balloon push the air?)
3 How could the set-up be improved to allow the balloon to travel further and faster?
4 What provided the force that accelerated the balloon?
5 If you inflate a balloon and then let it go when it isn't attached to a straw or the line, it flies all over the place in an unpredictable path. Why do you think this is?

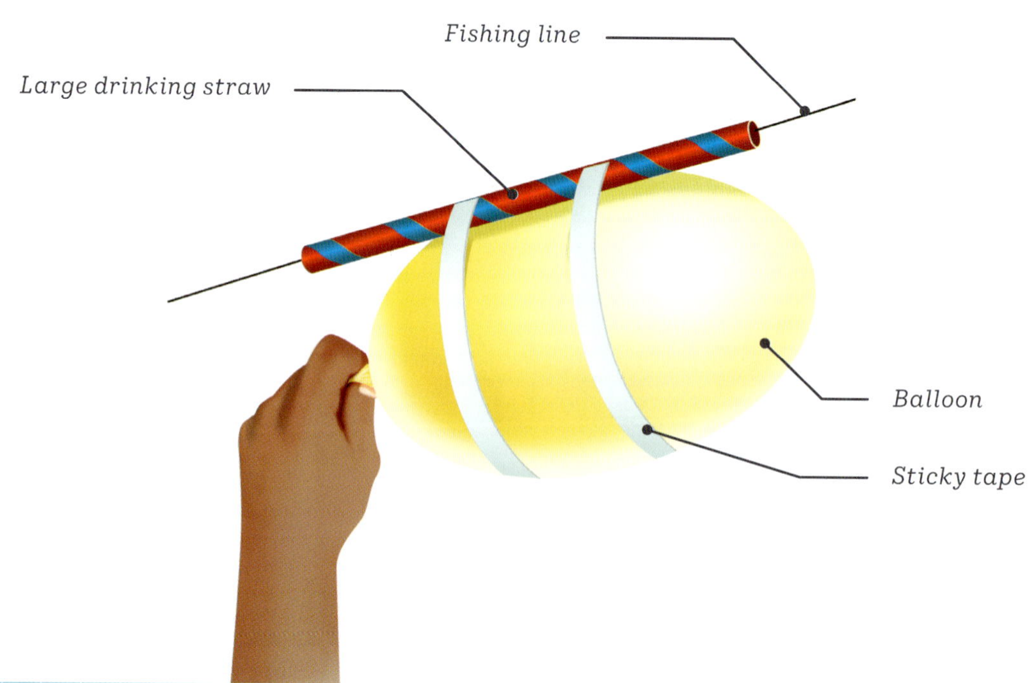

Investigation 2.5

Car safety

AIM

To investigate Newton's laws by using a model of a crash test dummy

MATERIALS

- plasticine
- dynamics trolley
- talcum powder
- ramp
- metre ruler
- brick

METHOD

1 Copy the results table into your notebook, adding a title.
2 Make three plasticine models – small, medium and large. These will represent a baby, a child and an adult.
3 Lightly dust the top of the dynamics trolley with talcum powder, and sit the 'baby' on top. Make sure it isn't stuck down.

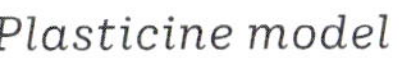

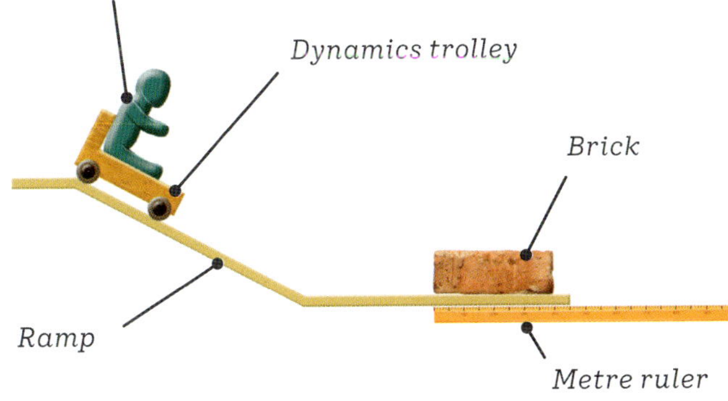

4 Set up the ramp and the brick as shown in the diagram. Place the metre ruler next to the brick so that the 0 cm mark is at the near edge of the brick.
5 Place the trolley with the plasticine 'baby' at the top of the ramp and then release it. The trolley should travel down the ramp and strike the brick at the bottom.
6 Record how far the 'baby' travels after the trolley strikes the brick.
7 Repeat steps 5 and 6 twice to obtain two more results. Calculate the average distance from the three trials.
8 Repeat steps 5–7 for the 'child' and 'adult' models.

DISCUSSION

1 On average, which sized person travelled the furthest after the trolley collided with the brick? Why do you think this is?
2 How could this experiment be improved to give more accurate results?
3 How does this experiment show that it is important to always wear a seatbelt in a car?
4 Describe how each of Newton's three laws can be demonstrated from one aspect of this experiment. (Hint: What happened to the motion of the brick and the cart when they collided?)

CONCLUSION

Copy and complete.

The results show that: (*respond to the aim*).

RESULTS TABLE I2.5

Plasticine model	Distance travelled (cm)			
	Trial 1	Trial 2	Trial 3	Average
Baby model				
Child model				
Adult model				

Investigation 3.1

Modelling a simple circuit

AIM

To model a simple circuit and the concepts of current and resistance

MATERIALS

- a piece of rope at least 4 metres long, tied in a loop
- two pairs of thick fabric gloves

METHOD

1 As a class, stand in a circle, holding the rope very loosely.

2 One student acts as the 'battery' and adds voltage to the circuit by pulling one end of the rope. Observe what happens to the rope (it should move in a circle).

3 The student increases the energy with which they pull the rope, just like a 'battery' adding more voltage. Observe what happens to the speed of the rope.

4 One or two students act as 'resistors'. They put on gloves and then hold the rope a little more tightly than the others. Observe what happens to the speed of the rope as resistance is added.

QUESTIONS

1 In what ways are models (like the one above) helpful in demonstrating scientific concepts?

2 What could you add or change in this model and role play in order to improve it?

3 Explain the role of the resistors in your own words.

4 How are resistors imporant in the conduction of electricity?

Investigation 3.2

Exploring Ohm's law

AIM

To investigate Ohm's law and determine the resistance of a resistor

MATERIALS

- fixed resistor (5–10 ohms)
- 0–12 V voltmeter
- 0–5 A ammeter
- variable DC power supply (2–12 V)
- 5 electrical leads

METHOD

1. Copy the results table into your notebook, adding a title.
2. Construct the circuit shown in the diagram but do not turn it on.
3. Set the power supply to 2 volts.
4. Close the switch. Note and record the readings on the ammeter and voltmeter in the results table.
5. Open the switch and increase the power supply by 2 volts. Note and record the readings on the ammeter and voltmeter.
6. Continue increasing the power supply by 2 volts, and recording the readings, until the results table is complete.

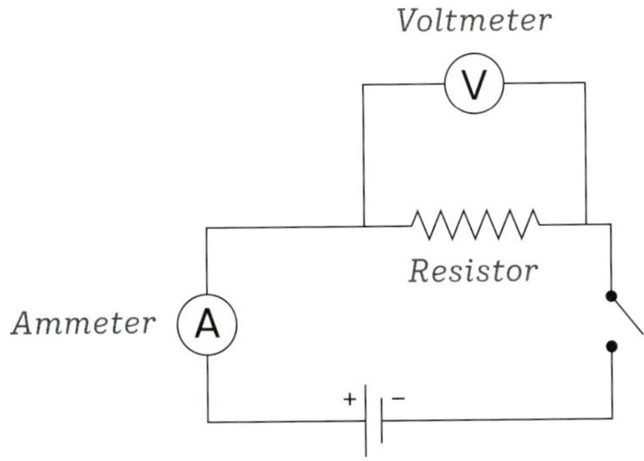

RESULTS

TABLE I3.2

Power (V)	Voltmeter reading (V)	Ammeter reading (A)
2		
4		
6		
8		
10		
12		

DISCUSSION

1. Identify the independent and dependent variables.
2. Draw a graph plotting current (y-axis) versus voltage (x-axis) using the data from your results table.
3. What does the slope of the line in the graph you drew equal?

CONCLUSION

Copy and complete.

The results show that: (*respond to the aim*).

Investigation 3.3

Series and parallel circuits

AIM

To investigate series and parallel circuits

MATERIALS

- 2 ammeters
- voltmeter
- 9 × electrical leads
- 2 × 10 ohm resistors
- power-pack

METHOD

1 Copy the results table into your notebook, adding a title.

PART A
SERIES CIRCUIT

2 Set up the circuit as shown in diagram 1.

3 Set the power-pack to 2 V and turn it on. Record the current and voltage across each resistor in the results table.

4 Repeat step 3, increasing the power supply by 2 V each time, until your results table is complete.

PART B
PARALLEL CIRCUIT

5 Set up the circuit as shown in diagram 2.

6 Set the power-pack to 2 V and turn it on. Record the current and voltage across each resistor in the results table.

7 Repeat step 6, increasing the power supply by 2 V each time, until your results table is complete.

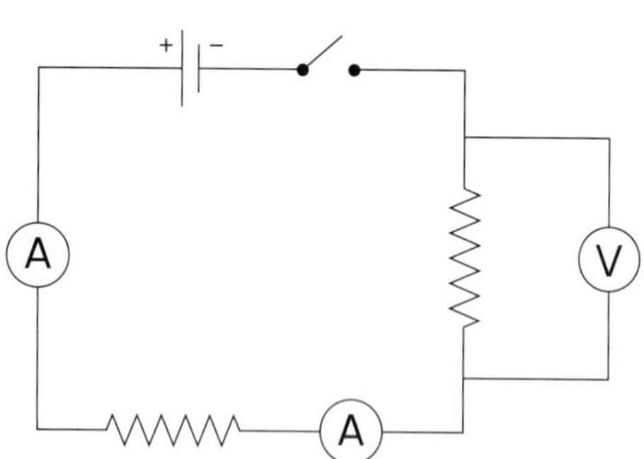

Diagram 1

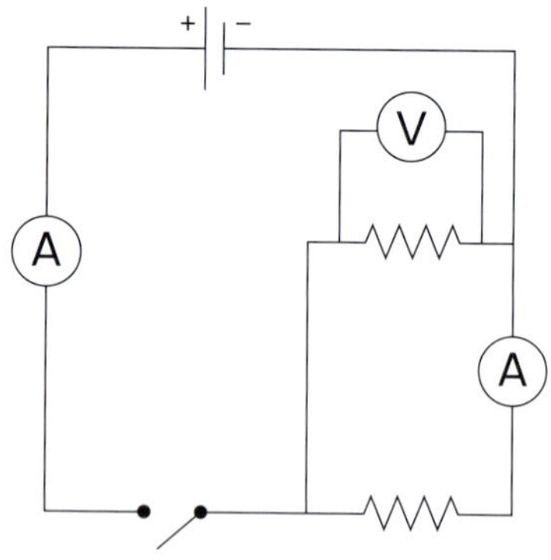

Diagram 2

RESULTS

TABLE I3.3

Power supply (V)	Part A		Part B	
	Voltmeter reading (V)	Ammeter reading (A)	Voltmeter reading (V)	Ammeter reading (A)
2				
4				
6				
8				
10				
12				

QUESTIONS

1 Consider the series circuit. What relationship exists between the voltage of the power-pack and the voltage across both resistors?

2 What do you notice about the voltage across each resistor in the parallel circuit?

3 What do you notice about the current through the ammeter of the series circuit compared to the parallel circuit?

Investigation 4.1

Galileo's pendulum

AIM

To investigate the law of conservation of energy by using a pendulum

MATERIALS

- metre ruler
- fishing line or poly string
- 50 g mass
- mass carrier (50 g)
- retort stand
- bosshead and clamp
- rubber stopper with hole
- pencil or pen

METHOD

1. Copy the results table into your notebook, adding a title.
2. Tie the mass carrier to the end of the fishing line, forming a pendulum.
3. Pass the other end of the fishing line through the rubber stopper and tie it off.
4. Insert the rubber stopper into a clamp on a retort stand as shown in the diagram.
5. Keeping the string taut, raise the mass up until the string is at about a 45° angle to the ground. Use the metre ruler to measure how far above the bench this is.
6. Release the mass, allowing it to swing backwards and forwards. Use the metre ruler to measure the height the mass returns to on its first swing. Record this height in your results table.
7. Interrupt the swing of the pendulum by placing a pencil in the path of the string. What height does the pendulum return to now?

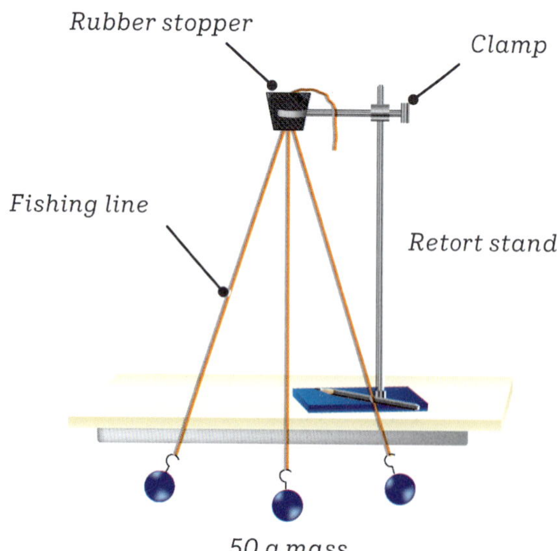

RESULTS TABLE I4.1

	Return height			
Starting height (cm)	Trial 1 (cm)	Trial 2 (cm)	Trial 3 (cm)	Average (cm)

DISCUSSION

1. Compare the return height to the starting height without the pencil in place. What did you notice?
2. Did the return height change when the pencil was placed in the path?
3. How could this experiment be improved?
4. How does this investigation demonstrate the conservation of energy?
5. Account for any energy losses in this investigation.

CONCLUSION

Copy and complete.
The results show that: (*respond to the aim*).

Investigation 4.2

Energy efficiency of bouncing balls

AIM

To investigate the energy efficiency of different types of balls

MATERIALS

- metre ruler
- sticky tape
- several different types of balls; e.g. tennis ball, golf ball, squash ball, basketball, superball, hi-bounce ball

METHOD

1 Copy the results table into your notebook, adding a title and rows as needed.
2 Tape the metre ruler vertically to the side of a desk or bench, so that the zero mark is at the bottom.
3 Take the first type of ball and hold it so that the 1-metre mark is just visible under the ball.
4 Drop the ball, measuring the maximum height the bottom of the ball reaches as it bounces back up. Record your results in your results table.
5 Repeat step 4 for two more trials of the same ball.
6 Repeat steps 3 and 4 for each type of ball.
7 Calculate the efficiency (%) of each type of ball by dividing the rebound height by the average drop height.

DISCUSSION

1 Which balls were the most and the least energy efficient?
2 Which ball lost the most energy?
3 How could this experiment be improved?
4 Is there a link between the loudness of the bounce and the energy efficiency of the ball?
5 Name two possible sources of energy loss for a bouncing ball.

CONCLUSION

Copy and complete.

The results show that: (*respond to the aim*).

RESULTS TABLE I4.2

Type of ball	Drop height (cm)	Rebound height				Efficiency $\left(\frac{\text{drop height}}{\text{average rebound height}} \times 100\right)$
		Trial 1 (cm)	Trial 2 (cm)	Trial 3 (cm)	Average (cm)	
Basketball	100					
Tennis ball	100					

Investigation 4.4

Effects of acid rain

Note: In this experiment you will use vinegar to represent acid rain. Vinegar has a similar pH to acid rain, but is more concentrated. Because of this, the effects seen will happen faster than they do in nature.

AIM

To investigate the effects of acid rain on organic and inorganic matter

MATERIALS

- 100–200 mL white vinegar
- 100–200 mL distilled water
- a selection of organic and inorganic materials; e.g. seashells, chalk, nails, flower petals, small pieces of apple, egg shells, green leaves. You will need two of each item.
- 2 large (500 mL) beakers

METHOD

1. Copy the results table into your notebook, adding a title and rows as needed.
2. Pour the white vinegar into one of the beakers, and the distilled water into the other. Label the beakers.
3. For each of the materials you are using, add one piece to the water beaker, and one to the vinegar beaker. If you only have one of an item, break it in half and add half to each beaker.
4. Record any immediate observations.
5. Leave the beakers for at least 1 day.
6. Compare the items that were in the water to those that were in the vinegar and record your observations.

RESULTS

TABLE I4.4

Item	Observations in water	Observations in vinegar

DISCUSSION

1. Describe the difference in appearance between the items in vinegar and those in water.
2. How could this experiment be improved?
3. Which item showed the greatest difference between being in water and being in vinegar?
4. Were any items unchanged?
5. What does this experiment show about the dangers of acid rain?
6. Consider the items you tested. Infer what sort of living and non-living things would be affected by acid rain. How would they be affected?

CONCLUSION

Copy and complete.

The results show that: (*respond to the aim*).

Investigation 5.1

Stargazing

AIM

To investigate the night sky

MATERIALS

- star map for the month and year you are viewing, or a stargazing app
- blank A4 paper
- pencil and highlighter
- red torch or torch with a red cellophane cover
- compass
- binoculars or telescope (optional)

METHOD

PART 1

EXPLORING THE NIGHT SKY

1. On a clear night, find a dark area where you will have a good view of the night sky.
2. Use the red torch to help you see your papers.
3. Use the compass to identify north, south, east and west in your location.
4. Orientate your star map.
5. Use your star map to identify:
   a. the Southern Cross and Pointers
   b. at least two other constellations
   c. a planet (if visible at the time you are observing).
6. Use the highlighter to highlight the objects on your sky map that you locate.
7. Compare what you observe with the naked eye to what you can see with a telescope or binoculars.

PART 2

LOCATE SOUTH USING THE SOUTHERN CROSS AND POINTERS

1. Draw an imaginary line with your finger from the top of the cross to the bottom and extending beyond it, as shown on the diagram.
2. Draw an imaginary line joining the two Pointers. Midway along this line, draw another line at right angles to it.
3. Where lines 1 and 2 meet is the South Celestial Pole. This is the point in the sky around which all the stars seen from the Southern Hemisphere rotate.
4. Locate south by dropping a vertical line from the South Celestial Pole to the horizon.

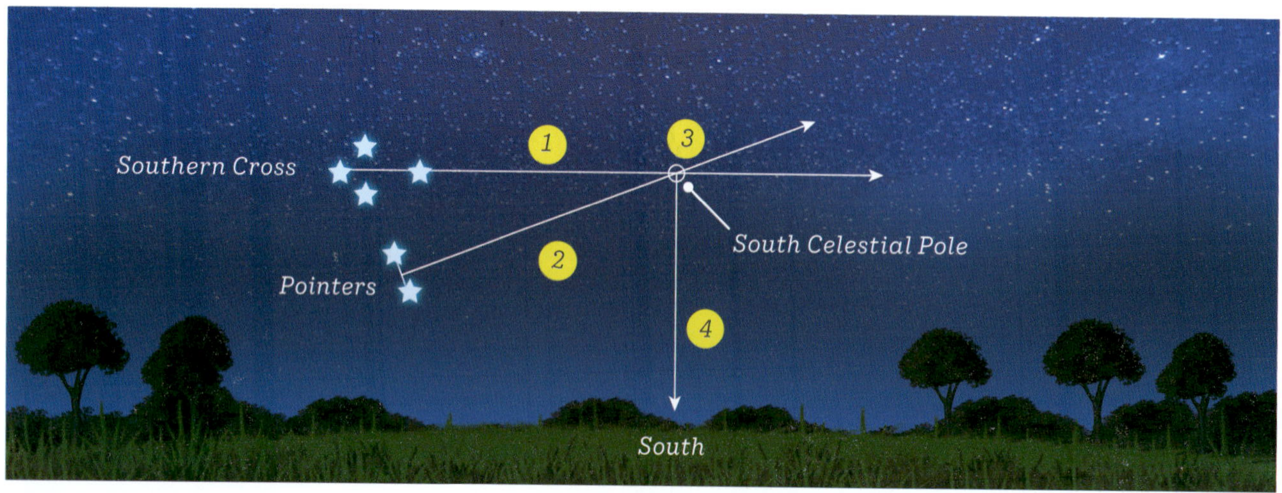

PART 3

OBSERVING THE MOTION OF THE STARS

1 Identify a bright star that will be easy for you to identify again. It should be close to an object such as a roof, tree or chimney.
2 Take note of where you are standing so that you can return to it later.
3 On your piece of paper, draw a silhouette of the object that you are comparing your bright star to, as shown in the diagram.
4 Mark the location of the bright star on your paper. Make a note of the time you make your observation.
5 For the period of your observations, return to the same spot every 30 minutes and note the location of the star.

QUESTIONS

PART 1

1 If you used binoculars or a telescope, describe the differences between the observations you made with them and those you made with the naked eye.

PART 2

2 Compare the technique using the Southern Cross to locate south with using the compass.

PART 3

3 Describe the movement of the star as time passed.
4 Identify the point that the movement is occurring around. Add this to your diagram if appropriate.
5 What causes this movement?

Investigation 5.3

Investigating orbits

AIM

To determine the relationship between orbit radius and orbital period

MATERIALS

- glass or plastic tubing 10–20 cm long
- 50 g mass
- mass carrier (50 g)
- string
- ruler
- rubber stopper
- stopwatch
- scissors
- marker pen

METHOD

1 Copy the results table into your notebook, adding a title.
2 Measure and cut 110 cm of string.
3 Tie one end of the string to the rubber stopper.
4 Thread the string through the tubing.
5 Tie the 50 g mass carrier to the other end of the string. Make sure there is 100 cm between the mass and the stopper.
6 Use the ruler and the marker to mark 10 cm intervals between the mass and the stopper.
7 With the stopper at a 50 cm distance from the tubing, hold the tube above your head and swing the stopper around until it is spinning steadily. The 50 g mass should hang in the same position. If it doesn't, you will need to increase or decrease the speed of spinning.
8 Use the stopwatch to time how long it takes for the stopper to complete 10 orbits of the tubing.
9 Record your results in the table and calculate the time taken for one orbit.
10 Repeat steps 7–9 for the other distances.
11 Draw a line graph of your results.

RESULTS

TABLE I5.3

Orbit radius (cm)	Time to complete 10 orbits (s)	Time to complete 1 orbit (s)
50		
40		
30		
20		
10		

QUESTIONS

1 Identify what represented the Sun and the planets in this model.
2 What was the relationship between distance from the tubing and time that it took the stopper to orbit?
3 What can you infer about the movement of the inner and outer planets from this model?
4 This table contains data for the distance from the Sun and orbital times. Create a line graph of this data.

Planet orbit distances and time to orbit

Planet	Distance from Sun (AU)	Time to orbit (Earth years)
Mercury	0.4	0.2
Venus	0.7	0.6
Earth	1.0	1.0
Mars	1.5	1.9
Jupiter	5.2	11.9
Saturn	9.5	29.5
Uranus	19.2	84.0
Neptune	30.2	164.8

5 Compare the trendline of your results with that of this second graph. How are they similar? How are they different?

Investigation 5.4A

Making a telescope

AIM

To use lenses to make a telescope

MATERIALS

- two convex lenses of different sizes
- ruler
- cardboard tube
- box cutter
- book, or sheet of paper with writing on it
- sticky tape
- Blu-Tack

METHOD

1 Hold the larger lens between you and the book or sheet of paper. The writing should look blurry when looking through the lens. This will be the primary or objective lens.
2 Hold the smaller lens between you and the larger lens. This will be the eyepiece lens. Move the lenses so that when you view the writing through the smaller lens it is in focus.
3 Record the distance between the first and second lenses.
4 Cut a slot in the cardboard tube about 2 cm from one end to hold the larger lens.
5 Cut a second slot for the smaller lens, the same distance away from the larger lens that you recorded in step 3.
6 Cut away any excess tubing, leaving about 2 cm.
7 Place the two lenses in the slots, holding them in place with tape or Blu-Tack.

QUESTIONS

1 Draw and label a scale diagram of your telescope. Include the objective lens, eyepiece lens and the measurements.
2 What is the purpose of the objective lens?
3 What is the purpose of the eyepiece lens?
4 What would you expect to happen with the detail you can see if you increase the size of these lenses? Explain.
5 What would you expect to happen with the length of the telescope if you increase the size of these lenses? Explain.

Investigation 5.4B

Lens diameter and resolution

AIM

To determine how lens diameter affects resolution

MATERIALS

- A4 piece of paper
- sticky tape
- marker pen
- ruler
- trundle wheel or long tape measure
- selection of binoculars and/or optical telescopes

METHOD

1 Copy the results table into your notebook, adding a title.
2 Measure and record the diameter of the lens of the binoculars and optical telescope and record it in the results table.
3 On the piece of paper, draw two thick lines 2 cm long and 2 cm apart. These will represent stars.
4 Use the sticky tape to tape the paper to a wall.
5 Position an observer at a point away from the wall where they observe the two lines as one.
6 The observer should walk towards the wall, stopping when they first observe the two lines as being distinct from one another.
7 Use the tape measure or trundle wheel to measure the distance between the observer and the wall. Record this in your results table.
8 Repeat steps 5–7 using the binoculars and telescopes that you have.
9 Construct a graph of your results.

QUESTIONS

1 What is the relationship between the diameter of the lens and the distance where the 'stars' are first observed as two separate objects?
2 Resolution is the ability to tell two separate objects apart, or the ability to see more fine detail. Which optical instrument had the best resolution and which had the worst?
3 Alpha Centauri is the brightest star in the constellation of the Pointers, and the third brightest star in the night sky. To the naked eye, Alpha Centauri appears as one star but it is a triple star system. Explain how resolution of an instrument is important for making these discoveries.
4 Compare the resolution of a telescope with the resolution of a microscope based on the ability to distinguish between objects that are close together.

RESULTS TABLE I5.4B

Optical instrument	Diameter of lens (mm)	Distance from 'stars' where they are first observed as two separate objects (m)
Human eye	10	
Binoculars		
Telescope		

Investigation 5.5

Modelling the expanding universe

AIM

To determine how a galaxy's distance from a reference point affects its speed

MATERIALS

- a round balloon
- 5 different coloured stick-on dots
- piece of string approximately 50 cm
- ruler
- stopwatch

METHOD

1. Copy the results table into your notebook, adding a title.
2. Blow up the balloon to about 150 mm and hold the nozzle closed (do not tie it up).
3. Stick the 5 dots (galaxies) onto the balloon (universe). Try to spread them evenly around the balloon.
4. Select one of the dots to be your home galaxy. Use the string to measure the distance between your home galaxy and the other galaxies, as shown in the diagram. Measure the string with the ruler to determine this distance in centimetres.
5. Record these distances in the results table. (Note: The distance to your home galaxy will be 0 cm.)
6. Fully inflate the balloon and use the stopwatch to time how long it takes. Tie the balloon and record the time taken in the results table.
7. Use the string and ruler to again measure the distance from your home galaxy to the other galaxies. Record these distances in the results table.
8. Calculate the change in distance by subtracting the first measurements from the second measurements.
9. Calculate the speed of each galaxy by dividing the change in distance (cm) by the time it took the balloon to inflate (s). (Note: The speed of your home galaxy will be 0 cm/s.)
10. Plot a line graph with speed (cm/s) on the y-axis and distance (cm) on the x-axis.

QUESTIONS

1. a Are the speeds of all the galaxies the same?
   b If not, what is the relationship between speed and the distance from the home galaxy?
2. Suggest what would happen to the results if you had used a different home galaxy.
3. Explain what the results of this investigation tell you about the way that the universe is expanding.
4. Use the results from this investigation to explain why we observe light from distant galaxies more redshifted than those closer to us.

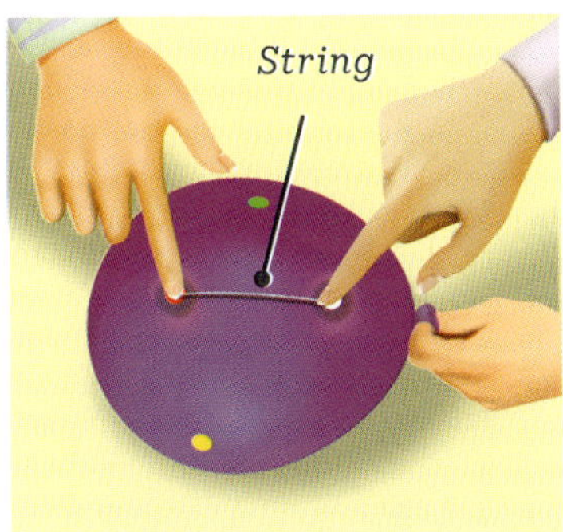

RESULTS TABLE 15.5

Colour of dot	Distance from home galaxy before inflating (cm)	Distance from home galaxy after inflating (cm)	Change in distance (cm)	Speed $\left(\frac{\text{distance}}{\text{time}}\right)$ (cm/s)
Red				
Green				
Blue				
White				
Yellow				
Time to fully inflate the balloon (s)				

Investigation 6.1

Modelling seafloor spreading

AIM

To model the process of seafloor spreading

MATERIALS

- 2 sheets of A4 paper
- pencil
- ruler
- coloured pencils or crayons
- sticky tape
- scissors

METHOD

1 Place a piece of A4 paper in landscape orientation. Use the ruler and pencil to draw a straight line down the centre of the paper as shown. Label the line 'Mid-ocean ridge'.

2 Carefully use the scissors to cut along the line so that there is a slit in the paper.

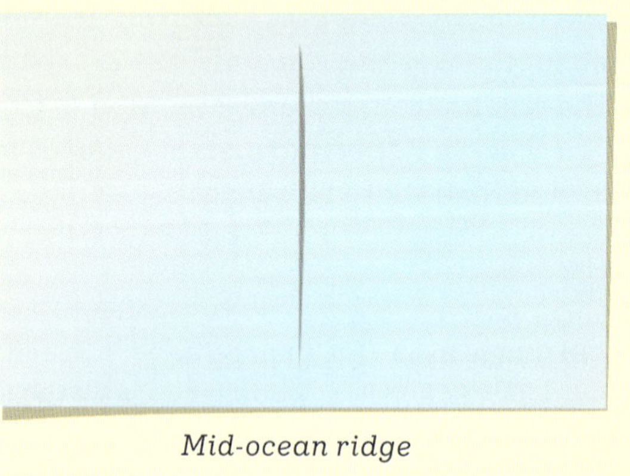

Mid-ocean ridge

3 Cut the second piece of A4 paper in half lengthwise as shown.

4 Tape the two pieces together along one of the short ends. This represents the oceanic crust being formed at the mid-ocean ridge.

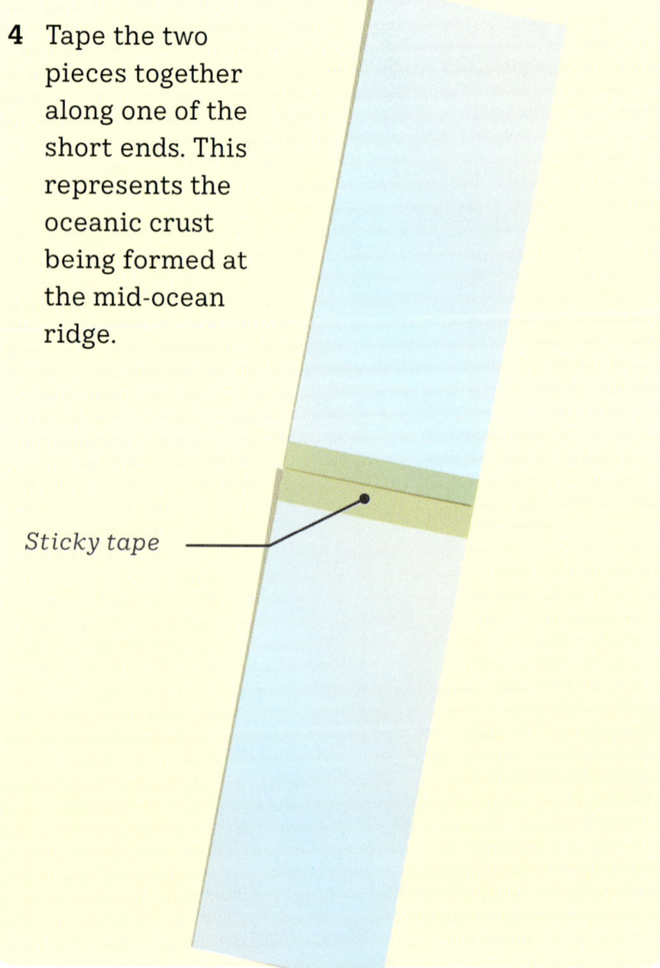

Sticky tape

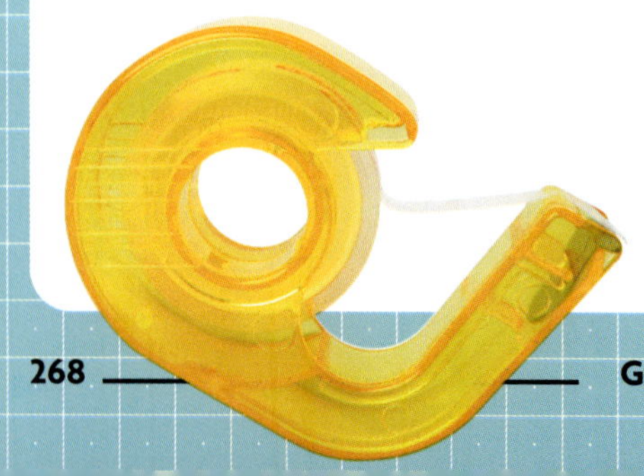

5 Carefully insert the two ends of this paper into the mid-ocean ridge slot from underneath. Separate the pieces so that one will move towards the left and the other towards the right. Leave about 2 cm exposed.

6 Write 'Continent' on both sides of the exposed paper and draw a line where the paper comes out of the mid-ocean ridge, as shown.

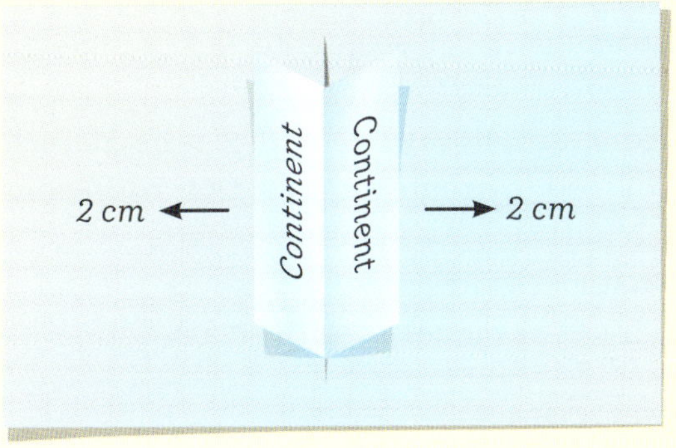

7 Evenly pull the paper outwards 1–3 cm and colour the white paper with the first colour.

8 Repeat step 7, using different colours until all of the paper has been pulled through the mid-ocean ridge.

9 Remove the coloured strip of paper for analysis.

QUESTIONS

1 What do the different coloured stripes on your model represent?

2 Describe any patterns that you observe in the stripes.

3 Your stripes may not all be of the same width. What does this represent?

4 Critique how well this model demonstrates seafloor spreading. How could it be improved?

Investigation 6.2

Modelling plate boundaries

AIM

To model the interactions happening at tectonic plate boundaries, using a Mars bar

MATERIALS

- paper towel or napkin
- Mars bar

METHOD

1. Wash your hands before you begin.
2. Unwrap the Mars bar. Hold it over the paper towel and use your fingernails to break the chocolate all around the middle of the bar.
3. Very slowly start to pull the Mars bar apart. Record your observations by drawing a diagram or taking a photograph.
4. Carefully place the two pieces of Mars bar back together again. While keeping them together, push each side in opposite directions. Record your observations by drawing a diagram or taking a photograph.
5. Squeeze the two pieces of Mars bar together from either end. Record your observations by drawing a diagram or taking a photograph.

QUESTIONS

1. Identify the types of plate boundary that you modelled in steps 2–4.
2. What does the top layer of chocolate represent?
3. What do the inside layers of the Mars bar represent?
4. Were you able to model subduction? Explain why or why not.

Investigation 6.3A

Modelling slab pull

AIM

To create a model that represents the force of slab pull on a tectonic plate

MATERIALS

- a variety of materials provided by your teacher

METHOD

1. Review the information on page 76 about how the force of slab pull moves a tectonic plate. You could also use the internet to undertake some further research.
2. Brainstorm how you could use the materials provided to create a model that represents slab pull moving a tectonic plate.
3. Construct your model.
4. Share your model with the class.

QUESTIONS

1. Explain how your model represents the force of slab pull moving a tectonic plate.
2. Critique your model. Is there anything that is inaccurate? Could it be improved?

Investigation 6.3B

Observing convection currents

AIM

To investigate the movement of convection currents

MATERIALS

- potassium permanganate crystals
- glass tube
- 500 mL beaker
- Bunsen burner
- tripod
- heatproof mat
- wire gauze
- matches

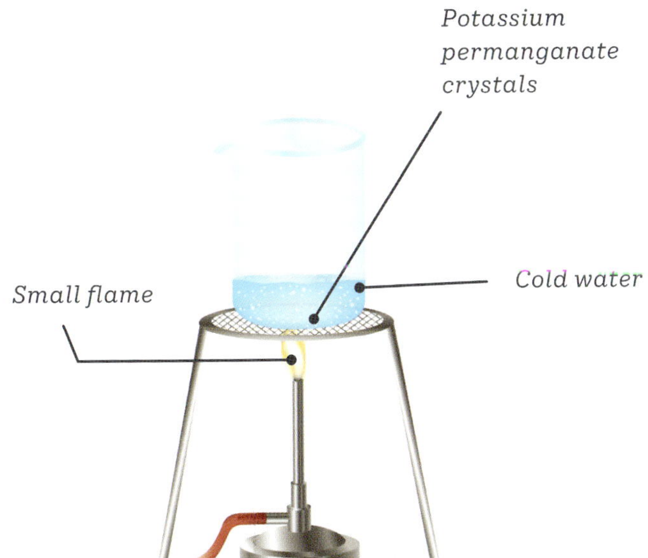

METHOD

1 Set up the Bunsen burner and heatproof mat.
2 Fill the beaker with cold water and place it on the tripod as shown in the diagram.
3 Use the glass tube to carefully place a couple of crystals of potassium permanganate on the bottom of one side of the beaker.
4 Light the Bunsen burner and place the flame under the crystals.
5 Observe how the coloured water moves.

QUESTIONS

1 Draw a diagram that illustrates the movement of the coloured water that you observed.
2 Explain what happened to cause the pattern of movement. Use the terms *heat* and *density* in your explanation.
3 Relate the movement of the coloured water in the beaker to the movement of rock in the mantle.

CONCLUSION

Copy and complete.
The results show that: (*respond to the aim*).

AN OPEN FLAME IS A HAZARD. TAKE CAUTION. IF YOU BURN YOURSELF, TELL YOUR TEACHER IMMEDIATELY AND PLACE YOUR SKIN UNDER COLD RUNNING WATER FOR 20 MINUTES.
DISPOSE OF WASTES APPROPRIATELY.

Investigation 6.4

Slinky waves

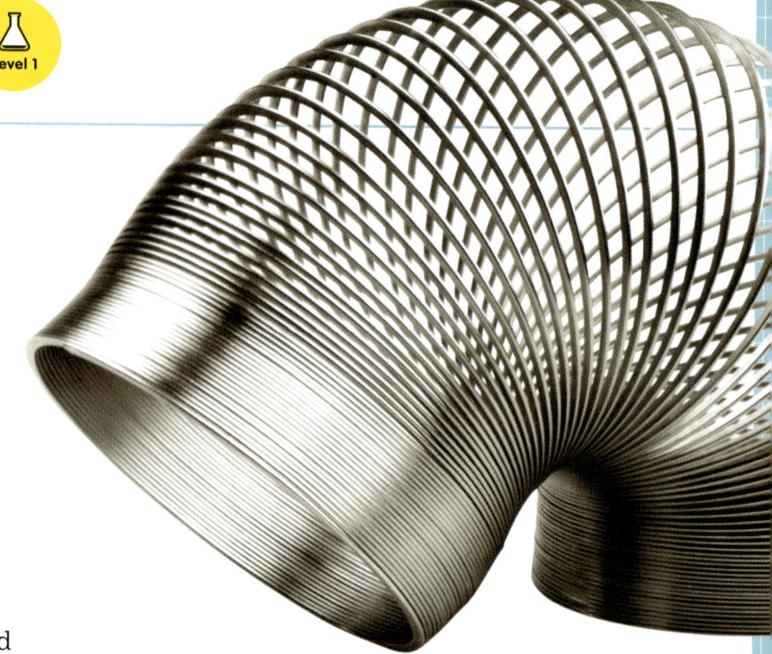

AIM

To use slinkies to model different seismic waves

MATERIALS

- 2 slinkies
- paperclip
- 5 cm piece of wool or string

METHOD

1 Copy the results table into your notebook, adding a title.
2 Have one person hold one end of the slinky on the ground. A second person holds the other end in the air above the first.
3 Tie the piece of wool or string in a bow around one of the coils.
4 The first person should pull some of the coils directly down and release them.
5 Observe how the wool moves and record your observations.
6 When the slinky is still again, the first person should move the bottom of the slinky from side to side.
7 Observe how the wool moves and record your observations.
8 Use the paperclip to create a hook to connect the end of one slinky to the middle of the other slinky. Have people hold each of the three ends. One slinky should be held vertically and the other horizontally.

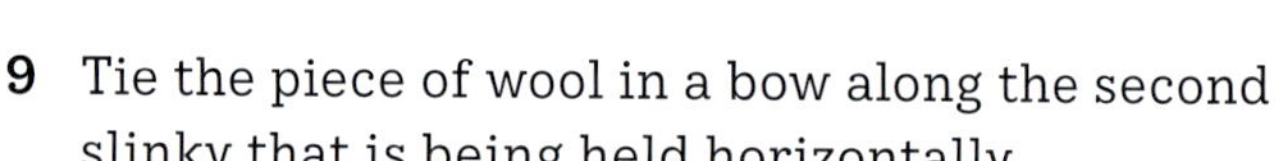

9 Tie the piece of wool in a bow along the second slinky that is being held horizontally.
10 The person holding the bottom of the vertical slinky should pull some of the coils down and to the side before releasing them.
11 Observe how the wool moves and record your observations.

QUESTIONS

1 Identify which movements relate to primary waves and secondary waves.
2 What were you modelling in steps 9–11?

RESULTS **TABLE I6.4**

Movement	Observations – description of how the string on the slinky moves
Compression	
Side to side	
Both	

Investigation 6.5A

Viscosity of lava

AIM

To investigate how the viscosity of a substance affects how well it flows over a distance

MATERIALS

- whipped cream
- thickened cream
- tray
- protractor
- 2 stopwatches
- 2 spatulas

METHOD

1 Copy the results table into your notebook, adding a title.

1 Use the spatula to place dollops about the size of a 20-cent coin of each type of cream at one end of the tray.

2 Start both stopwatches and immediately lift up that end of the tray up so that the tray is at a 45° angle.

3 Time how long it takes for each dollop of cream to reach the opposite end of the tray.

4 Record any other observations.

5 Clean and dry the tray and repeat steps 2–4 two more times.

QUESTIONS

1 Compare the flow rates of the two types of creams.

2 Explain how viscosity affected the time it took for the cream to travel down the length of the tray.

3 Relate the types of cream used in this investigation to lava from hot spot and strato volcanoes and the shape of volcano they form.

RESULTS **TABLE I6.5A**

	Time taken to travel the length of the tray (s)				Distance (cm)	Flow rate (distance/time) (cm/s)
	Trial 1	Trial 2	Trial 3	Average		
Whipped cream						
Thickened cream						

Investigation 6.5B

Wax volcano

AIM

To model a volcanic eruption using wax

MATERIALS

- coloured candle wax
- washed sand
- 500 mL beaker
- hot plate or Bunsen burner

METHOD

1 Melt the candle wax to form a 1 cm layer in the bottom of the beaker and allow it to set.
2 Cover the wax with a 1–2 cm layer of sand.
3 Carefully fill the beaker with water.
4 Place the beaker on the hotplate and turn on the heat.
5 Observe from a safe distance with written notes or by taking a short video.
6 Record the results of the eruption by drawing a diagram or taking a photograph.

QUESTIONS

1 Describe the eruption.
2 What structures of Earth do these layers represent in this model?
  a wax
  b sand
  c water
3 Critique this model. How does it reflect real volcanic eruptions? How is it different?

HOT WAX CAN BURN.
AN OPEN FLAME IS A HAZARD. TAKE CAUTION. IF YOU BURN YOURSELF, TELL YOUR TEACHER IMMEDIATELY AND PLACE YOUR SKIN UNDER COLD RUNNING WATER FOR 20 MINUTES.

Investigation 7.1

Observing interactions between the spheres

AIM

To investigate different interactions between the atmosphere, lithosphere, hydrosphere and biosphere

MATERIALS

- camera

METHOD

1 Copy the results table into your notebook, adding rows as required.
2 Find a space outside. This can be a natural or manufactured environment.
3 Sit and quietly observe your surroundings.
4 Record any interactions you can clearly observe between two or more of the four Earth spheres.

QUESTIONS

1 Describe the location where you made your observations.
2 What sphere(s) was easiest to observe? Why do you think this was the case?
3 What sphere(s) was hardest to observe? Why do you think this was the case?
4 Was there an observation you made that involved all of the spheres? If not, what was the maximum number of spheres that you observed interacting? Explain what the interaction was and how all of the spheres were involved.

RESULTS TABLE I7.1

Observation	Spheres that are involved				Explanation
	Atmosphere	Lithosphere	Hydrosphere	Biosphere	

Investigation 7.2A

Releasing dinosaur breath

AIM

To investigate the reaction between chalk and dilute hydrochloric acid

MATERIALS

- crushed chalk
- 1 mol/L acid
- 25 mL limewater (diluted calcium hydroxide)
- 100 mL conical flask
- 25 mL measuring cylinder
- boiling tube
- test-tube rack
- balloon

METHOD

1. Half-fill the boiling tube with limewater and place it in the rack.
2. Carefully add the crushed chalk to the conical flask.
3. Use the measuring cylinder to add 25 mL of 0.1 mol/L hydrochloric acid to the chalk.
4. Quickly and carefully place the opening of the balloon over the neck of the flask.
5. Once the reaction has slowed and the balloon is no longer inflating, pinch the balloon tightly around the neck as you remove it from the flask.
6. Place the neck of the balloon over the boiling tube and squeeze the balloon to force the gas in.

QUESTIONS

1. Chalk is the remains of tiny marine organisms with shells made of calcium carbonate ($CaCO_3$). Use your knowledge of the carbon cycle to identify where the carbon in the chalk came from. Where was it before then?
2. Write the word equation for the chemical reaction that occurred between the chalk and the acid.
3. Why could you say that 'dinosaur breath' was released from the chalk? Draw a diagram of the carbon cycle to help illustrate your response.

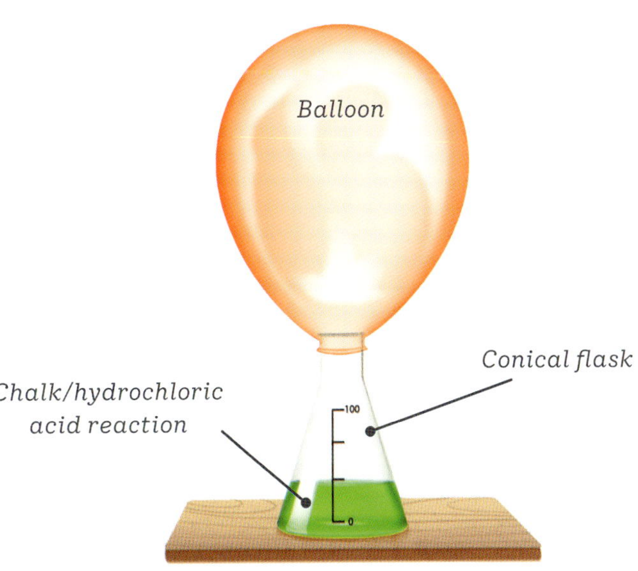

Investigation 7.2B

The effect of temperature on soil respiration

AIM

To determine how temperature affects the respiration rate of soil microbes (and therefore the rate at which they release carbon).

MATERIALS

- well-dried topsoil
- 1% sugar water solution
- 150 mL limewater (diluted calcium hydroxide)
- 3 × 600 mL plastic bottles
- 3 × 100 mL conical flasks
- 3 one-hole stoppers with glass tubes, to fit bottles
- 3 one-hole stoppers with glass tubes, to fit conical flasks
- 3 pieces of rubber tubing
- 25 mL measuring cylinder
- 50 mL measuring cylinder
- thermometer
- tape
- scissors
- safety blade
- small trowel or scoop
- marker pen
- ruler
- heat lamp

METHOD

1. Copy the results table into your notebook, adding a title.
2. Place a stopper in the mouth of a bottle. Use a ruler to measure and the pen to mark 1 cm from the base of the stopper. Remove the stopper.
3. Measure 4 cm from the top of the bottle and mark. Use the safety blade to cut about 80% of the way around the bottle. Carefully bend back the top.
4. Fill the bottle approximately one-third full with soil then use the measuring cylinder to add 20 mL of 1% sugar solution. Gently tap the bottle on the bench to compact and settle the soil.
5. Add more soil until the bottle is two-thirds full and use the measuring cylinder to add 20 mL of 1% sugar solution. Gently tap the bottle on the bench.
6. Fill the bottle with soil up to the cut. Use the measuring cylinder to add 20 mL of 1% sugar solution. Gently tap the bottle on the bench.
7. Flip the top back into place and seal the cut with tape.
8. Carefully add more soil until it reaches the mark 1 cm from the base of the stopper. Use the measuring cylinder to add 10 mL of 1% sugar solution. Gently tap the bottle on the bench.
9. Insert the stopper with glass tubing.
10. Repeat steps 2–9 for the other bottles.
11. Use a measuring cylinder to add 50 mL of limewater to each of the conical flasks.
12. Insert the stoppers with glass tubing into the conical flasks, ensuring the tubing is in the limewater.

13 Connect one soil bottle to each of with the conical flasks with the rubber tubing.
14 Place one apparatus under a heat lamp.
15 Place the second apparatus in a fridge.
16 Place the third apparatus in the classroom.
17 Record the air temperatures for each location.
18 Record your observations of the colour of the limewater over the next 3–5 days.

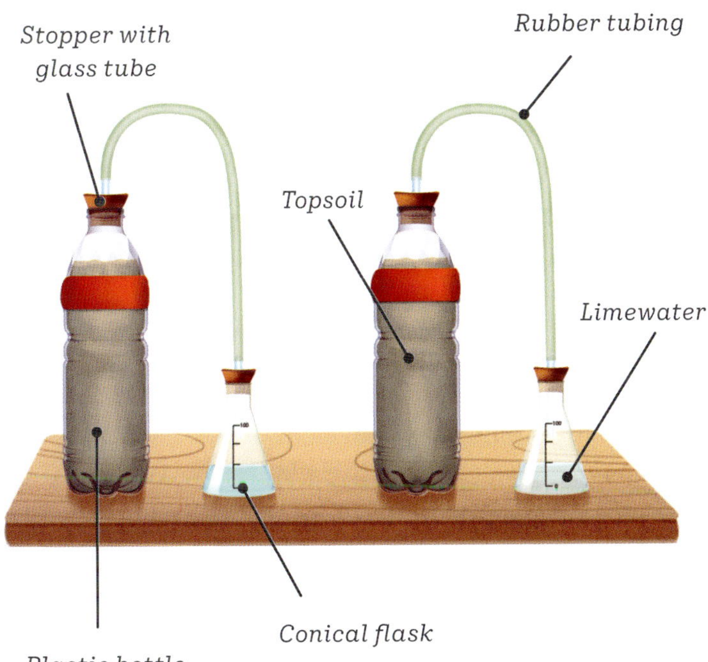

RESULTS **TABLE I7.2B**

Apparatus location	Temp. (°C)	Observations
Heat lamp		
Fridge		
Classroom		

DISCUSSION

1 Compare the colour in the limewater across the three flasks. What does a change in colour in the limewater indicate?
2 Microorganisms in the soil consume organic material, taking in carbon. They then release this carbon as carbon dioxide when they undergo cellular respiration. Compare the rates of cellular respiration in each flask, identifying which flask had the least and the most.
3 Describe the relationship that you observed between temperature and soil microbe respiration rate. Was this expected? Why or why not?
4 Identify the dependent, independent and controlled variables in this investigation.
5 Were there any variables that were not controlled? How could have they affected the results? Propose changes to the procedure to ensure that these variables are controlled.
6 Where does the carbon dioxide released by soil microbes usually go?
7 What respiration rate would you expect to observe in soils in a region that experiences a:
  a very cold climate?
  b tropical climate?
8 Is there any evidence from your investigation that a warming climate could change soil microbe respiration rates?

CONCLUSION

Copy and complete.
The results show that: (*respond to the aim*).

Investigation 7.5

The greenhouse effect

AIM

To investigate how a greenhouse affects air temperature

MATERIALS

- 3 glass microscope slides
- sticky tape
- 2 thermometers
- stopwatch
- Blu-Tack/plasticine

METHOD

1. Copy the results table into your notebook, adding a title and rows as needed.
2. Use the sticky tape to stick the long edges of the microscope slides together to form a triangular prism. This is your greenhouse.
3. Use the tape to seal shut one of the open ends.
4. Place the bulb of a thermometer through the other end of the prism before sealing the end with tape. Use Blu-Tack to support the thermometer so that the bulb does not touch the glass slides.
5. Place your greenhouse outside in direct sunlight. Place the second thermometer next to the greenhouse, using Blu-Tack or plasticine to support it. Make sure the thermometer bulb does not touch any surface.
6. Record the initial temperature of both thermometers.
7. Record the temperature of both thermometers every minute for 20 minutes.
8. Quickly cover or remove the greenhouse and thermometers from sunlight.
9. Continue to record the temperature every minute for a further 20 minutes.

DISCUSSION

1. Create a graph of your results.
2. Describe the trend shown on your graph.
3. Describe the difference between the temperature in the greenhouse and the outside air temperature when placed in sunlight to warm.
4. a Describe the difference between the temperature in the greenhouse and the outside air temperature out of the sunlight.
   b Explain how this is similar to the greenhouse effect caused by greenhouse gases.
5. Identify the independent, dependent and controlled variables in this investigation.
6. Identify any variables that should have been controlled. How would these have affected the results? Suggest changes to the method that would have ensured these were controlled.
7. Use the data gathered in this investigation to explain why horticulturists would grow plants inside greenhouses as opposed to in the open air.

CONCLUSION

Copy and complete.

The results show that: (*respond to the aim*).

RESULTS **TABLE I7.5**

In sunlight			Out of sunlight		
Time (min)	Temperature (°C)		Time (min)	Temperature (°C)	
	Inside greenhouse	External to greenhouse		Inside greenhouse	External to greenhouse

Investigation 7.6

Investigating sunscreens

AIM

To determine which material best blocks UV radiation

MATERIALS

- 5 sheets of sun-sensitive paper
- at least four different sun-protection materials (e.g. sunscreen, zinc cream, sunglasses, fabric)
- 5 ziplock bags
- tape
- stopwatch
- camera

METHOD

1. In a darkened room and away from UV light, label the back of a piece of sun-sensitive paper 'Control' and place it inside a ziplock bag.
2. Label the back of a second piece of sun-sensitive paper 'Sunscreen' and place it inside the ziplock bag. Apply the sunscreen to the front of the ziplock bag.
3. Repeat step 3 for the other sun-protection materials being tested, applying the cream or taping the material to the front of the ziplock bag.
4. Carefully cover the bags while you transport them outside.
5. Quickly set them up in direct sunlight and start the stopwatch.
6. Leave the bags for 5 minutes or as per the instructions on the sun-sensitive paper.
7. Bring the bags back inside.
8. Remove the paper from the bags and follow the instructions on the sun-sensitive paper to fix the image.
9. Photograph the results or record your observations.

QUESTIONS

1. Compare the papers that were 'protected' to the control. Were there significant differences?
2. **a** Which material was the most effective at blocking the UV radiation?
   b Which material was the least effective at blocking the UV radiation?
   c Why do you think this was the case?
3. Why is it important to block as much UV radiation as possible from hitting the skin?
4. Propose what you would expect to see from your results if the Montreal Protocol was not enacted.
5. Design an investigation that compares variations of one type of material; for example, different sunscreen brands, amount of sunscreen applied or different fabrics.

Investigation 8.1A

Sensory receptors

AIM

To determine which of three areas of the body has the highest concentration of pressure receptors

MATERIALS

- ruler
- 3 lengths of copper wire approximately 10 cm long

METHOD

NOTE: You will be testing three different areas to determine if the subject can tell the difference between being poked with one tip, two tips close together (5 mm apart), or two tips far apart (15 mm). You will need to be consistent with where you poke the subject (and be gentle!).

1 Copy the results table into your notebook, adding a title and rows as needed.
2 The test subject closes their eyes and holds out their dominant hand with the fingers together, palm up.
3 The tester GENTLY pokes the subject's forefinger with the sets of wires. You need to poke the subject 10 times with each set of wires (a total of 30 times), but don't do it in order. Mix up which wires you use and do it randomly so the tester can't guess which wire you are using.
4 Each time they are poked, the subject must say which set of wires were used. The tester records the subject's results in the table, using a tick for a correct statement and a cross if they were wrong. Do not tell the subject whether they are right or wrong.
5 Repeat steps 3 and 4 for the forearm and the back of the neck.

DISCUSSION

1 Which part of the body had the highest concentration of receptors (i.e. the subject got the most correct)? Why do you think this is?
2 Which part of the body was the least sensitive (i.e. the subject got the most incorrect)? Why do you think this is?
3 Compare your results with those of the rest of the class. Were they consistent?
4 Suggest two ways that you could improve this investigation in order to eliminate sources of error.

CONCLUSION

Copy and complete.

The results show that: (*respond to the aim*).

RESULTS TABLE I8.1A

Trial	Finger			Forearm			Back of neck		
	1 point	2 points 5 mm apart	2 points 15 mm apart	1 point	2 points 5 mm apart	2 points 15 mm apart	1 point	2 points 5 mm apart	2 points 15 mm apart
1									
2									
3									
4									
Total correct									

Investigation 8.1B

Reaction time

AIM

To determine whether a person's dominant or non-dominant hand has a faster reaction time

MATERIALS

- 30 cm ruler

METHOD

1 Copy the results table into your workbook, adding a title and rows as needed.
2 Have a partner sit at a table, with their arm resting on the table and their hand resting over the edge. Get them to hold their thumb and forefinger about 2 cm apart, so that they can catch the ruler in a pincer grip.
3 Hold the ruler upside down between their open thumb and forefinger but just above it, so that the 0 cm-mark on the ruler is aligned with the top of their thumb.
4 Without any obvious warning, drop the ruler for your partner to catch. Record the distance that they caught it at in the results table.
5 Repeat another 9 times for that hand. Then repeat the 10 trials for their other hand.
6 Calculate the average distance in centimetres that the ruler was caught in.
7 Use this value to calculate their reaction speed for each trial using the following formula:
where t = time to react in seconds
d = distance ruler dropped in cm
g = acceleration due to gravity, or 9.8 m/s^2.
8 Calculate the average reaction time.

DISCUSSION

1 Which hand had the faster reaction time? Was this what you expected? Explain why/why not.
2 Compare your results with those of the rest of the class.
3 Propose the series of actions and reactions that must occur in your body for you to react and catch the ruler.
4 Identify three variables that you kept consistent while undertaking this practical. Explain why it was important each of them was controlled.
5 Identify at least one variable that you should have kept consistent while undertaking this practical. Explain how you should have controlled it and how this may have affected the results.
6 Was there a relationship between the trial number and reaction distance? Suggest why this might have been the case.

CONCLUSION

Copy and complete.
The results show that: (*respond to the aim*).

RESULTS **TABLE I8.1B**

Trial	Distance dropped before ruler caught (cm)	
	Dominant hand	Non-dominant hand
1		
2		
3		
4		
Average reaction distance (cm)		
Average reaction time (s)		

Investigation 8.2

Sheep brain dissection

AIM

To investigate the structure of a mammalian brain

MATERIALS

- sheep brain
- dissection kit
- gloves
- dissecting board
- newspaper
- paper towel
- disinfectant

METHOD

1 Place some newspaper on the bench and the dissecting board on top of it.
2 Carefully place the sheep brain on the dissecting board. Identify the cerebrum, cerebellum and brain stem.
3 Draw a diagram of the brain, labelling these structures. (Alternatively, take a photograph.)
4 Notice that the cerebrum is made up of two halves, called hemispheres. Use the scalpel to carefully cut the brain in half, so that these hemispheres are separated.

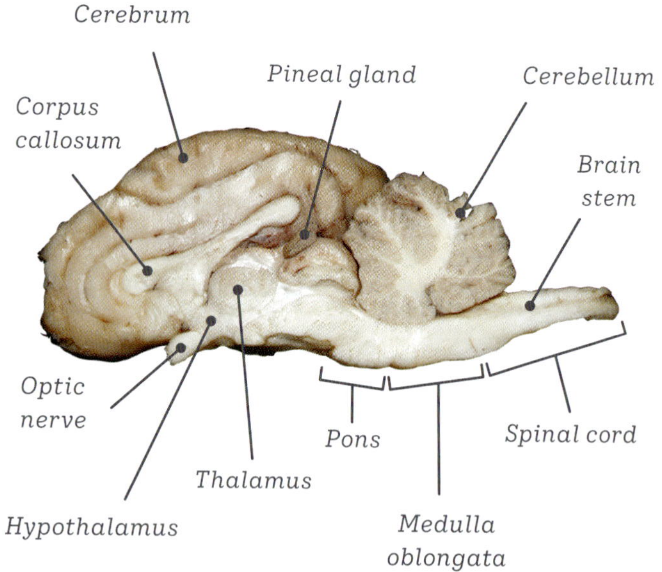

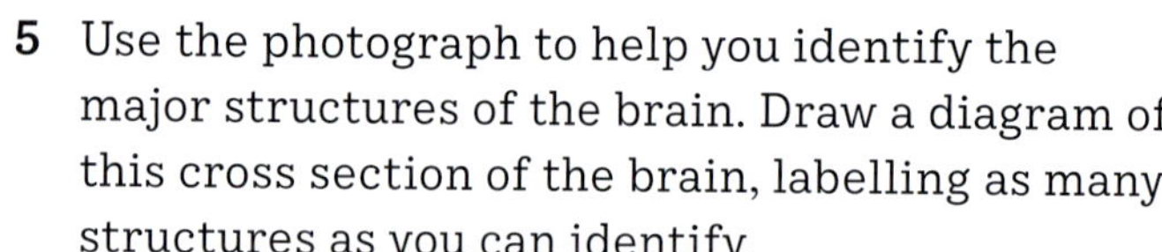

5 Use the photograph to help you identify the major structures of the brain. Draw a diagram of this cross section of the brain, labelling as many structures as you can identify.
6 Take one of the hemispheres and cut it across, perpendicular to the corpus callosum.
7 Notice that some parts of the cerebrum are white, whereas others are more pinky-grey. The white area is known as 'white matter', and the pinky-grey is the 'grey matter'.
8 Draw a diagram of this cross section, labelling the grey and the white matter.
9 When you have completed your dissection, carefully clean up and disinfect your work area.

QUESTIONS

Research to help you answer the following questions.

1 Find images that will allow you to compare the size and shape of a sheep brain and a human brain. What do you think influences these differences?
2 Find out more about the functions of the following brain structures.

Brain structure	Function
Cerebrum	
Cerebellum	
Brain stem (spinal cord, medulla oblongata, pons)	
Corpus callosum	
Hypothalamus	
Thalamus	

3 What is the difference between 'white matter' and 'grey matter'?

Investigation 8.3A

Heart dissection

AIM

To investigate the structure of the mammalian heart

MATERIALS

- sheep or cow heart
- dissection kit
- dissecting board
- newspaper
- disinfectant
- pins
- labels
- gloves

METHOD

1 Place some newspaper on the bench and the dissecting board on top of it.
2 Carefully examine the heart. The blood vessels around the outside are the coronary arteries and veins – those that take blood to and from the heart muscle.
3 Feel either side of the heart. One should feel thicker than the other. The thicker side is the left, and the smaller and thinner side is the right.
4 Look for the blood vessels on the top of the heart. Can you stick your fingers in to see where they lead to?
5 Place the heart on the cutting mat with the apex (pointed end) towards you, and the right side of the heart to your left.

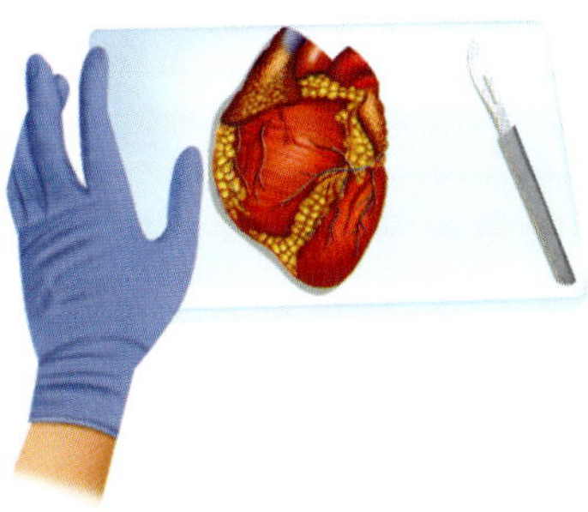

6 Carefully make a cut around the outside of the heart.

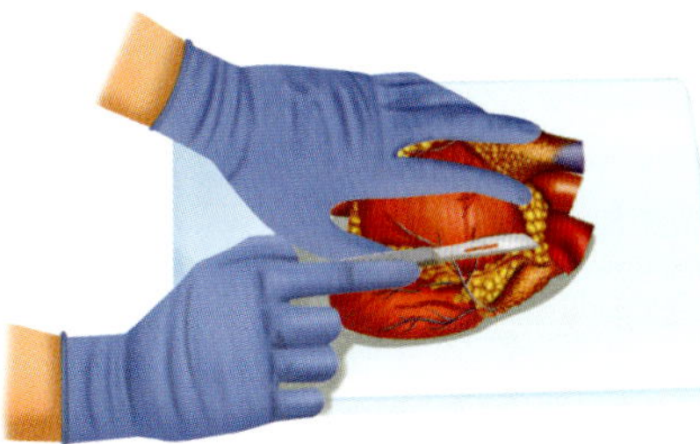

7 You should now be able to open the heart and see inside both the right and left sides.

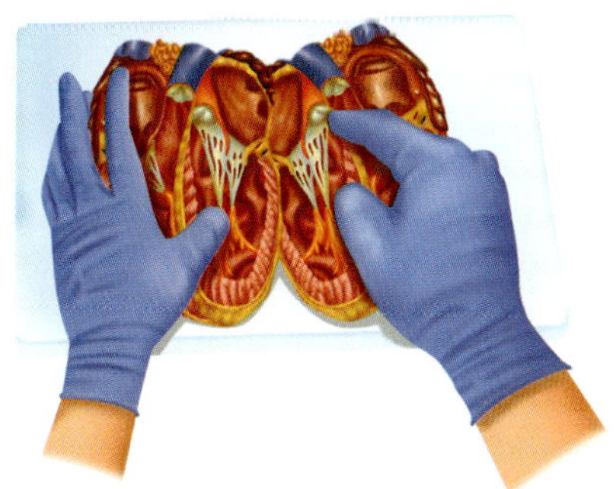

8 Carefully examine the inside of the heart. You should be able to identify the valves between the atrium and ventricles of each side. These are controlled by tendons – try pulling these.
9 Identify the septum – it is the structure that separates the left and right sides of the heart.
10 Use the pins and labels to label your heart with the correct structures.
11 Take a photo of your labelled, dissected heart.
12 When you have completed your dissection, carefully clean up and disinfect your work area.

QUESTIONS

1 Compare the size of the left and the right sides of the heart. Propose why there is such a size difference.
2 Compare the size of the atria and the ventricles. Propose why there is such a size difference.
3 What is the function of valves within the heart? What might occur if they did not function properly?
4 What is the function of the septum in the heart?

Investigation 8.3B

Heart rate, breathing rate and exercise

AIM

To investigate the effect of exercise on heart rate, breathing rate, perspiration and body temperature

MATERIALS

- skipping rope or other equipment to use to exercise
- heart rate monitor (if available)
- body temperature thermometer (if available)
- stopwatch

METHOD

1 Copy the results table into your workbook, adding a title.
2 Find an area that is suitable to exercise in.
3 Observe and record any redness in your partner's skin tone (face and arms). Use a scale of 1–4, where 1 is normal, 2 is slight redness, 3 is red and 4 is very red.
4 Observe and record your partner's level of perspiration. Use a scale of 1–5, where 1 is no sweating, 2 is slight perspiration, 3 is sweat building up under the armpits and 4 is extremely sweaty.
5 Measure and record your partner's heart rate (if using a pulse, measure for 15 seconds, then multiply by 4 to get beats per minute).
6 Measure and record your partner's breathing rate for 15 seconds. Multiply by 4 to get breaths per minute.
7 Measure and record your partner's body temperature.
8 Your partner will then undertake vigorous exercise (e.g. skipping or jumping jacks) for 5 minutes. Measure and record their skin redness, level of perspiration, heart rate and breathing rate every minute. Measure and record their temperature if you have a thermometer that allows you to do this.
9 Take these measurements again every minute for 5 minutes after they have stopped exercising.

RESULTS **TABLE 18.3B**

Time (min)	Skin redness (1–4)	Perspiration level (1–4)	Body temperature (°C)	Heart rate (beats in 15 s)	Heart rate (beats per min)	Breathing rate (breaths in 15 s)	Breathing rate (breaths per min)
0 (rest)							
1							
2							
3							
4							
5 STOP exercise							
6							
7							
8							
9							
10							

DISCUSSION

1. Construct line graphs of your data showing how the variables change over time.
2. How do the heart rate and breathing rate change? Why do they change?
3. Explain how the respiratory and circulatory systems work to maintain homeostasis during and after exercise.
4. What changes were observed in skin colour, perspiration and body temperature?
5. Explain how changes in skin colour, perspiration and body temperature during and after exercise are linked to homeostasis.
6. If you repeat an exercise routine regularly, you will become fitter and notice that your heart and breathing rate do not rise as high as they did before. Explain why.

CONCLUSION

Copy and complete.

The results show that: (*respond to the aim*).

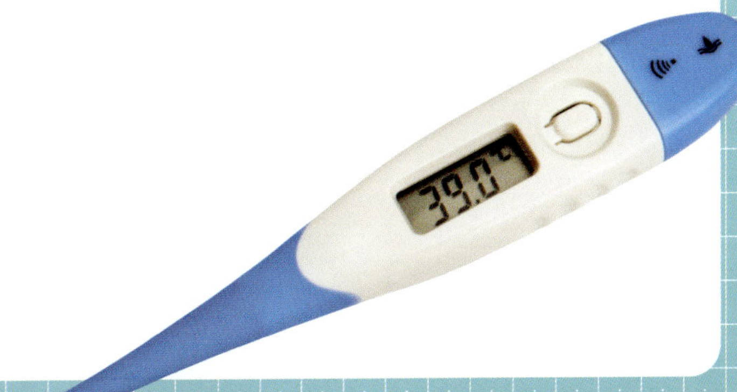

Investigation 9.1

Identifying community members in a terrestrial ecosystem

AIM

To investigate different species in a terrestrial ecosystem

MATERIALS

- magnifying glass
- gardening gloves
- species identification resources for your area
- camera

METHOD

1. Copy the results table into your notebook, adding a title and rows as required.
2. Select a small ecosystem in or around your school, no larger than 20 metres by 20 metres. Mark out an area if necessary, using witches hats or other physical markers.
3. Record all the different plants in your ecosystem. Photograph or draw the leaves of each plant.
4. Put on the gloves.
5. Look for small organisms living on the plants. Look on the underside of leaves, and in the branches and grasses.
6. Carefully turn over fallen leaves, and look under rocks and soil.
7. Sit quietly to record the different organisms you can see and hear in this ecosystem. Some may move in and out of the ecosystem or be nocturnal.
8. Record the organisms you find, including a photograph or drawing, and make note of any interesting behaviour or relationships with other organisms you observe.

QUESTIONS

1. Identify the organisms in your table as producers, consumers or decomposers.
2. Create a food web of the organisms you observed in your ecosystem. You may need to conduct some extra research to do this.
3. What types of relationships between organisms did you observe?

Location: __________ **Date:** __________ **Time:** __________

RESULTS **TABLE I9.1**

Organism	Photo/drawing	Number (approximate)	Observations of behaviour, including any relationships observed

Investigation 9.2

Measuring abiotic factors

AIM

To investigate the abiotic factors in a terrestrial ecosystem

MATERIALS

- soil multiprobe (pH, light, moisture)
- trowel
- thermometer

METHOD

1 Copy the results table into your notebook, adding a title and rows as required.
2 Select a small terrestrial ecosystem in and around your school.
3 Use the thermometer to measure the air temperature 1 metre above the ground in sunlight. Hold the thermometer in the air for several minutes until the reading stabilises.
4 Use the thermometer to measure the air temperature about 1 metre above the ground in shade.
5 Carefully place the end of the thermometer in the leaf litter/top of the soil. Use the trowel to dig a small hole if necessary. Leave for a few minutes to stabilise and record the value.
6 Carefully dig a small hole to bury the thermometer bulb about 5 cm deep. Leave for a few minutes to stabilise and record the value.
7 Switch the soil multiprobe to the 'moisture' setting. Carefully push the ends of the probe about 5 cm into the soil. Leave for a few minutes to stabilise and record the value.
8 Leave the multiprobe in the soil and carefully switch the setting to 'pH'. Leave for a minute to stabilise and record the value.
9 Leave the multiprobe in the soil and carefully switch the setting to 'light/LUX'. Leave for a minute to stabilise and record the value.
10 Collect a soil sample from about 5 cm deep in the specimen container. Fill it about half full.
11 Describe the soil colour and texture. Is it coarse, gritty, fine? Does it contain organic matter? You may wish to view the sample under a microscope.
12 Dispose of excess soil and return all equipment.

RESULTS TABLE I9.2

Air temperature, 1 m above ground in sun (°C)	
Air temperature, 1 m above ground in shade (°C)	
Soil temperature, surface (°C)	
Soil temperature, 5 cm deep (°C)	
Soil moisture, 5 cm deep	
Soil pH, 5 cm deep using multiprobe	
Light intensity (lux)	
Soil colour	
Soil texture	

QUESTIONS

1 Compare your results with those of your classmates. Were the groups in the same or different locations? How might this have affected the results each group obtained?
2 Propose how these results might differ if you were to repeat this experiment:
   a at night time
   b 6 hours earlier or later
   c in 6 months.
3 Research to find out about the temperature, light, water and soil requirements of at least two major plant species in your ecosystem. How do these correlate with your results?

Investigation 9.4

Algal balls

AIM

To investigate how different light levels affect the rate of photosynthesis, using algal balls

MATERIALS

- algal balls
- hydrogencarbonate indicator solution
- indicator colour chart
- 4 small sample jars with lids
- 10 mL measuring cylinder

METHOD

1 Copy the results table into your notebook, adding a title and rows as required.
2 Carefully place an equal number of algal balls in each sample jar.
3 Use the measuring cylinder to measure an equal amount of hydrogencarbonate indicator solution into each sample jar, ensuring the algal balls are covered.
4 Place the lids on the jars.
5 Place the jars in different places around your classroom that experience different levels of sunlight; for example, on the windowsill, on a shelf away from the window, in a cupboard.
6 Leave the jars until the next day.
7 Compare the colour of each solution to the indicator colour chart and record your results.

DISCUSSION

1 Consider the photosynthesis reaction. What:
  a are the reactants?
  b are the products?
  c else is required?
2 In which sample did the most photosynthesis take place? Which sample had the least? How do you know?
3 Design a procedure for an investigation using algal balls to test the effect of one of these variables on photosynthesis.
  a Wavelength of light (colour of light)
  b Temperature
  c Size of algal balls

RESULTS TABLE I9.4

Sample jar location	Colour	pH	Rank (most photosynthesis = 1, least photosynthesis = 4)

Investigation 10.2

Extracting DNA from strawberries

AIM

To investigate the process of extracting DNA from strawberries

MATERIALS

- strawberries
- liquid dishwashing detergent
- very cold ethanol (kept in ice or in the freezer)
- 10 mL measuring cyclinder
- plastic spoon
- salt
- water
- ziplock bag
- wooden skewer
- 10 cm squares of gauze
- elastic band
- plastic pipette
- funnel
- large test tube
- ruler
- test-tube rack
- measuring cylinder 50–100 mL
- electronic balance

METHOD

1 Copy the results table into your notebook, adding a title and rows as needed.

2 Make the DNA extraction buffer (this can also be pre-prepared by a lab assistant). Add 5 mL of detergent, 0.75 g of salt (weighed on electronic scales) and 45 mL of water to the measuring cylinder. Stir to mix. This will make enough for five 10 mL samples of extraction buffer.

3 Remove the green part of a strawberry and wash the strawberry carefully.

4 Place the strawberry in a ziplock bag with 10 mL of the extraction buffer and remove as much air as possible before sealing the bag tightly.

5 On the bench, crush and massage the strawberry inside the bag for a minute.

6 Set up the test tube in a rack on the bench.

7 Place the funnel in the test tube and then line the funnel with the gauze.

8 Pour the mixture from the ziplock bag into the funnel.

9 When the mixture has filtered (this will take some time), discard the gauze and strawberry pulp. Keeping only what filtered into the test tube.

10 Measure the amount of liquid in the test tube very roughly with a ruler.

11 Use the plastic pipette to add the same amount of ice-cold ethanol as the amount of liquid already in the test tube.

12 Observe what happens in the test tube. You should be able to see changes straight away.

13 Use the wooden skewer to gentle remove some DNA from the top of the test tube.

RESULTS

TABLE I10.2

Observations	Photo or diagram

QUESTIONS

1 a What errors could have occurred during this experiment?

b Outline how you could improve the method to ensure these errors don't occur.

2 What did the strawberry DNA look like?

3 Why do you think it was important to crush the strawberries before extracting the DNA?

4 What do you think the DNA extraction buffer did?

Investigation 10.3

Genetic trait survey

AIM

To investigate the different genetic traits that exist in a population

MATERIALS

- graph paper
- genetic trait survey
- at least 10 participants

METHOD

1. Copy the results table into your notebook, adding a title and rows as needed.
2. In the second column of the table, record whether you have the genetic trait listed in the first column.
3. Share your results with the rest of the class and create a class tally.
4. Copy the class tally into your table.

QUESTIONS

1. Graph your class results.
2. Which traits were the most prevalent in your class? Which were the least?
3. What other genetic traits could you have tested for?
4. Suggest why some traits are more common than others.
5. Genetic traits are passed down from parents on chromosomes. Describe the structure of a chromosome.

RESULTS TABLE I10.3

Trait	My response (yes or no)	Total number in class
Curly hair		
Brown eyes		
Attached earlobe		
Cleft chin		
Hitchhiker's thumb		
Ability to roll tongue		
Left-handedness		
Widow's peak hairline		

Investigation 10.4

DNA lolly model

AIM

To investigate the structure of DNA by making a model out of lollies

MATERIALS

- toothpicks
- red liquorice
- four of the same type of lollies in four colours, e.g. mini-marshmallows, jelly beans, gummy bears
- clean surface or cutting board

METHOD

1. Assign lolly colours to each of the four DNA bases – adenine, cytosine, guanine and thymine.
2. Use two pieces of red liquorice as the DNA backbone.
3. Pair your lolly bases: guanine with cytosine and adenine with thymine.
4. Thread the matching base pairs onto toothpicks.
5. Insert the ends of the toothpicks into the red liquorice DNA backbone.
6. Twist the red liquorice DNA backbone as you go, creating the DNA twisted double helix model.
7. Draw a diagram or take a photo of your finished DNA model.

QUESTIONS

1. Suggest why the use of models is important in science.
2. How could you make your model more accurate? (Hint: Think about what the backbone is made of.)
3. How could you alter this model to show DNA replication?
4. Outline the structure of DNA.
5. Which is bigger – DNA or a gene? Justify your answer.
6. Outline the steps this piece of DNA would go through in order to replicate.
7. What kind of mutations could occur to the DNA in your model?

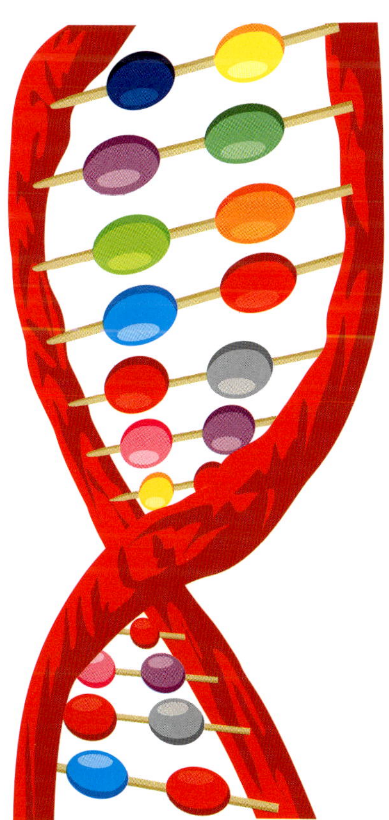

Investigation 11.1

Comparative anatomy

AIM

To investigate the comparative anatomy of a human, cat, whale and bat

MATERIALS

- images of human, cat, whale and bat appendages (or full skeleton/models if possible)

METHOD

1 Consider the diagram of the homologous structures of a human, a cat, a whale and a bat.
2 Copy the diagram into your notebook or obtain a photocopy from your teacher.
3 Colour each ulna the same colour on each of the four appendages.
4 Continue to colour the same bones on each appendage in matching colours.
5 Create a Venn diagram in your results section that shows the similarities and differences between each species appendage, based on your observations.

RESULTS

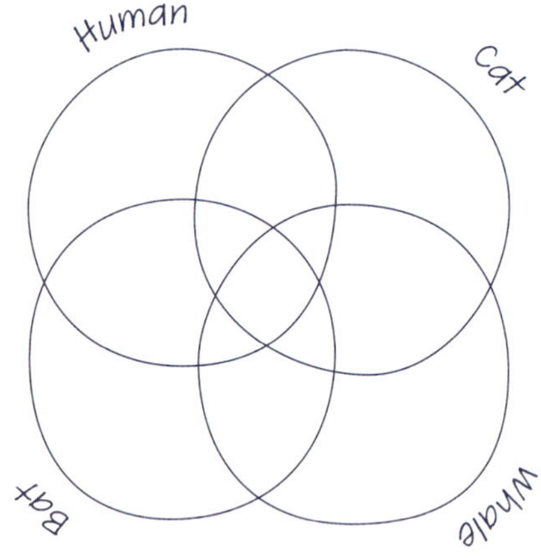

QUESTIONS

1 Identify the bones that are common across all four species.
2 Explain how comparative anatomy can be used as evidence of evolution.
3 Do your results suggest these species had a common ancestor? Why or why not?
4 What does the structure of the bones tell you about the organism and what it uses the appendage for?

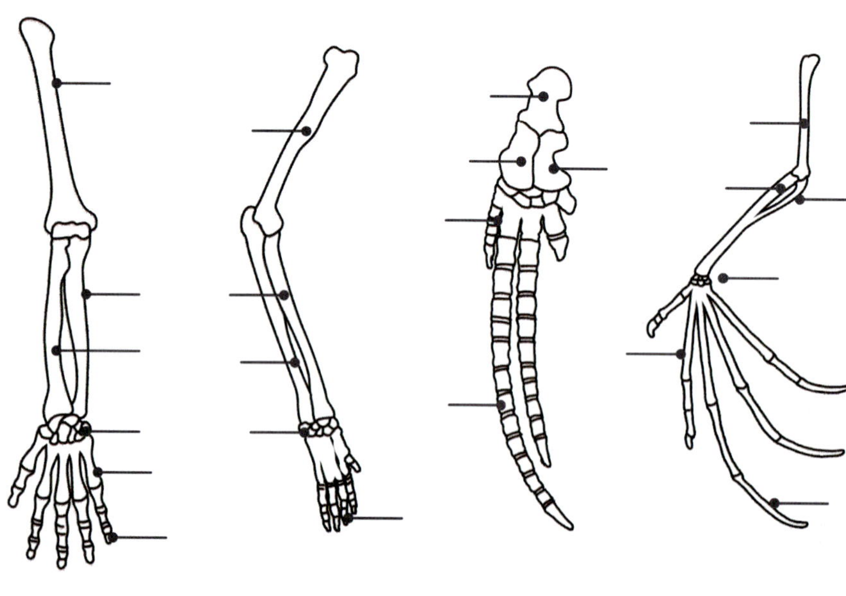

Investigation 11.2

Dating fossils

AIM

To investigate a model for absolute dating

MATERIALS

- 100 five- or ten-cent coins in a container

METHOD

1 Each coin represents an atom in the radioactive element carbon-14. You will be investigating how long the half-life of carbon-14 is. Scientists can calculate the age of fossils by identifying the half-lives of atoms surrounding the fossil.
2 Copy the results table into your notebook, adding a title.
3 Shake the container and carefully empty the coins onto the table. Spread them out so that none overlap.
4 Remove all the coins showing tails. These represent atoms that are no longer radioactive and have decayed.
5 Record the number of coins showing heads in your results table. These represent atoms that are still radioactive.
6 Put the coins showing heads back in the container. Shake the container, and carefully empty the coins onto the table. Repeat this process until all the coins are gone or you have completed 12 trials.
7 Draw a graph of your results.

RESULTS TABLE I11.2

Shake number	Number of coins showing heads (radioactive atoms)
1	
2	
3	
4	
5	
6	
7	
8	
9	
10	
11	
12	

QUESTIONS

1 How does this model relate to the technique used to identify the age of fossils?
2 What is the difference between relative and absolute dating?
3 How many shakes did it take for half of your coins to decay? This represents the half-life of your carbon-14.
4 The half-life of carbon-14 is actually 5730 years. If each coin shake is 5730 years, how many years would it take for carbon-14 to completely disappear?
5 What did your results tell you about absolute dating and the decay of radioactive isotopes?
6 What errors could have occurred during this investigation? Suggest ways to ensure they do not occur in the future.

Investigation 11.3

Natural selection in practice

AIM

To investigate the effectiveness of different 'beaks' in collecting different types of food

MATERIALS

- uncooked rice
- popcorn kernels
- string (cut into 3 cm pieces)
- marbles
- plastic tub – small to medium sized
- forceps
- crucible tongs
- kitchen tongs
- blunt pliers
- 4 paper cups
- stopwatch
- graph paper

METHOD

1. Copy the results table into your notebook, adding a title.
2. Place the marbles, rice, pieces of string and popcorn in the plastic tub.
3. Designate a different 'beak' (forceps, crucible tongs, kitchen tongs or blunt pliers) to each person in the group.
4. Choose a person to use the stopwatch to time 2 minutes.
5. When the two minutes starts, each participant should use their tool to pick up as many objects as possible and place them into their paper cup.
6. After 2 minutes, stop and count how many of each object is in each person's paper cup.
7. Record the totals in your results table.
8. Share your data so that each person can complete their results table.

QUESTIONS

1. What is natural selection?
2. How does this investigation demonstrate the principles of natural selection?
3. Which tools (or beaks) were best at picking up each type of food? Suggest why.
4. What errors could have occurred during this investigation? Suggest ways to prevent them in the future.

RESULTS TABLE I11.3

Tool	Beak type	Number of pieces of string	Number of grains of rice	Number of popcorn kernels	Number of marbles
Forceps	Long, thin, pointy				
Crucible tongs	Medium, long, pointy				
Kitchen tongs	Long, wide				
Blunt pliers	Short, thick, wide				

Investigation 11.4

Environmental and genetic factors affecting survival

AIM

To investigate the effect of colour on the survival of individuals in different habitats

MATERIALS

- 20 red toothpicks
- 20 yellow toothpicks
- 20 green toothpicks
- trundle wheel or tape measure

METHOD

1 Copy the results table into your notebook, adding a title.
2 You will need three different surfaces on which to conduct the investigation: carpet, concrete and grass.
3 Use the trundle wheel or tape measure to measure out a 2m × 2m square on the carpet.
4 Select one person in the group to be the predator. Ask the predator to turn away while the toothpicks (the prey) are scattered evenly over the measured square on the carpet.
5 The predator then has 10 seconds to pick up toothpicks.
6 Count how many toothpicks of each colour the predator has picked up and record your results in the table.
7 Repeat steps 3–6 for the concrete and the grass.
8 Calculate the percentage survival rate by dividing the number of survivors by 20 and then multiply that number by 100.

QUESTIONS

1 Which colours and surfaces had the highest survival rate? Suggest why.
2 Why did the predator find it harder to find prey in certain environments?
3 How does this investigation demonstrate the principles of natural selection?
4 What did your results tell you about the role of environmental and genetic factors on survival?
5 What errors could have occurred during this investigation? Suggest ways to ensure they do not occur in the future.

RESULTS **TABLE I11.4**

Surface	Toothpick colour	Number of prey caught	Number of survivors	Survival rate (%)
Carpet	Red			
	Yellow			
	Green			
Concrete	Red			
	Yellow			
	Green			
Grass	Red			
	Yellow			
	Green			

Investigation 12.2A

Making an atom

PART 1

THE BOHR MODEL

AIM

To produce a Bohr model of an atom and to better understand subatomic particles

MATERIALS

- paper plate
- three different type of lollies

METHOD

1 Copy and complete the results table into your notebook, adding a title.
2 Select an element, up to atomic number 6.
3 Work out the atomic number and atomic mass of the element by consulting a periodic table.
4 Calculate the number of protons, neutrons and electrons the atom has.
5 Select three different types of lollies.
6 On the paper plate, draw three circles. The inner circle represents the nucleus. The second and third circles represent electron orbitals. The second circle can only hold two ‘electrons’ and the third circle can hold the remaining ‘electrons’.

RESULTS TABLE I12.2A

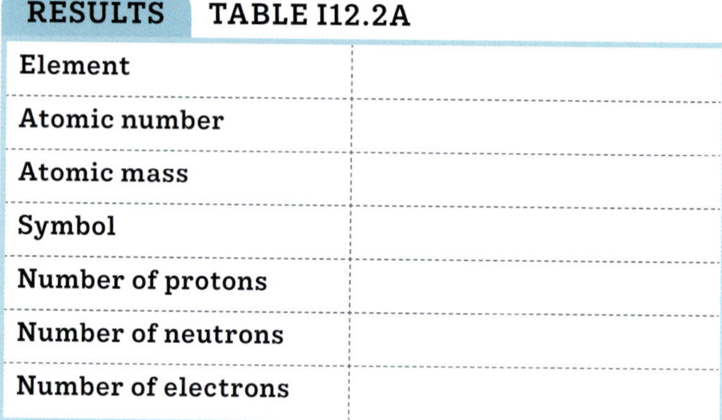

Element	
Atomic number	
Atomic mass	
Symbol	
Number of protons	
Number of neutrons	
Number of electrons	

PART 2

THE THOMSON MODEL

AIM

To produce Thomson's plum pudding model of an atom and to better understand subatomic particles

MATERIALS

- paper plate
- whipped cream
- one type of lolly

METHOD

1 Select an element, up to atomic number 10.
2 Work out the atomic number and atomic mass of the element by consulting a periodic table.
3 Calculate the number of electrons the atom has.
4 Select one type of lolly to represent electrons.
5 Use the whipped cream to represent a positively charged sphere on the paper plate.
6 Add the exact number of 'electrons' and spread them around on the whipped cream.

QUESTIONS

1 Describe the Bohr model that you constructed.
2 Describe the Thomson model that you constructed.
3 How are the two models similar and how are they different?
4 Describe the usefulness and limitations of using models in science.

Investigation 12.2B

Flame test

AIM

To investigate the distinctive colours produced by metallic ions in a flame test and use the results to identify an unknown metallic ion.

MATERIALS

- set of 0.1 mol/L metal chloride solutions ($NaCl$, $CuCl_2$, KCl, $CaCl_2$, $SrCl_2$, $LiCl$, $CoCl_2$, $BaCl_2$)
- 2 unknown solutions
- 8 popsticks
- matches
- Bunsen burner
- cobalt glass plate

METHOD

1. Copy the results table into your notebook, adding a title.
2. Light the Bunsen burner and adjust it to the blue flame.
3. Using a clean popstick, dip it into one of the solutions until it is saturated and then hold the soaked end in the hottest part of the flame, ensuring you do not set it on fire.
4. Observe the colour of the flame. Try not to let the popstick catch fire. Otherwise you won't be able to reuse it to double check the colour later.
5. Carefully record your observations in the results table. Be as accurate as possible – your description of the colour must be specific enough to distinguish this metal ion from the other ions tested.
6. Repeat steps 3–5s for each solution, using a new, clean popstick to test each solution.
7. When you examine the sodium and potassium ions, first look at their colour unaided, then look through a cobalt glass plate. Record both observations separately in your results table.
8. When you have tested all the known solutions and can accurately describe the colour of each metal ion, obtain two different unknown solutions from your teacher and determine which metal ions are present by performing a flame test and comparing this data to your previous data.

RESULTS TABLE I12.2B

Metal ion	Colour of flame
Lithium	
Strontium	
Calcium	
Barium	
Copper	
Cobalt	
Sodium	
Sodium (with cobalt glass)	
Potassium	
Potassium (with cobalt glass)	
Unknown 1	
Unknown 2	

Based on your observations, identify the two unknown ions.

Unknown ion 1 is ______________________

Unknown ion 2 is ______________________

AN OPEN FLAME IS A HAZARD. BE CAREFUL. IF YOU BURN YOURSELF, TELL YOUR TEACHER IMMEDIATELY AND PLACE YOUR SKIN UNDER COLD RUNNING WATER FOR 20 MINUTES.

DISPOSE OF WASTES APPROPRIATELY.

QUESTIONS

1 Which ions produce similar colours in the flame tests?
2 What was the purpose of the cobalt glass?
3 Explain how the colours observed in the flame tests are produced.
4 Did you correctly identify the two unknown solutions?
5 What factors contributed to your correct or incorrect identifications?
6 State at least three problems that may arise when using flame tests for identification purposes in a laboratory.

Investigation 12.3

Half-life coin experiment

AIM

To model radioactive decay, using coins

MATERIALS

- 100 coins and a large container with a lid

METHOD

1 Copy the results table into your notebook, adding a title.
2 Place 100 coins in the container and fasten the lid.
3 Shake the container several times and remove the lid. Carefully empty the coins onto a flat surface, making sure the coins don't roll away.
4 Remove all the coins that show heads.
5 Record the number of coins removed and the number of coins remaining in the results table.
6 Return the remaining coins to the container and repeat steps 3–5 until no coins are left in the container.
7 Draw a graph of your data. Label the x-axis 'Shake number' and the y-axis 'Number of coins remaining'.

RESULTS TABLE I12.3

Shake number	Number of coins removed	Number of coins remaining
1		
2		
3		
4		
5		
6		
7		
8		
9		
10		

QUESTIONS

1 Compare the decay curve of carbon-14 with your graph for the coins. Explain any similarities that you see.

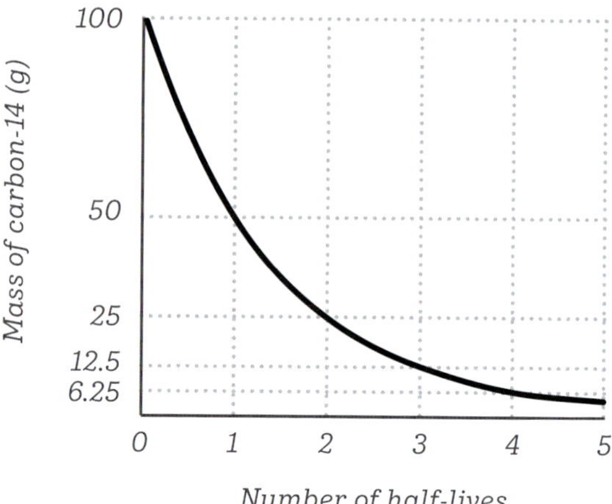

2 Recall that the probability of landing heads in a coin toss is $\frac{1}{2}$. Use this information to explain why the remaining number of coins is reduced by about half each time they are shaken and tossed.

Investigation 12.4

Nuclear energy debate

AIM

As a class, select one of the three critical and creative thinking strategies and use it to debate one of these issues about nuclear energy.

- The consequences of continuing to rely on fossil fuels are greater than those posed by a potential nuclear disaster.
- It is unethical to cure a patient of an illness using nuclear technology because it may cause long-term side effects for the patient.
- Australia must move to nuclear energy as our main source of electricity.

Use the information in spread 12.4 and other sources to help you form your arguments.

STRATEGY 1: LINE DEBATE

1. Divide the class evenly into two groups and form two lines, facing each other.
2. One line is ‘for’ the issue (the affirmative team), the other line is ‘against’ the issue (the negative team).
3. Any student from the affirmative team can put forward an argument to begin the debate.
4. If it is judged as a fair point (by your teacher, or a student in the adjudicator role), the affirmative team can ‘steal’ someone from the negative team to join their line.

5 The 'stolen' student now becomes a member of the affirmative team and must put forward an argument for the issue. If it is judged as a fair point, they can again steal someone from the negative team to join their line. If it is not a fair point, any student from the negative line is allowed to put forward an argument and then has the chance to steal an affirmative team member.
6 The debate continues in this fashion until the teacher calls 'time'. At this point, the line with the most students 'wins' the debate.

STRATEGY 2: FISH BOWL DEBATE

1 Two students sit facing each other in the centre of the room. One student is on the affirmative side and the other is on the negative side of the debate.
2 The rest of the class sits in a circle around them.
3 The two central students begin debating one of the issues, taking it in turns to put forward arguments one by one.
4 Students in the outside circle must listen carefully because they may be called upon to swap with one of the central people at any time in the debate. You could be called upon to join either the affirmative team or the negative team so you need to have arguments prepared for both sides of the issue.
5 Your teacher will nominate students to move from the outer circle into the centre periodically throughout the discussion. The goal is to keep the debate flowing.

STRATEGY 3: THINK-PAIR-SHARE DEBATE

1 Form a pair and select a topic to 'debate'. Allocate one person as 'affirmative' and one as 'negative'.
2 First, work independently. On your own, think about your response to the issue. Consider it in depth. Write notes if that helps you to organise your thoughts.
3 Now turn to your partner. Listen respectfully while your partner talks through their arguments for or against the issue, then offer your response. Consider the similarities and differences in your responses.
4 Share your opposing points of view on the issue with another pair or the whole class.

QUESTIONS

1 Reflect on the critical and creative thinking strategy that you used. How did the strategy help you to evaluate the use of nuclear energy?
2 Think about the way you prefer to learn. Develop a new critical and creative thinking strategy that your class could use next time you are called upon to evaluate an issue.

Investigation 13.3A

Physical properties of elements

AIM

To investigate and compare the physical properties of metals and non-metals

MATERIALS

- samples of various metals and non-metals, such as lead, zinc, aluminium, iron, magnesium, tin, sulfur and carbon
- power-pack, switch and electrical leads
- light bulb
- sand paper
- heatproof mat

METHOD

1 Copy the results table into your notebook, adding a title and rows as required.
2 Clean and polish each sample with sand paper and record the appearance after it's been cleaned. Record your observations.
3 Test for malleability by trying to bend the sample. Record your observations.
4 Using a power-pack, light bulb, switch and cables, construct a circuit and test each sample for electrical conductivity on the heatproof mat. Check if the bulb lights up, and record your observations.
5 Using a periodic table and your observations, determine whether the element you have tested is a metal or a non-metal.

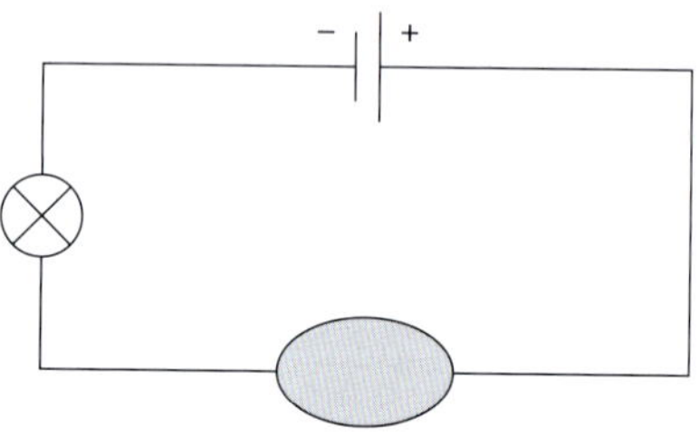

Tested material

QUESTIONS

1 List the physical properties of metals that you observed.
2 List the physical properties of non-metals that you observed.
3 Predict, using the elements tested and their properties, which other elements would have similar properties.

RESULTS TABLE I13.3A

Element	Appearance when polished	Bends or crumbles	Electrical conductivity	Metal or non-metal

Investigation 13.3B

Chemical properties of elements

AIM

To investigate and compare the chemical properties of metals and non-metals

MATERIALS

- 1 mol/L hydrochloric acid
- water
- 1 cm magnesium ribbon cut into three pieces
- iron filings
- copper turnings
- sulfur powder
- 4 test tubes
- 4 gas jars filled with oxygen
- 4 deflagrating spoons
- Bunsen burner
- heatproof mat
- matches
- safety glasses
- heatproof gloves

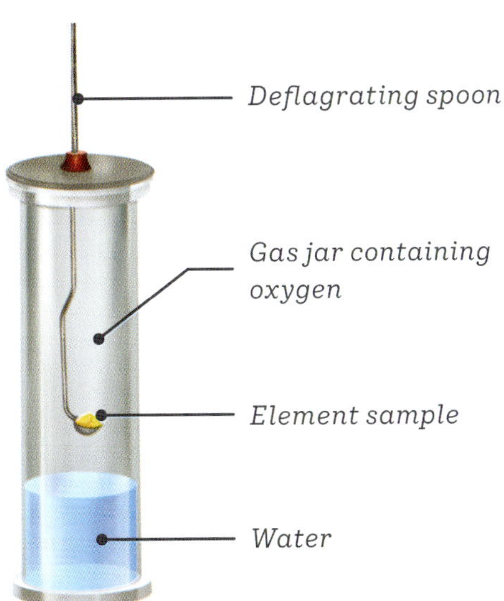

METHOD

1 Copy the results table into your notebook, adding a title and rows as required.
2 Put a small amount of copper turnings in a test tube. Add about 3 mL of hydrochloric acid. Record your observations.
3 Repeat step 2 with magnesium, iron filings and sulfur powder.
4 Place a small amount of copper in a deflagrating spoon and heat it. Then place it in a gas jar containing oxygen gas and a small amount of water. Record your observations. (Do not look directly into the flame.)
5 Repeat step 4 with the magnesium and iron filings. Record your observations.
6 Repeat step 4 with a small amount of sulfur. This must be performed in the fume cupboard.

RESULTS

TABLE I13.3B

Element	Reaction with acid	Reaction with oxygen

QUESTIONS

1 Describe the differences between the effects of acids on metals and non-metals.
2 Describe the effects of reacting metals and non-metals with oxygen.
3 Give reasons why the elements tested are in particular locations in the periodic table.

WEAR SAFETY GLASSES AND HEATPROOF GLOVES DURING HEATING. HEAT THE ELEMENTS WITH THE DEFLAGRATING SPOON. ENSURE YOU HEAT THE SULFUR IN A FUME CUPBOARD. DO NOT LOOK DIRECTLY AT THE FLAME.

Investigation 14.5A

Action of acids and bases – cleaning coins

AIM

To investigate the actions of acids and bases

MATERIALS

- 3 coins of the same size and type
- 20 mL of 1 mol/L hydrochloric acid
- 20 mL ammonia
- 20 mL water
- cling film
- 3 × 50 mL beakers
- 3 × 25 mL measuring cylinders

METHOD

1. Copy the results table into your notebook, adding a title.
2. Label the three beakers as hydrochloric acid, ammonia and water.
3. Add a coin to each 50 mL beaker.
4. Add 20 mL of hydrochloric acid in a measuring cylinder and add it to the first beaker.
5. Add 20 mL of ammonia in a measuring cylinder and add it to the second beaker.
6. Add 20 mL of water in a measuring cylinder and add it to the third beaker.
7. Cover all the beakers with cling film and leave them overnight. Record your observations in the results table.
8. Compare your results with those of other groups.

RESULTS TABLE I14.5A

Substance	Observations
Hydrochloric acid	
Ammonia	
Water	

DISCUSSION

1. List the controlled variables, independent variable and dependent variable.
2. What benefit is there to repeating the experiment?
3. Is the action of acids and bases with metal a chemical reaction?
4. What does this experiment tell you about cleaning metals?
5. Suggest what would happen to the coins if you left them in the solutions for a month.

CONCLUSION

Copy and complete.

The results show that: (*respond to the aim*).

ACIDS ARE CORROSIVE; WEAR SAFETY GLASSES.

AMMONIA GIVES OFF FUMES; DO NOT INHALE. USE IN A VENTILATED ROOM.

DISPOSE OF WASTES APPROPRIATELY.

Investigation 14.5B

Acid or base?

AIM

To prepare a natural indicator and use it to find out if some household substances are acids or bases

MATERIALS

- red cabbage
- solutions of household products: ammonia, baking soda, conditioner, detergent, lemon juice, oven cleaner, shampoo, lemonade, vinegar
- 0.1 mol/L hydrochloric acid
- 0.1 mol/L sodium hydroxide
- plastic pipette
- water
- 12 test tubes
- test-tube rack
- 3 × 10 mL measuring cylinders
- cutting board
- knife
- large beaker
- hot plate
- strainer

METHOD

1. Copy the results table into your notebook, adding a title and rows as required.
2. Chop the cabbage, put it in the large beaker and add enough water to cover it.
3. Put the beaker on the hot plate and bring it to a boil and then simmer for 5 minutes.
4. Once cooled, strain the cabbage. The resulting solution is a natural indicator.
5. Prepare three test tubes by adding 5 mL of hydrochloric acid to the first one, 5 mL of sodium hydroxide to the second one and 5 mL of water to the third one.
6. Add 5 drops of the red cabbage indicator to each test tube and record your results. Identify the colour changes in acid, base and neutral solutions.
7. To each of the remaining 9 test tubes, add 5 mL of one of the provided household solutions.
8. Add 5 drops of the red cabbage extract to each test tube. Shake the test tube and observe the colour.
9. Identify whether each household substance is an acid or a base and record your results in the table.

RESULTS TABLE I14.5B

Substance	Colour with red cabbage indicator	Acid or base
Hydrochloric acid		
Sodium hydroxide		
Water		
Ammonia		
Baking soda		

DISCUSSION

1. List the controlled variables, independent variable and dependent variable.
2. How can you improve the results of the experiment?
3. Is red cabbage an effective indicator? Explain.
4. Do you think the acidic and basic household substances are safe to use in the home? Explain.

THE HOUSEHOLD SUBSTANCES MAY IRRITATE YOUR SKIN AND EYES AND BE TOXIC. WEAR GLOVES AND SAFETY GLASSES. USE THE KITCHEN KNIFE WITH CARE. IF BURNT BY HOT LIQUIDS, RUN COLD WATER ON THE AFFECTED AREA.

DISPOSE OF WASTES APPROPRIATELY.

Investigation 14.5C

The effect of indicators on acids and bases

AIM

To observe the effect of indicators on acids and bases

MATERIALS

- 10 mL vinegar
- 10 mL of 0.1 mol/L hydrochloric acid
- 10 mL of 1 mol/L ammonia
- 10 mL of 0.1 mol/L sodium hydroxide
- universal indicator solution
- universal indicator chart
- bromothymol blue
- phenolphthalein
- methyl orange
- litmus paper
- 20 test tubes
- test-tube rack
- 4 × 10 mL measuring cylinders

METHOD

1 Copy the results table into your notebook, adding a title.
2 Add 5 mL of vinegar, hydrochloric acid, ammonia and sodium hydroxide to four test tubes.
3 Add 3 drops of universal indicator to each of the solutions. Use the universal indicator chart to identify each solution as acid or base.
4 Repeat the experiment with the rest of the indicator solutions – bromothymol blue, methyl orange, phenolphthalein, litmus paper (add a small piece of litmus paper to each test tube) – and observe their colours in acidic and in basic solutions.
5 Compare your results with those of other groups.

QUESTIONS

1 Identify the substances that are acidic and those that are basic.
2 If lemonade is acidic, what colour would it change bromothymol blue to?
3 What characteristics are useful in an indicator?
4 Would we be able to use bromothymol blue to identify if two unknown substances were acids or bases? Explain.
5 What limitations does the use of litmus paper as an indicator present?

RESULTS TABLE I14.5C

Substance	Acid or base according to indicator				
	Universal indicator	Bromothymol blue	Methyl orange	Phenolphthalein	Litmus paper
Vinegar					
Hydrochloric acid					
Ammonia					
Sodium hydroxide					

WEAR SAFETY GLASSES TO PROTECT YOUR EYES FROM CHEMICALS.

Investigation 14.6A

Reactions of acids with metals

AIM

To investigate the rate of reaction of hydrochloric acid with a range of metals

MATERIALS

- similar sized pieces of metals: aluminium, copper, iron, magnesium and zinc
- 50 mL of 2 mol/L hydrochloric acid
- 10 mL measuring cylinder
- 5 test tubes
- test-tube rack
- matches
- thermometer
- 5 rubber stoppers
- sandpaper

METHOD

1 Copy the results table into your notebook, adding a title.
2 Clean all the metals with sandpaper.
3 Add 5 mL of hydrochloric acid to each of the five test tubes.
4 Record the initial temperature of the acid in each test tube.
5 Add aluminium to one test tube and stopper it with a rubber stopper until the reaction stops.
6 Light a match, remove the rubber stopper and hold the match inside the test tube. If a pop sound occurs, it means that the gas produced was hydrogen.
7 Measure the final temperature of the contents of the test tube.
8 Repeat steps 5–7 with the rest of the metals.

QUESTIONS

1 List the controlled variables, independent variable and dependent variable.
2 What can you conclude about the bubbles produced?
3 Comment on the final temperature readings obtained.
4 Were these reactions exothermic or endothermic?
5 a Rank the metals from the most reactive to the least reactive.
  b Explain why you ranked the metals in that way.
6 Did all the reactions give a positive pop test? Explain.

ACIDS ARE CORROSIVE. IF YOU BURN YOUR SKIN, RINSE THE AFFECTED AREA UNDER COLD RUNNING WATER. WEAR SAFETY GLASSES TO PROTECT YOUR EYES FROM ACID OR METALS.

DISPOSE OF WASTES APPROPRIATELY.

RESULTS TABLE I14.6A

Metals	Initial temp. (°C)	Final temp. (°C)	Change in temp. (°C)	Observations
Aluminium				
Copper				
Iron				
Magnesium				
Zinc				

Investigation 14.6B

Reactions of acids with carbonate

AIM

To observe the reaction between hydrochloric acid and sodium carbonate

MATERIALS

- sodium carbonate
- 2 mol/L hydrochloric acid
- 10 mL limewater
- 2 × 10 mL measuring cylinders
- spatula
- 2 test tubes connected by a delivery tube with rubber stoppers
- retort stand
- bosshead and clamp

METHOD

1 Copy the results table into your notebook, adding a title.
2 Set up the equipment as shown in the diagram. Note that the delivery tube must go into the limewater.
3 Add 2 spatulas full of sodium carbonate to the clamped test tube (test tube 1) and have someone hold the test tube containing 10 mL of limewater (test tube 2).
4 Add 5 mL of hydrochloric acid to the test tube containing sodium carbonate and stopper both test tubes. Make sure they are connected by the delivery tube as shown in the diagram.
5 Observe what happens in both the test tubes and record your results.

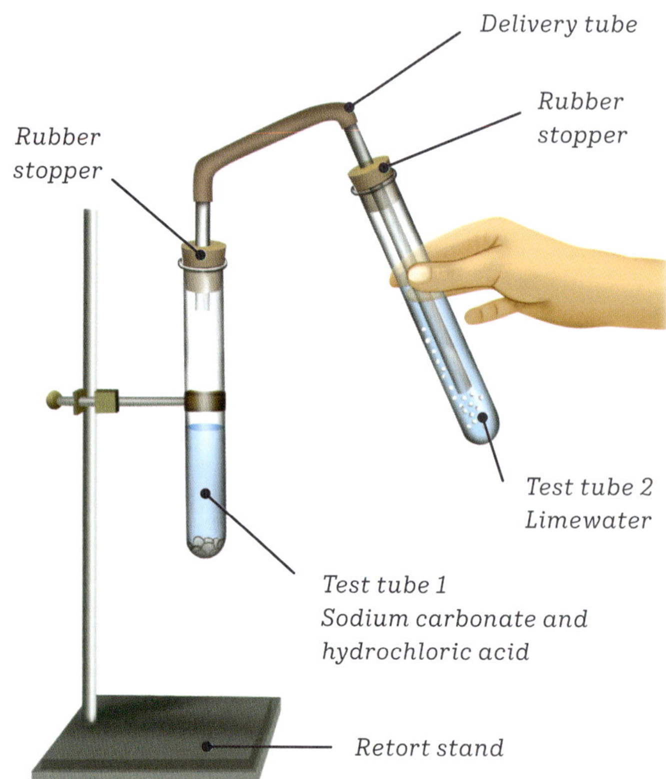

RESULTS

TABLE I14.6B

	Observations	Colour of limewater	Equation
Test tube 1			Sodium carbonate + hydrochloric acid → ____________

DISCUSSION

1 Did the test tube with the sodium carbonate and hydrochloric acid get hot?
2 What happened to the test tube containing limewater?
3 What colour did the limewater change to?
4 An endothermic reaction absorbs heat from the surroundings and the reaction container becomes cold to touch. An exothermic reaction releases heat to the surroundings and the reaction container becomes hot. Do you think this reaction was endothermic or exothermic?
5 Which gas do you think was released during this reaction?
6 Complete the sentence and equation:

   When ________ gas is bubbled through limewater, the limewater goes ________ in colour.

   Acid + carbonate → salt + ________ + water

CONCLUSION

Copy and complete.
The results show that: (*respond to the aim*).

HYDROCHLORIC ACID IS CORROSIVE. WEAR SAFETY GLASSES TO PREVENT IT FROM SPLASHING INTO YOUR EYES.

DISPOSE OF WASTES APPROPRIATELY.

Investigation 14.7A

Combustion of fuels

AIM

To find out which fuel – methanol, ethanol or propanol – will burn the fastest

MATERIALS

- 3 spirit burners – methanol, ethanol and propanol (50 mL each)
- water
- measuring cylinder
- thermometer
- stopwatch
- retort stand and clamp
- calorimeter
- insulating card or lid
- heatproof mat
- bosshead and clamp

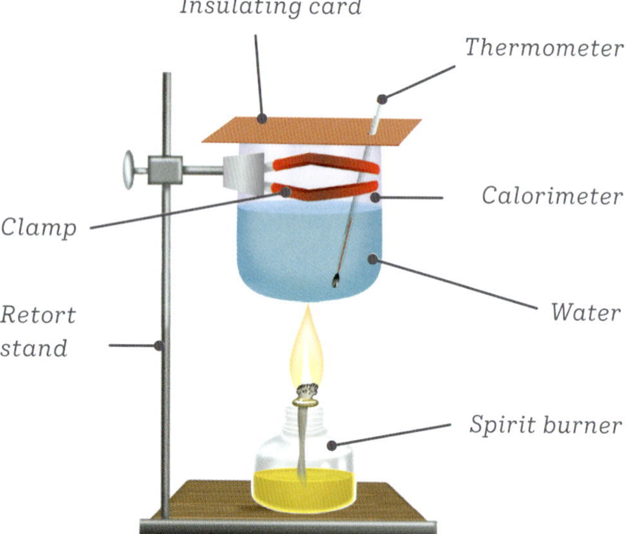

METHOD

1 Copy the results table into your notebook, adding a title.
2 Set up the equipment as shown in the diagram.
3 Fill the calorimeter with 100 mL of water. Fill the spirit burner with 50 mL of methanol.
4 Record the initial temperature of water.
5 Light the spirit burner and record the time taken for the temperature of water to increase by 10°C.
6 Extinguish the flame with the cap of the spirit burner.
7 Check and record if the bottom of the metal can or calorimeter has soot on it (black substance).
8 Repeat steps 3–7 with ethanol and propanol.
9 Record all your observations.

DISCUSSION

1 List the controlled variables, independent variable and dependent variable.
2 How would you make the results of your experiment reliable?
3 What are the products of combustion of these fuels?
4 Which was the best fuel? Explain.
5 List some assumptions (things you took for granted) that could have interfered with your results.
6 Can combustion occur without oxygen?

CONCLUSION

Copy and complete.
The results show that: (*respond to the aim*).

FUELS ARE FLAMMABLE. BE CAREFUL NOT TO SET THEM ALIGHT.

RESULTS **TABLE I14.7A**

Fuel	Initial temp. (°C)	Final temp. (°C)	Change in temp. (°C)	Time (s) for water temperature to increase by 10°C	Soot/no soot
Methanol					
Ethanol					
Propanol					

Investigation 14.7B

Complete and incomplete combustion reactions

AIM

To determine whether complete or incomplete combustion of gas will heat water to a higher temperature

MATERIALS

- water
- Bunsen burner
- matches
- thermometer
- tripod
- gauze mat
- retort stand
- clamp
- rubber stopper
- 250 mL beaker
- stopwatch
- heatproof mat
- bosshead and clamp

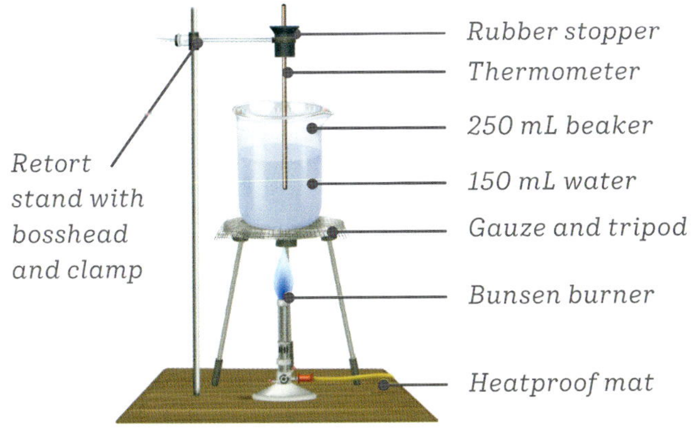

METHOD

1 Copy the results table into your notebook, adding a title.
2 Set up the equipment as shown in the diagram.
3 Record the initial temperature of water.
4 Light the Bunsen burner and, with the air hole almost closed, heat the water for 5 minutes.
5 Record the final temperature and find the change in temperature.
6 Observe the bottom of the beaker to see if any soot was formed.
7 Replace the beaker of water and repeat steps 3–5 with the air hole open.

DISCUSSION

1 What were your observations about the bottom of the beakers? Explain.
2 Identify the black particles you observed under one of the beakers.
3 Describe the difference in the colour of the flame in complete and incomplete combustion and explain why this was so.
4 What is the difference between complete and incomplete combustion?
5 Write a word and a formula equation for the complete combustion of methane.
6 Write a word and a formula equation for the incomplete combustion of methane.

CONCLUSION

Copy and complete.

The results show that: (*respond to the aim*).

RESULTS

TABLE I14.7B

	Flame colour	Soot/ no soot	Initial temp. (°C)	Final temp. (°C)	Temperature change after 5 minutes (°C)
Incomplete combustion					
Complete combustion					

Investigation 14.8A

Corrosion of iron

AIM

To investigate the conditions required for the rusting of iron

MATERIALS

- 5 ungalvanised iron nails
- 2 × 10 mL measuring cylinders
- tap water
- boiled water
- salt water
- oil
- drying agent such as calcium hydroxide
- 5 test tubes
- test-tube rack
- 5 rubber stoppers
- spatula
- paper towel
- electronic balance

METHOD

1. Copy the results table into your notebook, adding a title.
2. Read all the instructions before you begin this experiment.
3. Set 5 test tubes on a test-tube rack.
4. Label the first test tube 'Air only'.
5. Label the second test tube 'Drying agent'. Add one spatula of calcium hydroxide.
6. Label the third test tube 'Water and air'. Add 5 mL of water.
7. Label the fourth test tube 'Salt water'. Add 5 mL of salt water.
8. Label the fifth test tube 'Oil and boiled water'. Add 5 mL of boiled water.
9. Weigh the five nails separately. Note each mass and add a nail to each test tube.
10. Add oil to cover the boiled water in test tube 5.
11. Stopper all test tubes with the rubber stoppers.
12. Observe the test tubes every second day for a week.
13. Remove the nails, clean and dry them and weigh them.
14. Record all your results in the results table.

DISCUSSION

1. Rank the nails from the least rusted to the most rusted.
2. Should there have been a gain or a loss in mass of the nails? Explain.
3. Which environment causes the most rust? Research why this environment causes the most rusting of iron.
4. Why was it important to measure the mass of the iron nail before and after the experiment?

CONCLUSION

Copy and complete.

The results show that: (*respond to the aim*).

WEAR SAFETY GLASSES AND GLOVES WHILE DOING THIS EXPERIMENT TO PREVENT RUSTED METAL FROM ENTERING YOUR EYE OR TOUCHING YOUR SKIN.

RESULTS TABLE I14.8A

	Test tube				
	1 Air only	2 Drying agent	3 Water and air	4 Salt water	5 Oil and boiled water
Initial mass of nail (g)					
Final mass of nail (g)					
Change in mass (g)					
Observations					

Investigation 14.8B

Preventing corrosion

AIM

To investigate some methods of preventing corrosion

MATERIALS

- 4 ungalvanised iron nails
- 1 galvanised iron nail
- salt water
- magnesium ribbon
- paint
- oil
- 5 test tubes
- test-tube rack
- 5 stoppers to fit test tubes

METHOD

1 Copy the results table into your notebook, adding a title.
2 Prepare the nails.
   - Nail 1: leave as is for the control.
   - Nail 2: Paint it and let it dry.
   - Nail 3: Dip it in oil and let the oil drip off.
   - Nail 4: Twirl magnesium ribbon tightly around it.
   - Nail 5: The galvanised nail that is coated with zinc.
3 Label five test tubes 'Untreated', 'Painted', 'Oil coated', 'Magnesium twirled' and 'Galvanised'.
4 Half-fill all test tubes with salt water.
5 Add the nails to their appropriate test tubes.
6 Stopper the test tubes and leave them to stand for a week.
7 Record your observations and determine if the method used to prevent the corrosion of iron was effective.

DISCUSSION

1 List the controlled variables, independent variable and dependent variable.
2 What happened with the nail that had magnesium twirled around it? Explain why this happened.
3 What is a galvanised nail?
4 How does galvanising prevent corrosion?

CONCLUSION

Copy and complete.
The results show that: (*respond to the aim*).

METAL PARTS OR SOLUTIONS MAY ENTER YOUR EYES. WEAR SAFETY GLASSES TO PREVENT THIS.
WEAR GLOVES TO PREVENT THE PAINT OR OIL FROM TOUCHING YOUR SKIN.

RESULTS TABLE I14.8B

	Nail				
	1 Untreated (control)	2 Painted	3 Oil coated	4 Magnesium twirled	5 Galvanised
Observations					

Investigation 14.9

Precipitation reactions

AIM

To find out which reactions form a precipitate

MATERIALS

- 0.1 mol/L solutions in dropper bottles of: sodium sulfate, sodium carbonate, sodium chloride, potassium iodide, lead nitrate, silver nitrate, barium chloride, copper sulfate, sodium hydroxide
- spotting tile
- paper towel
- a piece of blank paper

METHOD

1 Copy the results table into your notebook, adding a title.
2 Put the spotting tile on the piece of blank paper. Label the cations on the left-hand side and the anions on the top of the spotting tile, as shown in the diagram. You may have to wash the spotting tile and re-use it for further tests.

Spotting tile

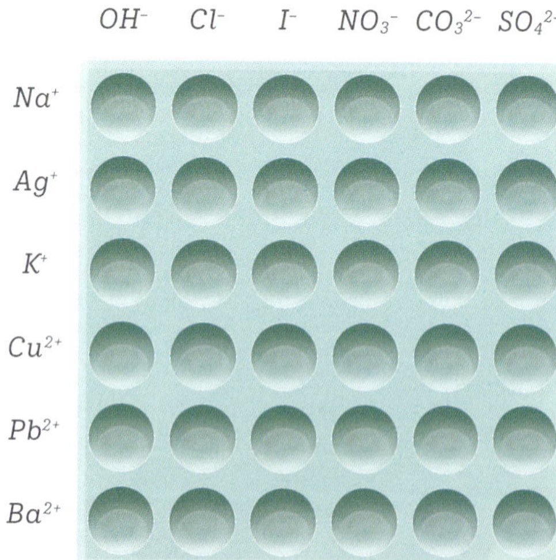

3 Add one drop of the appropriate anion and one drop of the appropriate cation into the same cavity and observe whether a precipitate has formed. Record your results in the table, by using 's' for soluble and 'p' for precipitate. Be careful not to contaminate the solutions.

RESULTS **TABLE I14.9**

Cations	Anions					
	OH^-	Cl^-	I^-	NO_3^-	CO_3^{2-}	SO_4^{2-}
Na^+						
Ag^+						
K^+						
Cu^{2+}						
Pb^{2+}						
Ba^{2+}						

DISCUSSION

1 Why did you use dropper bottles of solution instead of using solutions from a beaker?
2 Why did you use spotting tiles and not test tubes to carry out these reactions?
3 Write the net ionic equation for the formation silver chloride precipitate.
4 Use your results to predict whether the following salts will be soluble: barium sulfate, silver hydroxide, copper carbonate, silver chloride, lead iodide.
5 Explain how precipitation reactions occur.

CONCLUSION

Copy and complete.
The results show that: (*respond to the aim*).

CHEMICALS MAY ENTER YOUR EYES. WEAR SAFETY GLASSES TO PREVENT THIS. SILVER SALTS ARE TOXIC AND MAY STAIN YOUR SKIN AND CLOTHES. WEAR GLOVES TO PROTECT YOUR SKIN.

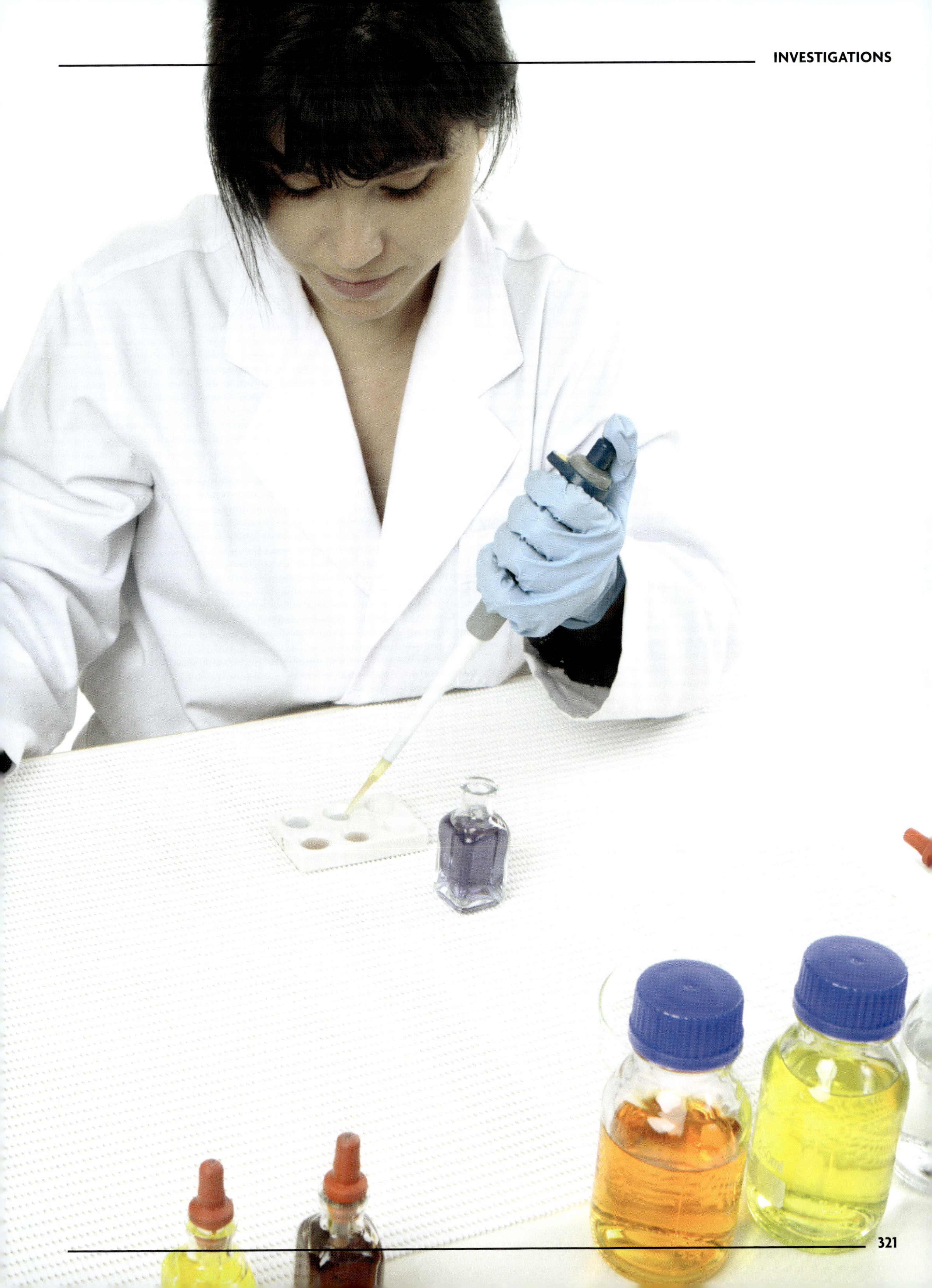

Investigation 15.1

Exothermic and endothermic reactions

AIM

To investigate whether a chemical reaction is exothermic or endothermic

MATERIALS

- vinegar
- sodium bicarbonate
- 3 cm piece of magnesium ribbon
- 1 mol/L hydrochloric acid (HCl)
- 1 mol/L copper sulfate solution
- 1 mol/L sodium hydroxide
- 10 mL measuring cylinder
- 1 mol/L sulfuric acid
- distilled water
- 5 test tubes
- stirring rods
- 3 spatulas
- steel wool
- watch glass or piece of wood
- thermometer

METHOD

Copy the results table into your notebook, adding a title.

PART A

1 Add 5 mL of vinegar to a test tube.
2 Use a thermometer to record the temperature of the vinegar.
3 Add a spatula of sodium bicarbonate to the test tube.
4 Mix gently with a stirring rod, being careful not to break the base of the test tube. After a minute, record the temperature of the mixture.
5 Record your results in the results table.

PART B

1 Add 5 mL of hydrochloric acid to a test tube.
2 Record the temperature of the hydrochloric acid.
3 Add 5 mL of 1 mol/L sodium hydroxide.
4 Mix gently with a stirring rod, being careful not to break the base of the test tube. After a minute, record the temperature of the mixture.
5 Record your results in the results table.

PART C

1 Add 5 mL of copper sulfate to a test tube.
2 Record the temperature of the solution.
3 Add a 3 cm piece of magnesium that has been cut in small pieces.
4 Record the temperature of the solution after 1 minute.
5 Record your results in the results table.

PART D

1 Add 5 mL of sulfuric acid to a test tube.
2 Record the temperature of the solution.
3 Add a piece of magnesium ribbon.
4 Record the temperature of the solution after 1 minute.
5 Put your results in the result table.

PART E

1 Add 10 mL of vinegar to a test tube.
2 Record the temperature of the vinegar.
3 Add a piece of steel wool so that it is completely submerged.
4 Record the temperature of the vinegar after 1 minute.
5 Record your results in the results table.

RESULTS TABLE I15.1

Part	Reactants	Initial temp. (°C)	Final temp. (°C)	Observations	Exothermic or endothermic
A	Vinegar + sodium bicarbonate				
B	Hydrochloric acid + sodium hydroxide				
C	Copper sulfate + magnesium				
D	Sulfuric acid + magnesium				
E	Vinegar + steel wool (iron)				

QUESTIONS

1 Which reactions were exothermic and which were endothermic? Give evidence to support your choices.

2 Copy and complete equations for all the reactions in the table. The first one has been done for you.

a acetic acid + sodium bicarbonate → sodium acetate + carbon dioxide + water

b ________ + ________ → sodium chloride + water

c ________ + ________ → magnesium sulfate + copper

d ________ + ________ → magnesium sulfate + hydrogen

e ________ + ________ → iron acetate + hydrogen

WEAR SAFETY GLASSES AND GLOVES WHILE DOING THIS EXPERIMENT TO PREVENT ACID FROM BURNING YOUR SKIN OR SPLASHING IN YOUR EYES, AND TO PREVENT METAL GETTING IN YOUR EYES.

DISPOSE OF WASTES APPROPRIATELY.

Investigation 15.3A

The effect of concentration on the rate of a reaction

AIM

To investigate how concentration affects the rate of chemical reactions

MATERIALS

- 0.15 mol/L sodium thiosulfate
- 2 mol/L hydrochloric acid
- 5 × 100 mL beakers
- 2 × 50 mL measuring cylinders
- 5 × 10 mL measuring cylinder
- black whiteboard marker pen
- stirring rod
- distilled water
- stopwatch

METHOD

1 Copy the results table into your notebook, adding a title.
2 Label the five beakers 1–5.
3 Mark the bottom of each beaker with an X.
4 Use the two 50 mL measuring cylinders to add sodium thiosulfate and distilled water in the volumes indicated in the results table.
5 Add 5 mL of hydrochloric acid solution to each 10 mL measuring cylinder and place them next to each beaker.
6 Carefully add the hydrochloric acid to the sodium thiosulfate solution in beaker 1. Stir the solution once with a stirring rod and immediately start timing.
7 Stop timing when the black X is no longer visible.
8 Record the reaction time in seconds in the results table.
9 Wash the stirring rod between each use. Repeat steps 6–9 with the remaining beakers.

Calculate $\frac{1}{\text{reaction time}}$ for each trial.

Plot concentration vs time and concentration vs $\frac{1}{\text{time}}$ on separate graphs.

DISCUSSION

1 List the controlled variables, independent variable and dependent variable.
2 Should the total volume of each beaker be the same or different? Explain your answer.

CONCLUSION

Copy and complete.

The results show that: (*respond to the aim*).

SODIUM THIOSULFATE IS HARMFUL IF INGESTED. DO NOT INGEST IT. WASH YOUR HANDS AFTER USING IT. HYDROCHLORIC ACID IS CORROSIVE TO EYES AND SKIN AND IS AN IRRITANT. IT MAY CAUSE BURNS. IF SPILT ON SKIN FLUSH THE AREA WITH COLD TAP WATER. WEAR SAFETY GLASSES. SULFUR DIOXIDE IRRITATES THE RESPIRATORY SYSTEM. MAKE SURE THE ROOM IS VENTILATED.

RESULTS TABLE I15.3A

Beaker	Sodium thiosulfate (mL)	Water (mL)	Sodium thiosulfate (mol/L)	Time (s)	$\frac{1}{\text{time}}$ (s^{-1})
1	50	0	0.15		
2	40	10	0.12		
3	30	20	0.090		
4	20	30	0.060		
5	10	40	0.030		

Investigation 15.3B

The effect of temperature on the rate of a reaction

AIM

To investigate the effect of temperature on the rate of a reaction

MATERIALS

- 3 Alka-Seltzer tablets
- distilled water at three different temperatures: cold, room temperature and warm (40°C)
- 3 × 250 mL beakers
- stopwatch
- thermometer

METHOD

1. Copy the results table into your notebook, adding a title.
2. Measure 200 mL each of warm water, room temperature water and cold water into three separate labelled beakers.
3. Measure the water temperature in each beaker and record it in the results table.
4. Drop an Alka-Seltzer tablet into each beaker. Use the stopwatch to immediately start timing how long it takes to dissolve entirely.
5. Stop the timer when the tablet has dissolved completely.
6. Record the reaction times (in minutes and seconds) in the results table.

RESULTS **TABLE I15.3B**

Type of water	Temperature (°C)	Time to dissolve (s)
Cold		
Room temperature		
Warm		

DISCUSSION

1. Graph the reaction time (minutes) on the *y*-axis against the water temperature (°C) on the *x*-axis.
2. List the controlled variables, independent variable and dependent variable.
3. How does reaction time change with temperature? Do the results support your hypothesis? Why or why not?
4. If some tablets used are whole and some are broken, would the temperature experiment still be valid? Why or why not?
5. If some students stirred the water while their tablets were dissolving, would the temperature experiment still be valid? Why or why not?
6. If everyone in the class used a different water temperature, could you average the results? Why or why not?

CONCLUSION

Copy and complete.

The results show that: (*respond to the aim*).

WEAR SAFETY GLASSES WHILE DOING THIS EXPERIMENT TO PREVENT ANY MATERIAL FROM ENTERING YOUR EYES.

Investigation 15.4A

The effect of a catalyst on the rate of a reaction

AIM

To investigate the effect of a catalyst on the decomposition of hydrogen peroxide

Hydrogen peroxide solution decomposes slowly at room temperature to water and oxygen:

$$\text{hydrogen peroxide} \rightarrow \text{oxygen} + \text{water}$$
$$2H_2O_2(aq) \rightarrow O_2(g) + 2H_2O(l)$$

The volume of oxygen given off can be measured by collecting it in an inverted measuring cylinder.

MATERIALS

- 3% hydrogen peroxide
- 0.5 g manganese dioxide
- conical flask with a delivery tube attached with a rubber stopper
- 50 mL and 100 mL measuring cylinders
- retort stand, boss head and clamp
- trough of water

METHOD

1 Copy the results table into your notebook, adding a title.
2 Using a 50 mL measuring cylinder, add 50 mL of hydrogen peroxide solution to a conical flask.
3 Fill the 100 mL measuring cylinder with water, seal the top with your fingers, and invert it into the water trough, as shown in the diagram.
4 Put the rubber stopper loosely into the conical flask and make sure the delivery tube connects to the inverted measuring cylinder.
5 Measure 0.5 g of the catalyst manganese dioxide.
6 Add the catalyst to the conical flask, put the stopper back into the flask and start the stopwatch.
7 Record the volume of gas given off every 10 seconds. Continue timing until no more oxygen appears to be given off. Record all your observations.

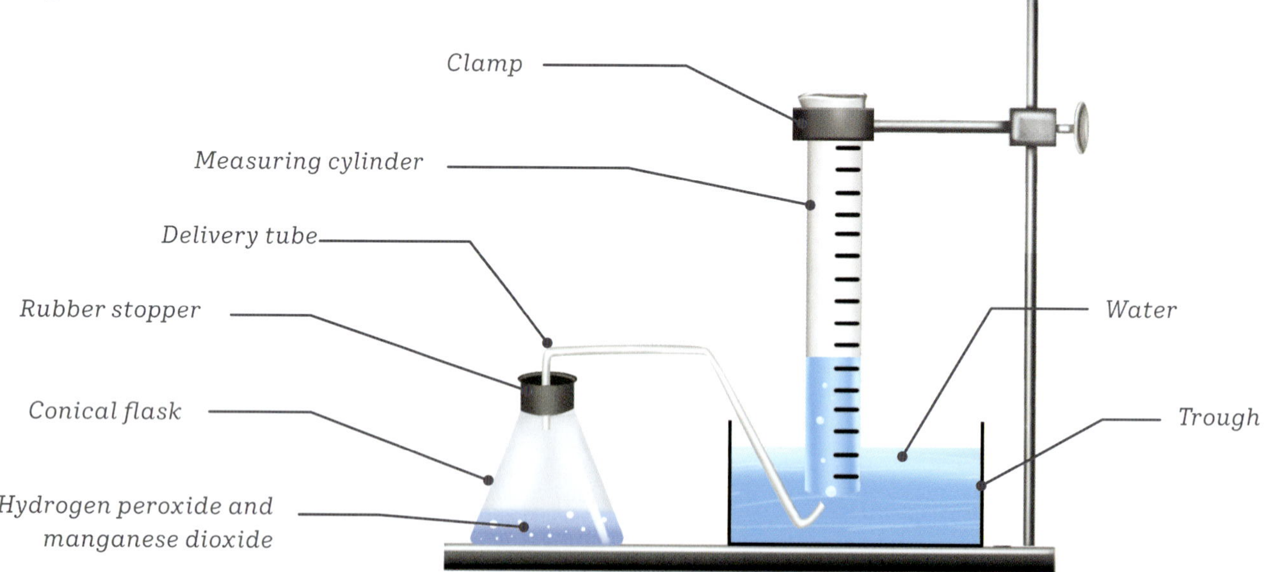

RESULTS TABLE I15.4A

Time (s)	Volume of $O_2(g)$ (mL) with catalyst	Volume of $O_2(g)$ (mL) without catalyst
10		
20		
30		
40		
50		
60		

DISCUSSION

1 Draw a graph of your results with time on the *x*-axis and volume of oxygen on the *y*-axis.
2 Comment on the trends you see on the graphs.
3 Identify the catalyst in this reaction.
4 What was the role of the catalyst in this experiment?
5 Did the temperature in the conical flask change?

CONCLUSION

Copy and complete.
The results show that: (*respond to the aim*).

HYDROGEN PEROXIDE CAN CAUSE SKIN IRRITATION. WEAR GLOVES AND WASH YOUR HANDS IMMEDIATELY IF HYDROGEN PEROXIDE GETS ONTO THEM.

MANGANESE DIOXIDE CAN IRRITATE IF INHALED, SO AVOID SPILLING IT.

DISPOSE OF WASTES APPROPRIATELY.

Investigation 15.4B

The effect of surface area on the rate of a reaction

AIM

To investigate the effect of increasing the surface area on the rate of a reaction

MATERIALS

- 4 Alka-Seltzer tablets
- water
- 4 × 100 mL beakers
- mortar and pestle
- 50 mL measuring cylinder

METHOD

1 Copy the results table into your notebook, adding a title.
2 Add 50 mL of water to each of the four beakers.
3 Of the four Alka-Seltzer tablets:
   - leave the first one whole
   - cut the second one in half
   - cut the third one into quarters
   - crush the fourth one with the mortar and pestle.
4 Add the whole tablet to the first beaker of water and time how long it takes to dissolve.
5 Add the pieces of the second tablet to the second beaker of water and time how long they take to dissolve.
6 Add the pieces of the third tablet to the third beaker of water and time how long they take to dissolve.
7 Add the crushed tablet to the fourth beaker of water and time how long it takes to dissolve.
8 Make your observations and record your results.

RESULTS TABLE I15.4B

Size of Alka-Seltzer tablet	Time taken for the tablet to dissolve (s)
Whole tablet	
Halved tablet	
Quartered tablet	
Crushed tablet	

DISCUSSION

1 List the controlled variables, independent variable and dependent variable.
2 Do you think your experiment was reliable? Why or why not?
3 Draw a column graph to show your results.
4 Explain what you think the rate of reaction depended on.
5 Explain how you changed the surface area of the tablets.

CONCLUSION

Copy and complete.
The results show that: (*respond to the aim*).

WEAR SAFETY GLASSES WHILE DOING THIS EXPERIMENT TO PREVENT ANY PARTICLES FROM ENTERING YOUR EYES.

GLOSSARY

abiotic non-living

abundant present in large amounts

acceleration a change in speed or direction of motion over time

accretion the process of matter collecting together into a bigger mass

action–reaction pair two equal and opposite forces exerted by two objects on each other

activation energy the energy required to start a reaction

adaptation a change in the structure or function of an organism over time that makes it better suited to its environment

adaptation a feature that enables an organism to survive in its environment

aerobic in the presence of oxygen

aerosol tiny droplets in the atmosphere

alkali metal an element in group 1 of the periodic table

alkaline earth metal an element in group 2 of the periodic table

allosome a sex chromosome (1 pair in humans)

alloy a substance made from mixing two or more metals

alpha particle a small positively charged particle

alpine ecosystem an ecosystem existing at high altitudes

amino acid a simple molecule that is the basic unit of a protein

ammeter a device that measures electric current

amplifier an electronic component that boosts electrical current

amplitude half the height of the wave

anaerobic in the absence of oxygen

angle of incidence the angle at which light hits a surface

angle of reflection the angle at which light reflects from a surface

antibiotic a medicine that kills microorganisms such as bacteria or stops them reproducing

antibody a protein that responds to a specific antigen

antigen a substance that triggers the production of antibodies

antimatter particles that have properties opposite to that of normal matter

asthenosphere the portion of Earth's mantle, underneath the lithosphere, that can flow

astronomer a scientist who studies space, stars and celestial objects

astronomical unit the average distance between Earth and the Sun (about 150 000 000 km)

astrophysicist a scientist who studies the physics of the universe

atmosphere the layer of gas that surrounds Earth and is 600 km thick

atmospheric nitrogen nitrogen that is found in the atmosphere

atom the smallest particle of matter, made up of electrons orbiting a nucleus of protons and neutrons

atomic mass the sum of the protons and neutrons in an atom

atomic number the number of protons in an atom

autosome a non-sex chromosome (22 pairs in humans)

balanced force a force acting on an object that is cancelled out by another force, so the net force is zero

biochemist a scientist who studies the chemical processes in living organisms

biodiversity the number of species in an ecosystem

biogeography the study of the past and present distribution of living organisms

biosphere all of the living things on Earth

biotechnology the use of living organisms to develop products

biotic living

GLOSSARY

carbohydrase a type of enzyme that digests carbohydrates

carbon capture the process of trapping carbon dioxide at its emission source so that it does not enter the atmosphere

carbon cycle the cycle that explains how carbon moves between Earth's spheres

carbon sink a place where carbon is stored in the carbon cycle

carbonate a substance containing the elements carbon and oxygen

catalyst substance that increases the rate of a chemical reaction without being used up

cathode ray a stream of electrons observed in a high-vacuum tube

caustic able to burn or corrode organic tissue through chemical action

cellular respiration the process that all living things use to produce cellular energy from glucose and oxygen

chemical formula chemical symbols showing the ratio of elements to one another

chemical property a property of a substance that is observed during a chemical reaction

chemical symbol one or two letters that represent an element

chromosome a tightly coiled strand of DNA

circuit a closed path containing a collection of components connected to a power source that allows charge to flow

coefficient a number placed before the chemical in a formula or chemical equation

collaborate work together

combustion a reaction that involves burning in the presence of oxygen to release heat

community a naturally occurring group of animals, plants and other organisms

comparative anatomy the study of features of different species to look for evidence of a common ancestor

compound substance formed when two or more elements are chemically bonded together

concentration the amount of substance in a volume of solution

conduction the transfer of heat through a substance

conductor a material that allows the movement of charge or transfer of heat

conservation of mass mass cannot be created or destroyed

conservation protection and maintenance

consumer an organism that gains energy by consuming other living organisms

contagious able to spread from one organism to another

continental drift the theory that the continents have moved position over time

convection the transfer of heat by movement of a liquid or gas

convergent boundary where two tectonic plates are moving towards each other

Coriolis force a force caused by the rotation of Earth

corrosive highly reactive and will damage or destroy another substance

covalent bond a bond in which two atoms share one or more pairs of electrons

current a measure of how fast charge moves in a circuit

cyclone an intense tropic storm that forms over warm oceans

Dalton's atomic theory the theory that states that all matter is made up of tiny particles

degradation the breaking down of a substance

dengue fever a mosquito-borne viral disease occurring in tropical and subtropical areas

deoxyribonucleic acid (DNA) found in the nucleus of a cell; the carrier of genetic information

diffuse to move from an area of high concentration to an area of low concentration

diploid the number of chromosomes found in most cells

displacement the distance an object is from its starting position

dissociate to split apart into ions

distance the amount of space between two points

dosimeter a device used to measure an absorbed dose of ionising radiation

double helix the structure of a DNA molecule; a double-stranded spiral

ductile can be drawn out into a wire

economy the system of how a country makes and spends money and provides goods and services

ecosystem a community of living things and their environment

effector a muscle, gland or organ that responds to a message sent by the nervous or endocrine system

efficient not wasteful

electrical charge a quantity of electricity caused by an imbalance of protons and electrons

electrically neutral having equal number of protons (positively charged) and electrons (negatively charged)

electrolysis a decomposition reaction using electricity

electromagnetic spectrum all the different electromagnetic waves

element a pure substance made up of one type of atom

epidemic widespread occurrence of an infectious disease in a community

endothermic a reaction that absorbs energy in the form of heat

energy a measure of the ability to do work

energy efficiency how much usable energy is produced compared to how much energy has been supplied

energy transfer the movement of energy from one place to another without changing form

energy transformation a change from one type of energy to another

enhanced greenhouse effect an increase in the greenhouse effect due to human greenhouse gas emissions

enzyme a biological catalyst that increases the rate of reactions in cells

epicentre the point on Earth's surface directly above the focus of an earthquake

equation a mathematical statement showing that one value is equal to another

erosion the natural process of wearing away by wind, water or other natural agents

ethical consideration a factor that takes into account what is right and what is wrong

euryhaline the ability to survive in water with a range of salinity levels, from fresh to very salty

eutrophication the process in which nutrient levels increase in a waterway, resulting in increased algal growth and decreased dissolved oxygen levels

evolution the change in characteristics of living things over time

evolve change over time to adapt to the environment

exoplanet a planet located outside our solar system

exothermic a reaction that releases energy in the form of heat

fault a break in Earth's surface where blocks of rock slide past each other

finite limited, won't last forever

focus the origin of an earthquake

fold mountain a mountain formed by the folding of continental crust when tectonic plates collide

force a push, pull or twist

formula a relationship expressed in symbols

fossil fuel a fuel that is formed from the decomposition of dead animals and plants over millions of years, e.g. oil and coal

fossil record a record of all fossils (found and not found) on Earth

fossil the geologically altered remains of an organism that was once alive

frequency the number of waves passing a point every second

friction a contact force that opposes motion, caused by objects rubbing against each other

frost line a boundary just inside Jupiter's orbit

galaxy a system of millions or billions of stars

gamete a sex cell – ova or sperm

gamma ray radiation emitted by radioactive material

gas exchange the exchange of oxygen and carbon dioxide between an organism and the environment

gene a segment of DNA, the basic functional unit of heredity

generate produce or make something

genetic variation the differences in genes within individuals of a population

genome an organism's entire set of DNA

genotype the genetic code for a gene or an organism's entire genome

GLOSSARY

gland tissue that secretes hormones

GM genetically modified; an organism that has had alterations to its DNA

GPS global positioning system

graft a piece of living tissue that is transplanted surgically

gravity the force that attracts objects to each other

greenhouse effect the trapping of the Sun's warmth by the atmosphere

greenhouse gas a gas that traps heat energy in the atmosphere

group a column in the periodic table

half-life the time taken for half of the radioactive material to decay

halogen an element in group 17 of the periodic table

haploid the number of chromosomes found in the sex cells (sperm and ova)

hereditary the passing of characteristics through the genes from one generation to the next

heredity the passing on of traits from parents to their offspring

homeostasis maintaining a stable balanced state within the body

homologous structures parts of organisms that show evidence of a common ancestor

hormone a chemical secreted by a gland that triggers a response in certain cells

hot spot volcano a volcano formed by magma upwelling underneath a tectonic plate

hydrocarbon a compound made up of carbon and hydrogen

hydrosphere all the water on Earth

igneous rock rock formed when molten rock cools and becomes solid

implication likely outcomes or consequences of an action

industrial chemist a scientist who converts raw materials to products on a commercial scale

inertia a property of matter that causes it to resist change in speed or direction (to remain at rest or in a state of uniform motion)

infrared light an electromagnetic wave with longer wavelength than red light

insulator a material that resists the movement of charge or transfer of heat

intensity a measure of the amount of destruction caused by an earthquake

inversely proportional when one value decreases at the same rate that the other increases

ion a charged particle

ionic bond a bond in which atoms gain and lose electrons, becoming ions

ionise move electrons from an atom or molecule

ionising radiation radiation made up of particles, X-rays or gamma rays with sufficient energy to cause cancer

isotopes atoms of the same element with the same number of protons but different numbers of neutrons

IVF (in vitro fertilisation) fertilisation of the sperm and egg that occurs outside the human body

joule the unit of energy

karyotype a picture of an individual's full set of chromosomes

keystone species a species that plays a crucial role in its ecosystem

kinetic energy energy of motion

lattice an interlaced structure or pattern

lava molten rock at Earth's surface

light-year the distance that light travels in one Earth year

lipase a type of enzyme that digests lipids (fats)

lithosphere Earth's rigid outer zone (crust and upper mantle), made up of tectonic plates

longitudinal running lengthwise rather than across

lymphocyte a type of white blood cell that produces antibodies in response to a pathogen

magma molten rock below Earth's surface

magnitude a measure of the energy released by an earthquake

malleable can be bent

mantle Earth's middle layer, made up of two layers

mass the amount of matter that something consists of

mass number the sum of the protons and neutrons in an atom

matter physical substance that has mass and takes up space

menstruation monthly discharge of blood and tissue from the lining of the uterus; also called a period

metallic bond a bond in which free electrons move around metal ions

metalloid an element with properties similar to both metals and non-metals

mid-ocean ridge a long chain of mountains under the ocean formed by plate tectonics

molecule the smallest unit of a chemical compound that can take part in a chemical reaction

moment magnitude scale a logarithmic scale used to compare the amount of energy released by earthquakes

mutation an error in DNA caused by a substitution, an insertion or a deletion of a base

mutualism a relationship in which both organisms benefit

natural selection the process where organisms that are better suited to their environment tend to survive and reproduce, passing their traits on to further generations

nebula a vast region of gas and dust

negative feedback a response that counteracts the stimulus

net force all the forces acting on an object added together

neuron a specialised cell that makes up the nervous system

neutralisation reaction a reaction involving an acid and a base

neutralise to make something chemically neutral

neutron star an extremely dense star left over after a supernova

newton the unit of measurement of force

nitrogenous containing nitrogen bonded with other elements

noble gas an element in group 18 of the periodic table

normal force the force of the floor or ground pushing an object upwards

nucleotide building block of DNA (adenine, guanine, thymine, or cytosine)

nucleus the central part of an atom containing protons and neutrons

Ohm's law a law that states that the current through a conductor between two points is directly proportional to the voltage across the two points

omnivore an organism that eats both plants and animals

opaque a material that doesn't allow any light to pass through

ovum the female sex cell, also known as the egg

oxidation a reaction taking place in the presence of oxygen

ozone a molecule made of three oxygen atoms bonded together

ozone-depleting substance a chemical that reacts with ozone and breaks the molecule down as well as stopping the molecule from re-forming

ozone layer a region of the stratosphere with a high concentration of ozone

palaeontologist a scientist who studies fossils

pandemic disease occurring within a whole country or around the world

parallel circuit a circuit in which all components are connected between the same points, so the current has more than one path to take

parsec 3.26 light-years

pathogen an agent that causes disease

penetrating power how well radioactive particles can pass through matter

period a row in the periodic table

permafrost frozen soil in the Arctic

pesticide a substance that controls pests, such as insects or weeds

petrification the process in which minerals replace organic matter and a dead organism turns into stone

pH a figure expressing the acidity or alkalinity of a solution

phagocyte a type of cell capable of engulfing and destroying bacteria

phenotype how the genotype is physically expressed

photosynthesis the process that plants use to make glucose from carbon dioxide and water

physical property a property of a substance that can be measured, such as colour, melting and boiling points, density and hardness

pollution a substance that enters the environment and has harmful or poisonous effects

polyatomic ion two or more atoms bonded together and acting as a single charged unit

GLOSSARY

population a group of the same species living in a specific area

precipitate to form an insoluble product

prefix a word or number placed before another word

producer an organism that makes its own food from the energy from the Sun

product a substance formed in a chemical reaction

protease a type of enzyme that digests proteins

pustule a small blister

pyroclastic flow a dense and fast-moving mass of extremely hot ash and gas

quarantine the isolation of people, plants, animals or objects that may have been exposed to biosecurity threats

radioisotope an isotope that emits radiation

radiotherapy a therapy that uses radiation to kill cancerous cells

reactant a substance that starts a chemical reaction

receptors specialised cells and organs that detect changes

reciprocal 1 divided by the number

red giant a star that has stopped fusing hydrogen in its core

redshift a change in light's wavelength towards the red end of the visible spectrum

reflect send back sound or light without absorbing it

refract bend light

resolution the ability to tell two separate objects apart

respiration exothermic reaction within cells that releases energy for the body's use

rift valley a valley formed when a continent is being pulled apart

scalar a quantity that has a magnitude but no direction

seismic wave a wave of energy that passes through Earth's layers and is caused by an earthquake

seismometer a scientific instrument that detects seismic waves

semen a protective fluid that contains the sperm

series circuit a circuit in which components are arranged in a chain, so the current has only one path to take

shell space around the nucleus where electrons circulate

singularity an infinitely dense point of matter that existed before the Big Bang

social consideration a factor that affects society and the people within society

sonar sound navigation and ranging

soot a black form of carbon formed by incomplete combustion

spectator ion an ion that does not take part in the reaction

spectral line a dark or bright line in an otherwise uniform spectrum

speed the distance an object travels divided by the time taken

sperm the male sex cell

spontaneous not planned

stationary not moving, still

stem cell a cell type that is unspecialised and can develop into other types of cell

strato volcano a volcano formed at a subduction zone

strong acid an acid that ionises completely in water

subduction when one tectonic plate moves underneath another

supernova an explosion of a massive star at the end of its life

surface area the area of the outermost layer of an object

surrogacy an arrangement in which a woman becomes pregnant and gives the baby to another person

surroundings a container or an external environment

suspended distributed throughout a fluid

sustainability long-term maintenance

sustainable able to be maintained long-term

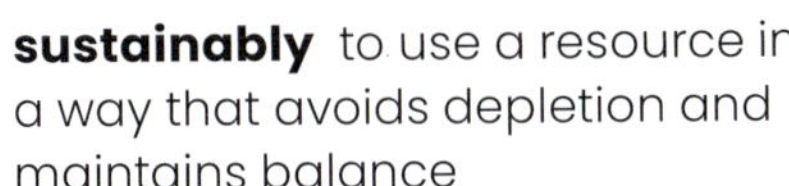

sustainably to use a resource in a way that avoids depletion and maintains balance

synapse the gap between the axon and dendrite of two neighbouring neurons

tectonic plate a section of Earth's lithosphere

transform boundary where two tectonic plates are sliding past one another

transform change from one form to another

transistor an electronic component that can act as a switch or an amplifier

transition metal an element in groups 3–12 of the periodic table

translucent a material that allows some light to pass through

transparent a material that allows light to pass through

transverse running across rather than lengthwise

trophic level the position an organism occupies in a food chain

tsunami a sea wave caused by the displacement of water as a result of an earthquake or other disturbance

unbalanced force a force acting on an object that is not cancelled out by another force, so a net force acts on the object

usable energy the type of energy that a device is designed to produce

ultraviolet radiation harmful rays from the Sun, which can cause cancers and harm animals and plants

vacuum empty space

valency an element's power to combine with other atoms when it forms molecules

vector a quantity that has both a magnitude and a direction

velocity a measure of how quickly displacement changes

vestigial structure a part of an organism that has lost some or all of its original function

volcano a point in Earth's crust where lava erupts

voltage the difference in potential energy between two points in a circuit

voltmeter a device that measures potential difference (voltage)

wavelength the distance from the peak of one wave to the next

weak acid an acid that partially ionises in water

weight force the force generated by gravity pulling an object downwards

white dwarf a small, very dense star formed at the end of a small star's lifetime

X-ray a high-energy ray that can penetrate materials

zone of tolerance the range of an abiotic factor that an organism can survive in

INDEX

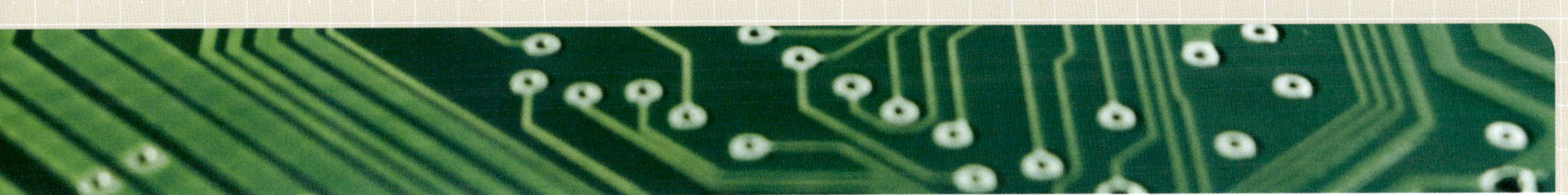

INDEX

INDEX

ACKNOWLEDGEMENTS

The author and publisher are grateful to the following for permission to reproduce copyright material:

PHOTOGRAPHS

AAP/ Luke Aikins, **17** (top), /Great Barrier Reef Marine Park Authority, **101**, **108**, /Matt Dunham/AP, **147** (top), /Ron Siddle, Antelope Valley Press/AP, **158** (bottom); Alamy Stock Photo/Custom Life Science Images, **46** (bottom)/, dpa picture alliance, **19** (bottom), /Juice Images, **175** (top), /Martin Siepmann/imageBROKER, **91** (bottom), Prisma by Dukas Presseagentur GmbH, **168**, (right), /A. T. Willett, **230**, **240** (top right), /AB Forces News Collection, **69** (bottom), /Aleksandr Belugin, **5**, Anton Lebedev, **52** (bottom, right), /Archive PL, **103**, /Auscape, **142**, /Bernd Settnik/dpa-Zentralbild/dpa, **111** (bottom),/Blue Planet Archive-PNE, **129** (top), /Bob Kreisel, **195**, **200** (bottom right), /cbstockfoto, **128** (top), /David Forster, **16** (bottom), /David Shale/NPL, **170** (top), /Ekin Yalgin, **111** (top), /Gerry Pearce, **140** (bottom), /Historic Images, **200** (bottom centre right), /INTERFOTO, 198, **200** (bottom centre), /Jaroszpilewski, **30**(top), /Kerstin Langenberger, **84**, /Laura Chiesa/Pacific Press, **31** (top), /Nigel Spiers, **80** (left), /Paul Faith/PA Images, **122**, /Peter Hermes Furian, **10**, /Photo Researchers/Science History Images, **124**, **154** (bottom), /Pictorial Press Ltd , **186** (centre right), /PJF Military Collection, **226** (top right), /Science History Images, **200** (bottom centre left), /Science Picture Co, **120**, / sciencephotos, **216** (bottom), /Seaphotoart, **129** (bottom), /Sebastian Kaulitzki, **158** (top), /Susan E. Degginger, **51** (top), / tom pfeiffer, **274–5**; Auscape/Mike Gillam, **140**, /Wayne Lawler, **138**, /David Wachenfeld, **141**, **144** (bottom left); CERN/Anna Pantelia, **205**; CSIRO/David McClenaghan, **63**, Dr Louise Morin, **123**, 126 (bottom); Event Horizon Telescope Collaboration, **57**; ESA/ATG medialab, (CC BY-SA 3.0) IGO, **61** (bottom), ESA, Hubble, NASA, and S. Beckwith (STScI) and the HUDF Team, **55** (top); ESO/M. Kornmesser, **55** (bottom), **66** (top, left), H. Drass et al., **56**, **330–35**; European Chemical Society (CC BY ND), **189** (top); Flickr/Lennart Halling, 1960-11 (CC BY 2.0), **189** (bottom); Fotolia/Georgios Kollidas, **186** (top centre right); Getty Images, **217**, /BANANASTOCK/1996-98 AccuSoft Inc, **203** (bottom), /Ulrich Baumgarten, **90** (centre), /BEDI/AFP, **227** (bottom), /Bettmann, **186** (bottom centre right), /Monica Bertolazzi, **58**, **66** (top, right), /Ben Curtis/PA Image, **156** (top), **160** (bottom centre), /C12, **95**, /De Agostini, **69** (top), /DNY59, **136**, /Daniel Eskridge/Stocktrek Images, **163** (bottom), /Encyclopaedia Britannica, **80** (right), /Elliot & Fry/Hulton Archive, **186** (bottom right), /foto-ruhrgebiet/iStock, **200** (top right), /Heritage Images/Hulton Archive, **186** (top right), /Hero Images, **306**, /Icon Sportswire, **18** (bottom), /Jxfzsy, **110** (top), /Knowlesgallery, **176**, /Kyodo News, **50**, /Pamela Moore/istock, **222**, /Morsa Images/Digital Vision, **146** (bottom), /Tomohiro Ohsumi/AFP, 175 (bottom), /PhotoDisc, **24** (bottom), /QAI Publishing/Universal Images Group, **82** (top), /Jack Scott, **143**, /Schroptschop, **97**, /Sebastian Crespo Photography, **68** (bottom, right), /Sena Vidanagama/AFP, **86**, **88** (bottom, left), /Cameron Spencer, **26**, /Stuart C. Wilson , **39** (top), /The Washington Post, **104**, /Digital Vision/Klaus Vedfelt, **110** (bottom), /Westend61, **12**, /Vera Zurkova, **261**; /iStockphoto/Aj_OP, **51** (bottom), /f3rd14n, **48**, /1xpert, **90** (bottom), /alanphillips, **147** (bottom), /angel_nt, **236**, **240** (bottom right), /antpkr, **297** (bottom),/ Ar Ducha Misfa'i, **213**, /Arsty, **35** (bottom), **40** (bottom), /artisteer, iv, **6**, **272**, /AzmanL, **202** (centre), /Betka82, **297** (top centre), /bjdlzx, **65** (top), /bonchan, **237**, /BookyBuggy, **297** (bottom centre), /Matt_Brown, **256**, /bymuratdeniz, **235**, /JensenChua, **162** (centre), /CribbVisuals, **82** (right), /Dean_Fikar, **331** (left), /Design Cells, **165**, /Dovapi, **viii–xi**, **2**, **242–3**, /DrPAS, **9**, /dszc, **231** (bottom), /eAlisa, **11** (top), **242**, / ekremguduk, **273** (left), **273** (right), /Evgeniy Fedorcov, **4**, /FactoryTh, **30** (left), /fatihhoca, iv, **16** (left), /fcafotodigital, **301**, /fotostorm, **2** (top), /frentusha, **299**, /GaryAlvis, **323**, /gilaxia, **14**, /GlobalP, **vi**, **162** (left), /goir **19** (top, left), /graphicassault, **226** (bottom left), /PeterHermesFurian, **74** , /hudiemm, **191**, /ImageMediaGroup, **45**, /Inner_Vision, **38** (bottom), **40** (centre bottom), /inus12345, **33**, **40** (top), /izusek, **8** (bottom), /jaroon, **185**, /jeridu, **297** (top), /jlazouphoto, **96** (top), /joggiebotma, **24** (top), /johan63, **iv–vii**, **54** (centre), /kamilpetran, **174** (bottom), /Kanawa_Studio, **232** (top left), /koya79, **39** (bottom), /Lemonan, **232** (top right), /lovleah, **162** (bottom), /lusia83, **52** (bottom, left), /David Maki, **174** (top), /mevans, **42** (top), /mgokalp, **42** (bottom), /MicroStockHub, **46** (top), /MsLightBox, **128** (centre), /mvp64, **231** (top), /ShaneMyersPhoto, **68** (top, left), /natrot, **8** (top), /Krystian Nawrocki, **52** (top), /normaals, **78**, **88** (top left), /olga_sweet, **286** (right), /oversnap, **54** (top), /paulafrench, **172** (bottom centre left), /pic_studio, 188 (centre), /pioneer111, 38 (centre), **40** (centre left), /PJPhoto69, **14**, /Priyono, **90** (top),

ACKNOWLEDGEMENTS

/Emilija Randjelovic, **174** (left), /Rawpixel, **28**, /rentusha, **156** (top), **160** (centre), /robynmac, **305**, /romrodinka, **128** (bottom), /rustyfox, **68** (bottom, left), /s-c-s, **172** (bottom left), /sharply_done, **42** (left), Sjo, **43** (top), /skodonnell, **226** (top left), /smirkdingo, **146** (centre), /CarolinaSmith, v, **54** (left), /solidcolours, 2**29**, 2**40** (top centre), /South_agency, **25**, / spxChrome, **226** (centre left), /sweetmoments, **23**, /t:Grafner, **49**, /Talaj, **37**, /ThamKC, **208**, /TheCrimsonMonkey, **216** (top), /tsvibrav, **142**, **144** (bottom right), /ttsz, **85**, /ufuka, **16** (centre), /vencavolrab, **168** (left), /vgajic, **44**, /VIDOK, **52** (centre), **333**, /Vitus, **72**, 2**18**, /VvoeVale, **83**, /wabeno, **19** (top, right), /WachiraS, **259**, /Wavebreakmedia, **16** (top), /Kitsada Wetchasart, **14**, /yothinpi, **188** (bottom), /YsaL, **3** (bottom), /yurazaga, **30** (centre); NASA, **22** (bottom), **62** (left), **66** (bottom, left), **54** (bottom), /NASA/WMAP Science Team, **65** (bottom), **65** (centre), **66** (bottom, centre), /Photo courtesy NASA JSC's Gateway to Astronaut Photography of the Earth, **98**, /NASA, ESA and the Hubble Heritage Team (STScI/AURA) – Hubble/ Europe Collaboration; Acknowledgement: H. Bond (STScI and Pennsylvania State University), **17** (bottom), /NASA/BARREL/ David Milling **107**, /NASA/NOAA GOES Project, **91** (top); Photo courtesy of the National Park Service, **31** (bottom); Nature Picture Library/Dave Watts, **170** (bottom), **172** (top centre left), /Felicity Lanchester, **3** (top); Newspix/Melvyn Knipe, **171**; Science Photo Library/TONY CAMACHO, **172** (top left), /Martyn F. Chillmaid, **209**, **215**, **220** (bottom), /Frank Fox, 11(bottom), /Steve Gschmeissner, **163** (top left), **163** (bottom right), **163** (top right), /Rob Judges/Oxford University Images, **159**, **160** (centre left), /Andrew Lambert Photography, **220** (top), /Arno Massee, **239** (bottom), /Science Photo Library, **43** (bottom), **199** (bottom), **203** (top), /SCIEPRO, **163** (bottom left); Science Source/A. Barrington Brown, **154** (top), **160** (bottom right), / Pascal Goetgheluck, **42** (centre), /Francis Leroy, Biocosmos, **148**, /Science Photo Library, **196**, **200** (top left), /Turtle Rock Scientific, **219** (bottom), **233** (top), **233** (bottom), **302**, /Charles D. Winters, **227** (top); Shutterstock/Africa Studio, **226** (bottom right), /arka38, **132**, /blurAZ, **21**, /BNK Maritime Photographer, **87**, /Ellen Bronstayn, **68** (top, right), /Susan Bughdaryan, **2** (bottom), /Linda Bucklin, **110** (centre), /cigdem, **286** (left), /Donna Ellen Coleman, **150**, /concept w, **192–3**, / Dotted Yeti, **61** (top), /Damsea, **96** (bottom), /Everett Historical, **190** (left), /sonya etchison, Sanit Fuangnakhon, **139**, **144** (centre), **146** (left), /kai hecker, **328**, /JIANG HONGYAN, **135**, /Jiang Dao Hua, **112**, **126** (top), /ID1974, **202** (top), /Katarina S, **226** (centre right), Kateryna Kon, **153**, ktsdesign, **2** (left), **13**, Jatupol Leaupathanasuk, **18** (top), leungchopan, **30** (bottom), / Lipskiy, **287**,
/Oleksandr Lytvynenko, **219** (top), /Magic mine, **202** (bottom), /MaraZe, **202** (left), /Kelly Marken, **134**, /Maximusmeridi, **162** (top), /michaelheim, **110** (left), /mirzavisoko, **238**, **240** (bottom left), /MrGoSlow, **210**, /Ortis, **254** (right), / osmandarinas, **197**, /Elena Pavlovich, **160**, /Pixel-Shot, **232** (bottom left), /posteriori, **90** (left), /ra3rn, **35** (top), /Rakic, **188** (left), /Serorion, **34**, /Yurchanka Siarhei, **146** (top), /Tracy Starr, **296**, /Cristian Storto, **38** (top), **40** (top left), /stockphotograf, **281**, /Stocksnapper, **254** (left), /Bodnar Taras, **22** (top), /theLIMEs, 2**65**, /Leah-Anne Thompson, **321**, / VectorMine, **71** (top), vectortatu, **59**, **66** (bottom, right), Marc Ward, **47**, Ken Weinrich, **239** (top), Steve Wilkins, **62** (right), / Mark Yarchoan, **284**, /Victor Yong, **128** (left), /Yok_onepiece, **184**, /YuRi Photolife, **166** (top), /3Dalia, **190** (right), /3Dsculptor, **228**, **240** (top right); stock.adobe.com/Juulijs, **186** (centre right), /Montri , **125**; Stocksy United/Song Heming, i; The Ocean Agency, **133**; U.S. Navy/Shannon E. Renfroe, **20**; Wellcome Collection (CCBY **4.0**), **198** (top), **200** (bottom left), **199** (top), **200** (bottom centre right).

OTHER MATERIAL

NSW Science K–10 Syllabus © Copyright 2012 NSW Education Standards Authority,
viii–xi, 4, 6, 8, 10, 12, 18, 20, 22, 24, 26, 32, 34, 36, 38, 44, 46, 48, 56, 58, 60, 62, 64, 70, 72, 74, 76, 78, 82, 86, 92, 94, 98, 100, 102, 106, 112, 114, 116, 118, 120, 122, 124, 130, 132, 134, 136, 138, 140, 142, 148, 150, 152, 154, 156, 158, 164, 166, 168, 170, 176, 178, 182, 184, 190, 194, 196, 198, 204, 206, 208, 210, 212, 214, 2016, 218, 220, 222, 228, 230, 232, 234, 236, 238.